Structural Integrity

Volume 10

The *Structural Integrity* book series is a high level academic and professional series publishing research on all areas of Structural Integrity. It promotes and expedites the dissemination of new research results and tutorial views in the structural integrity field.

The Series publishes research monographs, professional books, handbooks, edited volumes and textbooks with worldwide distribution to engineers, researchers, educators, professionals and libraries.

Topics of interested include but are not limited to:

- Structural integrity
- Structural durability
- Degradation and conservation of materials and structures
- Dynamic and seismic structural analysis
- Fatigue and fracture of materials and structures
- Risk analysis and safety of materials and structural mechanics
- Fracture Mechanics
- Damage mechanics
- Analytical and numerical simulation of materials and structures
- Computational mechanics
- Structural design methodology
- Experimental methods applied to structural integrity
- Multiaxial fatigue and complex loading effects of materials and structures
- Fatigue corrosion analysis
- Scale effects in the fatigue analysis of materials and structures
- Fatigue structural integrity
- Structural integrity in railway and highway systems
- Sustainable structural design
- Structural loads characterization
- Structural health monitoring
- Adhesives connections integrity
- Rock and soil structural integrity

Springer and the Series Editors welcome book ideas from authors. Potential authors who wish to submit a book proposal should contact Dr. Mayra Castro, Senior Editor, Springer (Heidelberg), e-mail: mayra.castro@springer.com

More information about this series at http://www.springer.com/series/15775

Zengtao Chen · Abdolhamid Akbarzadeh

Advanced Thermal Stress Analysis of Smart Materials and Structures

Springer

Zengtao Chen
Department of Mechanical Engineering
University of Alberta
Edmonton, Canada

Abdolhamid Akbarzadeh
Department of Bioresource Engineering
McGill University
Island of Montreal, QC, Canada

Department of Mechanical Engineering
McGill University
Montreal, QC, Canada

ISSN 2522-560X ISSN 2522-5618 (electronic)
Structural Integrity
ISBN 978-3-030-25203-8 ISBN 978-3-030-25201-4 (eBook)
https://doi.org/10.1007/978-3-030-25201-4

This Springer imprint is published by the registered company Springer Nature Switzerland AG
The registered company address is: Gewerbestrasse 11, 6330 Cham, Switzerland

Preface

Advanced smart materials and structures are under rapid development to meet engineering challenges for multifunctionality, high reliability, and high efficiency in modern technology. Advanced manufacturing technology calls for high-energy, non-contact, laser-forming materials in a sophisticated spatial and temporal environment. Thermal stress analysis of advanced materials offers a viable tool for optimized design of advanced, multifunctional devices and for accurate modelling of advanced manufacturing processes.

In classical thermal analysis, thermal stress is caused by the constrained deformation when a temperature variation occurs in an elastic body. How the material reaches the final temperature from the initial temperature will not affect the calculation of the steady-state or quasi-static thermal stresses. Classical Fourier heat conduction theory is widely used in the thermal stress analysis leading to perfect and trustworthy results. In transient and high-temperature gradient cases, when materials experience sudden changes in temperature within an extremely short time, the reaction to this ultrafast, temporal temperature changes or heat flux would be expected to have a time delay since heat propagation takes time to occur. This delay might not be felt for most metallic materials as relaxation time of metals is in the range of 10^{-8}–10^{-14} s, opposed to soft and organic materials with a relaxation time between 1 s and 10 s, where the delay is not negligible and the subsequent thermal wave propagation is evident. Non-Fourier, time-related heat conduction models have been proposed to compensate the effect of this delay in a heat transfer process. A natural outcome of the non-Fourier heat conduction models is the wave-like heat conduction equation, where a thermal wave is required to spread the heat. Thermal stress analysis in advanced materials based on the non-Fourier heat conduction theories has become a popular topic in the past decades. The authors of this book are among the researchers who initiated and continued working on promoting the research in the area of thermal stresses in advanced smart materials. Although the literature in this area is booming in recent years, a single-volume monograph summarizing the recent progress in implementing non-Fourier heat conduction theories to deal with the multiphysical behaviour of smart materials and structures is still missing in the literature.

This monograph is a collection of the research on advanced thermal stress analysis of advanced and smart materials and structures mainly written by the authors of this book, and their students and colleagues. This book is organized into seven chapters. Chapter 1 provides a brief introduction to the non-Fourier heat conduction theories, including the Cattaneo–Vernotte (C–V), dual-phase-lag (DPL), three-phase-lag (TPL) theory, fractional phase-lag, and non-local heat conduction theories. Chapter 2 introduces the fundamental of thermal wave characteristics by reviewing different methods for solving non-Fourier heat conduction problems in representative homogenous and heterogeneous advanced materials. Chapter 3 provides the fundamentals of smart materials and structures, including the background, application, and governing equations. In particular, functionally graded smart structures are introduced as they represent the recent development in the industry; a series of uncoupled thermal stress analyses on one-dimensional smart structures are also presented. Chapter 4 presents coupled thermal stress analyses in one-dimensional, homogenous and heterogeneous, smart piezoelectric structures considering alternative coupled thermopiezoelectric theories. Chapter 5 introduces a generalized method to deal with plane crack problems in smart materials and structures based on classical Fourier heat conduction. Thermal fracture analysis of cracked structures based on non-Fourier heat conduction theories is presented in Chap. 6. Finally, Chap. 7 lists a few perspectives on the future developments in non-Fourier heat conduction and thermal stress analysis.

We sincerely thank our students, colleagues, and friends for their contributions in preparation of this book. We are indebted to our families for their sacrifice, patience, and constant support during the composition of this book.

Edmonton, Canada — Zengtao Chen
Montreal, Canada — Abdolhamid Akbarzadeh
April 2019

Contents

1 Heat Conduction and Moisture Diffusion Theories 1
1.1 Introduction 1
1.2 Heat Conduction 2
1.2.1 Fourier Heat Conduction 2
1.2.2 Non-Fourier Heat Conduction 7
1.3 Moisture Diffusion 16
1.3.1 Fickian Moisture Diffusion 16
1.3.2 Non-Fickian Moisture Diffusion 18
References 20

2 Basic Problems of Non-Fourier Heat Conduction 23
2.1 Introduction 23
2.2 Laplace Transform and Laplace Inversion 23
2.2.1 Fast Laplace Inverse Transform 24
2.2.2 Reimann Sum Approximation 25
2.2.3 Laplace Inversion by Jacobi Polynomial 25
2.3 Non-Fourier Heat Conduction in a Semi-infinite Strip 26
2.4 Nonlocal Phase-Lag Heat Conduction in a Finite Strip 31
2.4.1 Molecular Dynamics to Determine Correlating Nonlocal Length 37
2.5 Three-Phase-Lag Heat Conduction in 1D Strips, Cylinders, and Spheres 42
2.5.1 Effect of Bonding Imperfection on Thermal Wave Propagation 46
2.5.2 Effect of Material Heterogeneity on Thermal Wave Propagation 48
2.5.3 Thermal Response of a Lightweight Sandwich Circular Panel with a Porous Core 50
2.6 Dual-Phase-Lag Heat Conduction in Multi-dimensional Media . . . 53

2.6.1 DPL Heat Conduction in Multi-dimensional Cylindrical Panels 53
2.6.2 DPL Heat Conduction in Multi-dimensional Spherical Vessels 58
References 62

3 Multiphysics of Smart Materials and Structures 65
3.1 Smart Materials 65
3.1.1 Piezoelectric Materials 67
3.1.2 Magnetoelectroelastic Materials 69
3.1.3 Advanced Smart Materials 73
3.2 Thermal Stress Analysis in Homogenous Smart Materials 74
3.2.1 Solution for the a Thermomagnetoelastic FGM Cylinder 76
3.2.2 Solution for Thermo-Magnetoelectroelastic Homogeneous Cylinder 82
3.2.3 Benchmark Results 86
3.3 Thermal Stress Analysis of Heterogeneous Smart Materials 90
3.3.1 Solution Procedures 92
3.3.2 Benchmark Results 97
3.4 Effect of Hygrothermal Excitation on One-Dimensional Smart Structures 98
3.4.1 Solution Procedure 101
3.4.2 MEE Hollow Cylinder 107
3.4.3 MEE Solid Cylinder 108
3.4.4 Benchmark Results 110
3.5 Remarks 114
References 115

4 Coupled Thermal Stresses in Advanced Smart Materials 119
4.1 Functionally Graded Materials 119
4.2 Hyperbolic Coupled Thermopiezoelectricity in One-Dimensional Rod 120
4.2.1 Introduction 121
4.2.2 Homogeneous Rod Problem 122
4.2.3 Solution Procedure 125
4.2.4 Results and Discussion 131
4.3 Hyperbolic Coupled Thermopiezoelectricity in Cylindrical Smart Materials 134
4.3.1 Introduction 135
4.3.2 Hollow Cylinder Problem 135
4.3.3 Solution Procedure 139
4.3.4 Results and Discussion 142
4.4 Coupled Thermopiezoelectricity in One-Dimensional Functionally Graded Smart Materials 146

4.4.1 Introduction 146
4.4.2 The Functionally Graded Rod Problem 147
4.4.3 Solution Procedures 150
4.4.4 Results and Discussion 156
4.4.5 Introduction of Dual Phase Lag Models 161
4.4.6 Results of Dual Phase Lag Model Analysis 163
4.5 Remarks 168
References 169

5 Thermal Fracture of Advanced Materials Based on Fourier Heat Conduction 171
5.1 Introduction 171
5.2 Extended Displacement Discontinuity Method and Fundamental Solutions for Thermoelastic Crack Problems 171
5.2.1 Fundamental Solutions for Unit Point Loading on a Penny-Shaped Interface Crack 174
5.2.2 Boundary Integral-Differential Equations for Interfacial Cracks 185
5.2.3 Stress Intensity Factor and Energy Release Rate 194
5.3 Interface Crack Problems in Thermopiezoelectric Materials 197
5.3.1 Basic Equations 198
5.3.2 Fundamental Solutions for Unit-Point Extended Displacement Discontinuities 200
5.3.3 Boundary Integral-Differential Equations for an Interfacial Crack in Piezothermoelastic Materials 204
5.3.4 Hyper-Singular Integral-Differential Equations 206
5.3.5 Solution Method of the Integral-Differential Equations 210
5.3.6 Extended Stress Intensity Factors 219
5.4 Fundamental Solutions for Magnetoelectrothermoelastic Bi-Materials 220
5.5 Fundamental Solutions for Interface Crack Problems in Quasi-Crystalline Materials 225
5.5.1 Fundamental Solutions for Unit-Point EDDs 228
5.6 Application of General Solution in the Problem of an Interface Crack of Arbitrary Shape 234
5.7 Summary 235
References 236

6 Advanced Thermal Fracture Analysis Based on Non-Fourier Heat Conduction Models 243
6.1 Introduction 243
6.2 Hyperbolic Heat Conduction in a Cracked Half-Plane with a Coating 243
6.2.1 Basic Equations 245
6.2.2 Temperature Field 247
6.2.3 Temperature Gradients 251
6.2.4 Numerical Results 252
6.3 Thermoelastic Analysis of a Partially Insulated Crack in a Strip 255
6.3.1 Definition of the Problem 256
6.3.2 Thermal Stresses 259
6.3.3 Asymptotic Stress Field Near Crack Tip 264
6.3.4 Numerical Results and Discussions 266
6.4 Thermal Stresses in a Circumferentially Cracked Hollow Cylinder Based on Memory-Dependent Heat Conduction 269
6.4.1 Problem Formulation 270
6.4.2 Thermal Axial Stress in an Un-cracked Hollow Cylinder 271
6.4.3 Thermal Stress in the Axial Direction 274
6.4.4 Stress Intensity Factors 279
6.4.5 Results and Discussion 283
6.5 Transient Thermal Stress Analysis of a Cracked Half-Plane of Functionally Graded Materials 284
6.5.1 Formulation of the Problem and Basic Equations 287
6.5.2 Solution of the Temperature Field 290
6.5.3 Solution of Thermal Stress Field 293
6.5.4 Numerical Results and Discussion 296
6.6 Summary 298
References 298

7 Future Perspectives 303
7.1 Heat Conduction Theories 303
7.2 Application in Advanced Manufacturing Technologies 304

Chapter 1
Heat Conduction and Moisture Diffusion Theories

1.1 Introduction

The design of high performance micro/macro-scale composite structures working at high temperature and humidity environmental conditions needs an accurate heat and moisture transfer analysis through the solid structure. The hygrothermal deformation, developed by temperature and moisture distribution, under adverse operating conditions degrades the structural integrity and results in lowering the structural stiffness and strength. While increasing temperature could majorly induce thermal stresses, more fluid or moisture could be absorbed into voids and microscopic defects in the solids, specifically in composites with polymeric matrices, which results in additional moisture induced stresses [1, 2].

Heat is defined as the energy transport within a body from hotter regions to cooler regions according to the second law of thermodynamics. This energy could be provided by the constituent particles such as atoms, molecules, or free electrons. While the heat flow cannot be measured, there exists a measurable and macroscopic quantity known as *temperature*. Temperature can be interpreted as the combined effect of all kinetic energies of a large number of molecules in solid, liquid, or gaseous state. *Conduction*, *convection*, and *radiation* are three distinct modes of heat transfer. In a conduction process, the heat passes through the materials by microscopic diffusion and collision of constituent particles. The process of transferring heat by a moving fluid and a relative motion of a heated body is called convection [3, 4]. The energy could also be emitted directly between the distant portions of a body via the electromagnetic radiation. Since the convection and radiation are usually negligible in solids, the heat conduction is merely considered in this book for thermal analysis except for the boundary conditions, which could be any type of heat transfer modes. Since the process of moisture transfer is basically the same as the heat transfer, the Fourier and non-Fourier heat conduction theories are first introduced in detail in Sect. 1.2 and then some of the moisture diffusion models are reviewed in Sect. 1.3.

Z. T. Chen and A. H. Akbarzadeh, *Advanced Thermal Stress Analysis of Smart Materials and Structures*, Structural Integrity 10,
https://doi.org/10.1007/978-3-030-25201-4_1

1.2 Heat Conduction

Heat conduction is encountered in many engineering applications, such as electronic packaging, casting, food processing machines, biomedical devices, and thermal shield design for aerospace vehicles [5]. The relation between the heat flux vector q and temperature gradient ∇T is called the constitutive relation of heat flux. The heat flux, q [W/m^2], is defined as the heat flow Q per unit time and per unit normal vector of the area of an isothermal surface [6]. The heat conduction equation is then established by using the constitutive relation of heat flux and the energy conservation equation or first law of thermodynamics. The classical Fourier and non-Fourier heat conduction theories are two main categories of thermal analysis discussed in this chapter.

1.2.1 Fourier Heat Conduction

The earliest constitutive relation of heat flux was proposed by Fourier in 1807, which based on experimental observations, assume that the heat flux and temperature gradient occur at the same instant of time. The Fourier heat conduction specifies the proportionality between the heat flux and temperature gradient as follows [7]:

$$q(x,t) = -k^T \nabla T(x,t) \tag{1.1}$$

where k^T [W/(m K)] is thermal conductivity coefficient and x and t, respectively, stand for the general coordinate of a material point and time. While k^T is a positive scalar quantity for isotropic materials, k^T should be replaced by the components of a second-order tensor of thermal conductivity for anisotropic media [6]. The state theorem and second law of thermodynamics prove that k^T should be positive definite and a function of pressure and temperature (two independent properties) [8, 9]. In a homogeneous medium, k^T is constant through the body, while it is a function of position in a heterogeneous material ($k^T(x)$).

To derive the differential equation of heat conduction, the first and second laws of thermodynamics should be utilized. As elucidated in [10, 11], the summation of the first variation of absorbed heat by a medium (Q) and the work done on the medium (W) is equal to the differential of internal energy change (U):

$$\delta Q + \delta W = dU \tag{1.2}$$

where the first variation operator δ is path dependent in contrast with the total derivative operator d. The internal energy of a medium could be obtained by the summation of the kinetic energy K and the intrinsic energy I as:

$$U = I + K \tag{1.3}$$

Whilst the first law of thermodynamics quantitatively describes the energy transport process, the second law of thermodynamics defines the direction of energy transfer. When a thermodynamic system accomplish a cycle, the Clausius inequality reads according to the second law of thermodynamics [10, 12]:

$$\int \frac{\delta Q}{T} \leq 0 \tag{1.4}$$

where T [K] stands for the absolute temperature. For an irreversible process, the left hand side of the inequality is always negative, while it is zero for a reversible process. Two independent properties of temperature and entropy define the thermodynamic state of a system. Using Eq. (1.4), the entropy change of a reversible cycle is defined as [13, 14]:

$$dS = \frac{\delta Q}{T} \tag{1.5}$$

The first and second law of thermodynamics are reduced to the following energy and entropy equations for a reversible thermodynamic process [10]:

$$-\nabla.q(x,t) + \rho r_T(x,t) = \rho T_0 \frac{\partial s(x,t)}{\partial t} \tag{1.6a}$$

$$\rho s(x,t) = \frac{\rho c}{T_0}(T(x,t) - T_0) \tag{1.6b}$$

where r_T [W/kg] and s [J/kg K] are the heat generation per unit time per unit mass and entropy per unit mass, respectively. The density, specific heat, initial temperature, and divergence operator are, respectively, represented by ρ [kg/m^3], c [J/(kg K)], T_0, and ∇. It is worth mentioning that the Fourier heat conduction Eq. (1.1) and the first and second laws of thermodynamics in Eq. (1.6) need modifications for the heat transport analysis of thermoelectrics. For thermoelectricity analysis, the heat loss from thermoelectric couples to interstitial gas is considered along with the Peltier effect, which is the contribution of electric current density in heat flux and the first law of thermodynamics [15, 16].

Differential equation of transient Fourier heat conduction is achieved by substituting Eqs. (1.1) and (1.6b) into Eq. (1.6a):

$$\nabla.\left(k^T \nabla T(x,t)\right) + \rho r_T(x,t) = \rho c \frac{\partial T(x,t)}{\partial t} \tag{1.7}$$

For steady-state analysis, the right-hand side of Eq. (1.7) is zero and the heat generation rate r_t is omitted. Moreover, for a homogenous isotropic material, when the thermal conductivity is assumed independent of temperature, Eq. (1.7) is simplified as [17, 18]:

$$\nabla^2 T(x,t) + \frac{\rho r_T(x,t)}{k^T} = \frac{1}{\alpha} T(x,t) \tag{1.8}$$

where $\alpha = \frac{k^T}{\rho c}$ $[\mathrm{m}^2/\mathrm{s}]$ is thermal diffusivity and ∇^2 represents the Laplacian operator. The typical mathematical operators used in the multiphysics analysis are given in Table 1.1 for different coordinate systems.

The boundary and initial conditions should be specified along with the differential equation of heat conduction to obtain the temperature distribution within a medium. The number of boundary conditions on the spatial domain and number of initial condition in the time domain are equal to the order of the highest derivative of a variable in the governing equation with respect to the space and time domain [10, 20].

1.2.1.1 Thermal Boundary Conditions

The typical thermal boundary conditions, encountered in thermal analysis, could be categorized as: (a) prescribed temperature, (b) prescribed heat flux, and (c) heat transfer by convection and irradiation [4, 10, 17]:

(a) The temperature along the boundary surface (*S*) of a medium is prescribed as:

$$T(x,t)|_{x=S} = \overline{T}(S,t) \tag{1.9}$$

where $\overline{T}$ is generally a known function of position and time.

(b) The heat flux across the boundary surface is known as a function of position and time:

$$k^T \nabla T(x,t)|_{x=S} = \pm q_S(S,t) \tag{1.10}$$

where q_S is the prescribed heat flux transferred toward the boundary surface. The plus or minus signs on the right-hand-side of Eq. (1.10) depends on the direction of the heat flux transferred from or to the surface. An insulated boundary condition is a special form of Eq. (1.10) where $q_S = 0$.

Table 1.1 Mathematical operators in different coordinates [19]

Coordinate	Operator
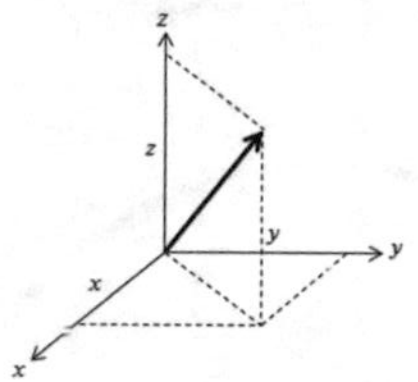 Cartesian	Gradient: $\nabla = \frac{\partial}{\partial x} e_x + \frac{\partial}{\partial y} e_y + \frac{\partial}{\partial z} e_z$
	Laplacian: $\nabla^2 = \frac{\partial^2}{\partial x^2} + \frac{\partial^2}{\partial y^2} + \frac{\partial^2}{\partial z^2}$
	Divergence: $\nabla . A = \frac{\partial A_x}{\partial x} e_x + \frac{\partial A_y}{\partial y} e_y + \frac{\partial A_z}{\partial z} e_z$
	Curl: $\nabla \times A = \left(\frac{\partial A_z}{\partial y} - \frac{\partial A_y}{\partial z}\right) e_x + \left(\frac{\partial A_x}{\partial z} - \frac{\partial A_z}{\partial x}\right) e_y + \left(\frac{\partial A_y}{\partial x} - \frac{\partial A_x}{\partial y}\right) e_z$
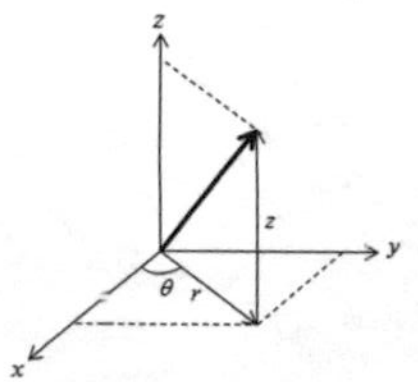 Cylindrical	Gradient: $\nabla = \frac{\partial}{\partial r} e_r + \frac{1}{r}\frac{\partial}{\partial \theta} e_\theta + \frac{\partial}{\partial z} e_z$
	Laplacian: $\nabla^2 = \frac{1}{r}\frac{\partial}{\partial r}\left(r \frac{\partial}{\partial r}\right) + \frac{1}{r^2}\frac{\partial^2}{\partial \theta^2} + \frac{\partial^2}{\partial z^2}$
	Divergence: $\nabla . A = \frac{1}{r}\frac{\partial}{\partial r}(A_r r) + \frac{1}{r}\frac{\partial A_\theta}{\partial \theta} + \frac{\partial A_z}{\partial z}$
	Curl: $\nabla \times A = \left(\frac{1}{r}\frac{\partial A_z}{\partial \theta} - \frac{\partial A_\theta}{\partial z}\right) e_r + \left(\frac{\partial A_r}{\partial A_z} - \frac{\partial A_z}{\partial r}\right) e_\theta + \left(\frac{1}{r}\frac{\partial (A_\theta r)}{\partial r} - \frac{1}{r}\frac{\partial A_r}{\partial \theta}\right) e_z$
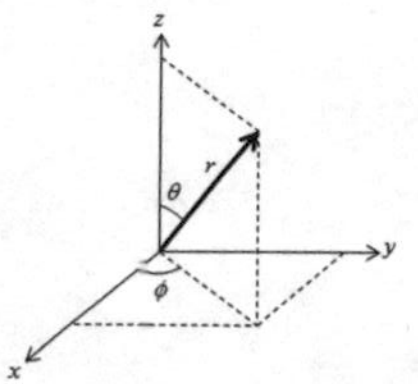 Spherical	Gradient: $\nabla = \frac{\partial}{\partial r} e_r + \frac{1}{r}\frac{\partial}{\partial \theta} e_\theta + \frac{1}{r \sin\theta}\frac{\partial}{\partial \phi} e_\phi$
	Laplacian: $\nabla^2 = \frac{1}{r^2}\frac{\partial}{\partial r}\left(r^2 \frac{\partial}{\partial r}\right) + \frac{1}{r^2 \sin\theta}\frac{\partial}{\partial \theta}\left(\sin\theta \frac{\partial}{\partial \theta}\right) + \frac{1}{r^2 \sin\theta}\frac{\partial^2}{\partial \phi^2}$
	Divergence: $\nabla . A = \frac{1}{r^2}\frac{\partial \left(r^2 A_r\right)}{\partial r} + \frac{1}{r \sin\theta}\frac{\partial (A_\theta \sin\theta)}{\partial \theta} + \frac{1}{r \sin\theta}\frac{\partial A_\phi}{\partial \theta}$
	Curl: $\nabla \times A = \frac{1}{r \sin\theta}\left[\frac{\partial \left(A_\phi \sin\theta\right)}{\partial \theta} - \frac{\partial A_\theta}{\partial \phi}\right] e_r + \frac{1}{r \sin\theta}\left[\frac{\partial A_r}{\partial \phi} - \sin\theta \frac{\partial \left(r A_\phi\right)}{\partial r}\right] + \frac{1}{r}\left[\frac{\partial (r A_\theta)}{\partial r} - \frac{\partial A_r}{\partial \theta}\right] e_z$

(c) The convection heat transfer with ambient and radiation heat exchange with a radiator happens at the boundary surface:

$$k^T \nabla T(x,t)|_{x=S} = \pm\left(h(T(x,t) - T_\infty) + \varepsilon\sigma\left(T(x,t) - T_r^4\right)\right)\big|_{x=S} \tag{1.11}$$

where h, T_∞, and T_r are convection heat transfer coefficient, ambient temperature, and the temperature of radiative body, respectively. The Stefan-Boltzmann constant σ has the value of $\sigma = (5.67040 \pm 0.00004)10^{-8}$ [W/m^2 K^4] and the emissivity of ε is always lower or equal to one ($\varepsilon \leq 1$) [20, 21].

1.2.1.2 Thermal Initial Conditions

Time derivatives of temperature appear in the differential equation of transient heat conduction. Therefore, the initial conditions are required to be specified for time-dependent thermal analysis problems. Regarding the first-order derivative of temperature in the Fourier heat conduction Eq. (1.7), the initial condition could be specified as:

$$T(x, t = 0) = T_0(x) \tag{1.12}$$

where $T_0(x)$ is a known function of the spatial coordinate x. The initial conditions for higher-order time derivatives of temperature should also be specified for non-Fourier heat conduction.

1.2.1.3 Thermal Interfacial Conditions

When a medium is composed of bonding the layers of dissimilar materials, thermal interfacial conditions should be also specified along with the boundary and initial conditions. The bonding of two materials typically contains imperfection, such as

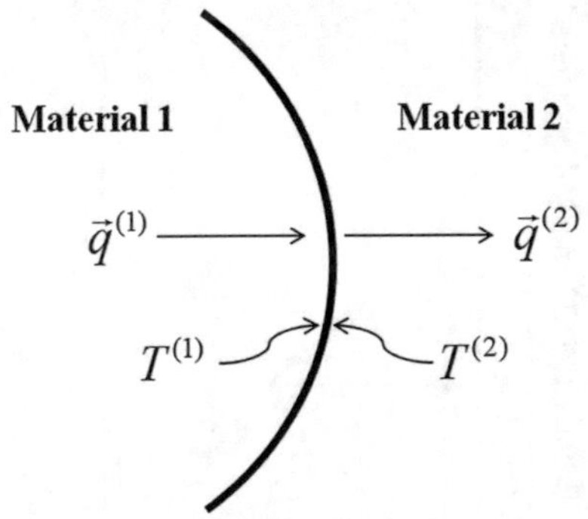

Fig. 1.1 Thermal boundary conditions at the interface of two dissimilar materials

small voids and defects. The following thermal interfacial conditions need to be considered at the interface of two layers with thermally weak conduction (Fig. 1.1) [22, 23]:

$$\left(T^{(1)}(x,t) - T^{(2)}(x,t)\right)\Big|_{x=S} = R^T q^{(1)}(S,t) \tag{1.13a}$$

$$q^{(1)}(S,t) = q^{(2)}(S,t) \tag{1.13b}$$

where R^T is thermal compliance constant or thermal contact resistance for the imperfect interface. The perfectly bonded interfaces could achieved by substituting $R^T = 0$ in Eq. (1.13a).

1.2.2 Non-Fourier Heat Conduction

The conventional heat conduction theory based on the classical Fourier law admits an infinite speed for thermal wave propagation due to the parabolic-type partial differential equation of Fourier heat conduction [Eq. (1.7)]. Fourier law assumes an instantaneous thermal response and a quasi-equilibrium thermodynamic condition, which implies that a thermal disturbance is felt instantaneously at all spatial points within a medium. The classical diffusive-like theory has been widely used in macroscopic heat transfer problems; however, the heat transmission has been observed to be a non-equilibrium phenomenon with a finite thermal wave speed for applications involving very low temperature, extremely high temperature gradient, ultrafast laser heating, ballistic heat transfer, and micro temporal and spatial scales [16, 24, 25]. For example, it has been observed that the measured surface temperature of a slab immediately after an intense thermal shock is 300 °C higher than the temperature obtained by Fourier's law [26, 27]. As a result, several non-Fourier heat conduction theories have been developed to eliminate the drawbacks of Fourier heat conduction while predicting the thermal wave behaviour, referred as second sound, and incorporate the microstructural effects on the heat transport.

1.2.2.1 Cattaneo-Vernotte Heat Conduction

The simplest hyperbolic non-Fourier heat conduction theory was proposed by Cattaneo [28] and Vernotte [29] (C-V model) to achieve a finite thermal wave speed for heat propagation. The thermal relaxation time was introduced in Fourier's law of heat conduction as follows:

$$q(x,t) + \tau_q \frac{\partial q(x,t)}{\partial t} = -k^T \nabla T(x,t) \tag{1.14}$$

where τ_q [s], a non-negative parameter, is the time delay called the thermal relaxation time. The thermal relaxation time could be mathematically interpreted as the time delay between heat flux vector and temperature gradient in the fast-transient transport, or micro-structurally interpreted as the time when the intrinsic length scales in diffusion and the thermal wave becomes equal [16, 24]. It could be calculated in terms of thermal diffusivity and the speed of sound (v_s) as [16]:

$$\tau_q = \frac{3\alpha}{v_s^2} \tag{1.15}$$

The value of thermal relaxation time varies between 10^{-14} s and 10^3 s for metals, organic tissues, and materials with microstructural non-homogeneity, such as polyethylene/graphite nanosheets [22, 24, 27, 30].

The C-V heat conduction differential equation is derived by omitting the heat flux q between the constitutive relation of heat flux (1.14) and the energy Eq. (1.6). For a homogenous isotropic material, when the thermal properties are assumed independent of temperature, the C-V heat conduction could be simplified as:

$$\nabla^2 T(x,t) = \left(1 + \tau_q \frac{\partial}{\partial t}\right)\left(\frac{1}{\alpha}\frac{\partial T(x,t)}{\partial t} - \frac{\rho r_T(x,t)}{k^T}\right) \tag{1.16}$$

Equation (1.16) is a hyperbolic-type differential equation, in contrast to the parabolic-type differential equation of Fourier's law [Eq. (1.7)], which depicts the propagation of temperature disturbance as a wave with thermal damping. Comparing Eq. (1.16) with wave equation, the finite thermal wave speed of C-V model $\left(C^T_{C-V}\right)$ is obtained as [16, 31]:

$$C^T_{C-V} = \sqrt{\frac{\alpha}{\tau_q}} \tag{1.17}$$

In the absence of thermal relaxation time, C^T_{C-V} reaches infinity and Eqs. (1.14) and (1.16) reduces to the classical Fourier heat conduction. The C-V model predicts a finite thermal wave speed based on the consideration of phonon collision in microstructural heat transport and results in the thermal shock formation and thermal resonance phenomenon, which cannot be observed by the conventional Fourier heat conduction. However, the heat transport by the dispersion of phonon collisions and phonon-electron interaction is overlooked. The C-V model does not take into account the relaxation times among electrons and the atomic lattice due to the macroscopic considerations. In addition, the C-V model presumes an instantaneous heat flow due to the immediate response between the temperature gradient and the energy transport [16]. Some unusual physical solutions introduced by the C-V model have also been reported. Therefore, the applicability of C-V model in superior conductors and fast-transient heat transport is debatable [16, 32–34].

1.2.2.2 Dual-Phase-Lag Heat Conduction

While the C-V model assumes an average macroscopic thermal behaviour over grains of a medium, microstructural effects and delayed thermal responses become pronounced in the fast transient process of heat transport as well as the induced delayed responses by low-conducting pores in sand media and inert behaviour of molecules at low temperatures [16]. Dual-phase-lag (DPL) heat conduction was introduced by Tzou [35, 36] to remove the assumptions made in the C-V model and to take into account the effect of relaxation time between electrons and atomic lattices in the transient process of heat conduction. Since the heat transport process needs a finite amount of time to take place on the macroscopic level due to the interactions conducted on the microscopic level, Tzou presented the following constitutive equation for the delayed thermal responses and provided the experimental supports for the DPL formulation [16, 37]:

$$q(x, t + \tau_q) = -k^T \nabla T(x, t + \tau_T) \tag{1.18}$$

where τ_q [s], like the C-V model, is the phase lag of the heat flux or thermal relaxation time and τ_T [s] represents the phase lag of temperature gradient. Furthermore, the phase lags τ_q and τ_T are positive and intrinsic properties of materials [16]. It is worth mentioning that the three characteristic times should be distinguished in the DPL heat conduction, which are: time t for the onset of the heat transport, time $t + \tau_q$ at which heat flows through the material, and $t + \tau_T$ for the occurrence of temperature gradient within the medium.

While τ_q could be defined as the relaxation time due to the fast transient effects of thermal inertia, τ_T could be interpreted as the time delay caused by the microstructural interactions such as phonon-electron or phonon scattering [16, 38]. In quantum mechanics, phonons determine the energy states of a metallic lattice.

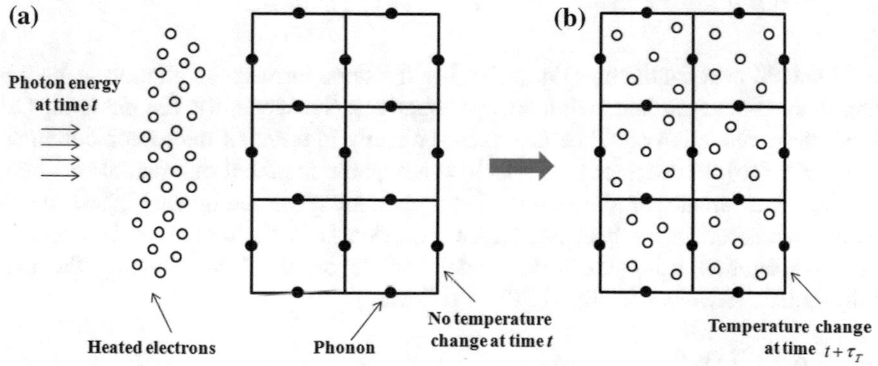

Fig. 1.2 The delayed thermal responses caused by phonon-electron interaction: **a** electron gas heating by photons and **b** metal lattice heating by phonon-electron interactions. [Reproduced from [16] with permission from John Wiley and Sons]

The delayed thermal responses caused by phonon-electron interactions during the short-pulse laser heating of metallic media is depicted in Fig. 1.2 [16]. The electron gases are first heated up by the photons of laser beams at time t when no temperature change can be detected in the metallic lattice (Fig. 1.2a). The phonon-electron interactions then cause the energy transport from the heated electrons to phonons which leads to the temperature rise in the lattice at time $t + \tau_T$. However, as asserted by Tzou, a detailed understanding of the delayed thermal responses caused by microstructural interactions needs a profound knowledge of quantum mechanics and statistical thermodynamics [16, 39, 40].

As seen in Eq. (1.18), temperature gradient or heat flux vector could precede the other depending on the value of τ_q and τ_T. For instance, if $\tau_T > \tau_q$, the temperature gradient within a medium is established as a result of the heat flux vector, which means that the heat flux vector is a cause and the temperature gradient is an effect [16]. In order to omit the heat flux between Eq. (1.18) and energy equation [Eq. (1.6], where the thermo-physical phenomena occur at the same time, Taylor series expansion of Eq. (1.18) with respect to time t is used. To develop a model equivalent to the microscopic hyperbolic two-step heat conduction model [35, 40], Tzou assumed small values for τ_q and τ_T so that the third- and higher-order terms of τ_q and second- and higher-order terms of τ_T are negligible. As a result, the second-order Taylor series expansion of Eq. (1.18) for τ_q and the first-order expansion for τ_T lead to:

$$\left(1 + \tau_q \frac{\partial}{\partial t} + \frac{\tau_q^2}{2} \frac{\partial^2}{\partial t^2}\right) q(x,t) = -k^T \left\{ \nabla T(x,t) + \tau_T \frac{\partial}{\partial t} [\nabla T(x,t)] \right\} \tag{1.19}$$

Eliminating the heat flux q between Eq. (1.19) and energy Eq. (1.6) results in the DPL heat conduction differential equations for isotropic materials with temperature-independent material properties:

$$\left(1 + \tau_T \frac{\partial}{\partial t}\right) \nabla^2 T(x,t) = \left(1 + \tau_q \frac{\partial}{\partial t} + \frac{\tau_q^2}{2} \frac{\partial^2}{\partial t^2}\right) \left(\frac{1}{\alpha} \frac{\partial T(x,t)}{\partial t} - \frac{\rho r_T(x,t)}{k^T}\right) \tag{1.20}$$

The DPL heat conduction Eq. (1.20) has the same form as the phonon scattering and phonon-electron interaction energy equations. Subsequently, the phase-lags of DPL heat conduction could be determined directly in terms of microscopic thermal properties [16]. Furthermore, Eq. (1.20) is a hyperbolic partial differential equation. As the characteristics of temperature distribution are governed by the highest-orders of differentiation in the heat conduction equation [16], thermal wave behavior in heat propagation using the DPL model can be observed by isolating the two third-order derivatives in Eq. (1.20) as follows:

$$\frac{\partial}{\partial t}\left(\frac{\partial^2 T(x,t)}{\partial t^2} - \left(C_{DPL}^T\right)^2 \nabla^2 T(x,t)\right) + lower - order\ terms = 0 \tag{1.21}$$

where the finite thermal wave speed of DPL model (C_{DPL}^T) is [41]:

$$C_{DPL}^T = \frac{\sqrt{2\alpha\tau_T}}{\tau_q} = C_{C-V}^T\sqrt{\frac{2\tau_T}{\tau_q}} \tag{1.22}$$

Consequently, this type of DPL heat conduction could be named as wave-like DPL or hyperbolic-type DPL model. Equation (1.22) shows that the DPL thermal wave speed depends on the phase-lag of heat flux and phase-lag of temperature gradient as well as the thermal diffusivity. In addition, it is found that τ_q and τ_T have opposite effects on thermal wave speed. While increasing τ_T, enhances the thermal wave speed, increasing τ_q decreases the speed of thermal wave. Depending on the ratio of $\frac{\tau_T}{\tau_q}$, Eq. (1.22) shows that C_{DPL}^T could propagate faster than C_{C-V}^T, e.g. in most metals ($\tau_T < \tau_q$), or slower, e.g. in superfluid liquid helium ($\tau_T > \tau_q$) [16].

On the other hand, in the absence of Taylor series expansion term $\frac{\tau_q^2}{2}$, Eq. (1.19) leads to a parabolic differential Eq. (1.20), which reveals a diffusive thermal behavior for heat flow. This type of DPL model is called diffusive-like DPL or parabolic-type DPL. It is also worthwhile mentioning that expanding both τ_q and τ_T in Eq. (1.18) up to the second-order leads to a parabolic differential equations and results in the nonlinear thermal lagging behavior which could be employed to describe the heat transport phenomena in biological systems with multiple energy carriers [42]. Moreover, in the absence of the phase-lag of temperature gradient, $\tau_T = 0$, and the second-order Taylor series expansion term, Eqs. (1.19) and (1.20) reduce to the C-V heat conduction equations [38].

1.2.2.3 Three-Phase-Lag Heat Conduction

The aforementioned Fourier and non-Fourier heat conduction theories have been employed to describe classical and generalized thermoelasticity models, respectively. Among generalized thermoelasticity models, Lord and Shulman (L-S) [43], Green and Lindsay (G-L) [44], Chandrasekhariah and Tzou (C-T) [24, 36], and Green and Naghdi (G-N) [45] could be mentioned. However, the G-N model cannot be obtained by the C-V and DPL heat conduction theories. As a result, three-phase-lag (TPL) heat conduction theory was proposed by Choudhuri [46] to describe all generalized thermoelasticity models and encompass all previous theories for non-Fourier heat conduction at the same time. The utilization of phase-lags of heat flux, temperature gradient, and thermal displacement gradient in the TPL theory is significant for understanding several physical phenomena, such as bioheat transfer and laser heating in living tissues, exothermic catalytic reactions, and harmonic plane wave propagation [47, 48].

As an extension to the G-N thermoelastic model [45, 49], where thermal displacement gradient υ was introduced along with the temperature gradient in the heat flux constitutive equations, Choudhuri proposed the following generalized constitutive equation to describe the lagging responses:

$$q\left(x,t+\tau_q\right) = -\left[k^T \nabla T(x,t+\tau_T) + k^{*T} \nabla \upsilon(x,t+\tau_\upsilon)\right] \tag{1.23}$$

where $\upsilon = T$ and $k^{*T} \geq 0$ is the TPL material constant (the rate of thermal conductivity). The third phase-lag τ_υ [s] introduced in TPL could be interpreted as the phase-lag of thermal displacement gradient $(0 \leq \tau_\upsilon < \tau_T < \tau_q)$. For $k^{*T} = 0$, Eq. (1.23) reduces to the DPL constitutive equation given in Eq. (1.18). Furthermore, the TPL heat conduction theory is reduced to the C-V theory by assuming $k^{*T} = \tau_T = 0$ in Eq. (1.23). Finally, the TPL theory reduces to the conventional Fourier heat conduction by setting $k^{*T} = \tau_q = \tau_T = 0$. In the absence of phase lags τ_q, τ_T, and τ_υ, Eq. (1.23) reduces to the one considered by Green and Naghdi, where the damped thermal wave behavior is admitted (G-N type III):

$$q(x,t) = -\left[k^T \nabla T(x,t) + k^{*T} \nabla \upsilon(x,t)\right] \tag{1.24}$$

For $k^T \ll k^{*T}$, Eq. (1.24) leads to the G-N model of type II without energy dissipation. Several types of Taylor series expansion could be considered to develop the TPL heat conduction equations. The second-order Taylor series expansion for τ_q and the first-order Taylor series expansion for τ_T and τ_υ in Eq. (1.23) result in:

$$\begin{aligned} &\left(1 + \tau_q \frac{\partial}{\partial t} + \frac{\tau_q^2}{2} \frac{\partial^2}{\partial t^2}\right) q(x,t) \\ &= -\left\{\left(k^T + k^{*T}\tau_\upsilon\right) \nabla T(x,t) + k^T \tau_T \frac{\partial}{\partial t}[\nabla T(x,t)] + k^{*T} \nabla \upsilon(x,t)\right\} \end{aligned} \tag{1.25}$$

The TPL heat conduction differential equation for isotropic materials with temperature-independent material properties is obtained by taking the divergence and time derivative of Eq. (1.25) and using $\upsilon = T$, taking the time derivative of energy Eq. (1.6), and then eliminating $q(x,t)$ between the resulting equations:

$$\begin{aligned} &\left[\frac{k^{*T}}{k^T} + \left(1 + \frac{k^{*T}\tau_\upsilon}{k^T}\right) \frac{\partial}{\partial t} + \tau_T \frac{\partial^2}{\partial t^2}\right] \nabla^2 T(x,t) \\ &= \left(1 + \tau_q \frac{\partial}{\partial t} + \frac{\tau_q^2}{2} \frac{\partial^2}{\partial t^2}\right) \left(\frac{1}{\alpha} \frac{\partial^2 T(x,t)}{\partial t^2} - \frac{\rho}{k^T} \frac{\partial r_T(x,t)}{\partial t}\right) \end{aligned} \tag{1.26}$$

The TPL heat conduction Eq. (1.26) is a hyperbolic partial differential equation. As a result, this type of TPL is named wave-like TPL or hyperbolic-type TPL model. Thermal wave behavior in the TPL heat conduction equation could be observed, similar to the DPL model, by isolating the two fourth-order derivatives in Eq. (1.26) [50]:

$$\frac{\partial^2}{\partial t^2}\left(\frac{\partial^2 T(x,t)}{\partial t^2} - \left(C_{TPL}^T\right)^2 \nabla^2 T(x,t)\right) + lower - order\ terms = 0 \tag{1.27}$$

where the finite thermal wave speed of TPL model (C_{TPL}^T) is:

$$C_{TPL}^T = C_{DPL}^T = \frac{\sqrt{2\alpha\tau_T}}{\tau_q} = C_{C-V}^T\sqrt{\frac{2\tau_T}{\tau_q}} \tag{1.28}$$

Equation (1.28) reveals that the DPL and TPL predict the same thermal wave speed; however, the characteristics of transient thermal responses of DPL and TPL are dissimilar due to the difference between the heat conduction equation of DPL [Eq. (1.20)] and TPL [Eq. (1.26)] [50]. The other types of TPL heat conduction can also be developed by Taylor series expansion of Eq. (1.23). For instance, the first-order or second-order Taylor series expansion of all the phase-lags of τ_q, τ_T, and τ_υ leads to a parabolic differential equation for heat conduction which does not show the wave-like behavior for thermal response. These types of TPL models are called diffusive-like TPL or parabolic-type TPL models. Furthermore, another type of wave-like TPL model could also be developed by second-order Taylor series expansion of τ_q and τ_υ and the first-order Taylor series expansion of τ_T which predicts a higher thermal wave speed for TPL compared to Eq. (1.28):

$$C_{TPL}^T = \frac{\sqrt{2\alpha\tau_T + \alpha^*\tau_\upsilon^2}}{\tau_q} \tag{1.29}$$

where $\alpha^* = \frac{k^{*T}}{\rho c_p}$.

1.2.2.4 Fractional Phase-Lag Heat Conduction

Another type of TPL model can be developed by taking the Taylor series expansion of time-fractional order α_F on the both sides of Eq. (1.23) [47]. The fractional calculus differentiation and integration of arbitrary order has been employed to modify the existing formulation for physical process in chemistry, biology, electronics and signal processing, wave propagation, viscoelasticity, and chaos/fractals. Ezzat et al. [47, 54] retained terms up to $2\alpha_F$-order for τ_q and up to α_F-order for τ_T and τ_υ in fractional Taylor series expansion of Eq. (1.23) as follows:

$$\left(1+\frac{\tau_q^{\alpha_F}}{\alpha_F!}\frac{\partial^{\alpha_F}}{\partial t^{\alpha_F}}+\frac{\tau_q^{2\alpha_F}}{(2\alpha_F)!}\frac{\partial^{2\alpha_F}}{\partial t^{2\alpha_F}}\right)q(x,t)$$
$$=-\left\{\left(k^T+\frac{k^{*T}\tau_\upsilon^{\alpha_F}}{\alpha_F!}\frac{\partial^{\alpha_F-1}}{\partial t^{\alpha_F-1}}\right)\nabla T(x,t)\right. \tag{1.30}$$
$$\left.+\frac{k^T\tau_T^{\alpha_F}}{\alpha_F!}\frac{\partial^{\alpha_F}}{\partial t^{\alpha_F}}[\nabla T(x,t)]+k^{*T}\nabla\upsilon(x,t)\right\}\quad 0\leq\alpha_F<1$$

where the fractional derivative is defined as [54, 55]:

$$\frac{\partial^{\alpha_F}}{\partial t^{\alpha_F}}f(x,t)=\begin{cases}f(x,t)-f(x,0) & \alpha_F\to 0\\ I^{\alpha_F-1}\frac{\partial f(x,t)}{\partial t} & 0<\alpha_F<1\\ \frac{\partial f(x,t)}{\partial t} & \alpha_F=1\end{cases} \tag{1.31}$$

where the Riemann–Liouville fractional integral I^{α_F} could be written in a convolution-type form as:

$$I^{\alpha_F}f(x,t)=\int_0^t\frac{(t-\zeta)^{\alpha_F-1}}{\Gamma(\alpha_F)}f(x,\zeta)d\zeta \tag{1.32}$$
$$I^0f(x,t)=f(x,t)$$

where $\Gamma(\ldots)$ represents the Gamma function. Eliminating heat flux $q(x,t)$ between Eqs. (1.6) and (1.30) provides the following fractional TPL heat conduction equation:

$$\left[\frac{k^{*T}}{k^T}+\left(1+\frac{k^{*T}\tau_\upsilon^{\alpha_F}}{k^T\alpha_F!}\right)\frac{\partial^{\alpha_F}}{\partial t^{\alpha_F}}+\frac{\tau_T^{\alpha_F}}{\alpha_F!}\frac{\partial^{\alpha_F+1}}{\partial t^{\alpha_F+1}}\right]\nabla^2T(x,t)$$
$$=\left(1+\frac{\tau_q^{\alpha_F}}{\alpha_F!}\frac{\partial^{\alpha_F}}{\partial t^{\alpha_F}}+\frac{\tau_q^{\alpha_F}}{(2\alpha_F)!}\frac{\partial^{2\alpha_F}}{\partial t^{2\alpha_F}}\right)\left(\frac{1}{\alpha}\frac{\partial^2T(x,t)}{\partial t^2}-\frac{\rho}{k^T}\frac{\partial r_T(x,t)}{\partial t}\right) \tag{1.33}$$

For $\alpha_F=1$, Eq. (1.33) reduces to the TPL heat conduction given in Eq. (1.26). Other types of fractional TPL could be developed by considering different fractional order Taylor series expansion, for example, up to $2\alpha_F$-order for phase lags of τ_q, τ_T, and τ_υ. Another form of fractional TPL model could also be developed by the generalization of the fractional C-V heat conduction, proposed by Youssef [56], as follows:

$$q(x,t+\tau_q)=-\left[k^TI^{\alpha_F-1}\nabla T(x,t+\tau_T)+k^{*T}I^{\alpha_F-1}\nabla\upsilon(x,t+\tau_\upsilon)\right]\quad 0<\alpha_F\leq 2 \tag{1.34}$$

The associated fractional TPL heat conduction could be derived by Eqs. (1.6) and (1.34) as:

$$\begin{aligned} &\left[\frac{k^{*T}}{k^T} I^{\alpha_F-1}\left(1+\tau_\upsilon\frac{\partial}{\partial t}\right)+I^{\alpha_F-1}\left(1+\tau_T\frac{\partial}{\partial t}\right)\frac{\partial}{\partial t}\right]\nabla^2 T(x,t) \\ &= \left(1+\tau_q\frac{\partial}{\partial t}+\frac{\tau_q^2}{2}\frac{\partial^2}{\partial t^2}\right)\left(\frac{1}{\alpha}\frac{\partial^2 T(x,t)}{\partial t^2}-\frac{\rho}{k^T}\frac{\partial r_T(x,t)}{\partial t}\right) \end{aligned} \tag{1.35}$$

Implied by Kimmich [57] and Youssef [56], Eq. (1.35) could be called as a generalized anomalous heat conduction which describers various heat transport phenomena in media with weak conductivity $(0<\alpha_F<1)$, normal conductivity $(\alpha_F = 1)$, and superconductivity $(1<\alpha_F\leq 2)$. Subdiffusive, e.g. dielectrics and semiconductors, and superconductive, e.g. porous glasses and polymer chains, media are examples of materials that could exhibit anomalous diffusion and heat conduction [51–53, 58].

1.2.2.5 Nonlocal Phase-Lag Heat Conduction

To accommodate the effect of thermomass, the distinctive mass of heat, of dielectric lattices in the heat conduction, Tzou [59, 60] has also included the nonlocal behavior, in space, in addition to the thermal lagging, in time. The nonlocal (NL) TPL constitutive equation could be derived as an expansion of the nonlocal DPL model proposed by Tzou in the following form:

$$q(x+\lambda_q, t+\tau_q) = -\left[k^T\nabla T(x+\lambda_T, t+\tau_T)+k^{*T}\nabla\upsilon(x+\lambda_\upsilon, t+\tau_\upsilon)\right] \tag{1.36}$$

where λ_q, λ_T, and λ_υ are correlating nonlocal lengths of heat flux, temperature gradient, and thermal displacement gradient in the heat transport constitutive equation [60]. The Taylor series expansion of Eq. (1.36) with respect to either nonlocal lengths and/or phase-lags could lead to a number of local or nonlocal heat conduction constitutive equations. However, only the constitutive equations with coordinate independent property must be considered. Eliminating the heat flux $q(x,t)$ between Eqs. (1.6) and (1.35) provides the nonlocal TPL heat conduction equation. Since the nonlocal behavior in the heat flux is only related to existing microscale heat transfer models, the nonlocal length λ_q is involved hereafter. Recently, Akbarzadeh et al. [61] has also introduced a heat conduction model, called nonlocal fractional three-phase-lag (NL FTPL), to take into account the size-dependency of thermophysical properties, subdiffusion or superdiffusion of heat transport, and phonon-electron interaction in ultrafast heat transport.

$$\begin{aligned} q(x+\lambda_q, t+\tau_q) = &-\left[k^T I^{\alpha_F-1}\nabla T(x, t+\tau_T)\right. \\ &\left.+k^{*T} I^{\alpha_F-1}\nabla\upsilon(x, t+\tau_\upsilon)\right] \quad 0<\alpha_F\leq 2 \end{aligned} \tag{1.37}$$

Among the abovementioned non-Fourier heat conduction theories, C-V, DPL, NL C-V, TPL, and fractional TPL models are considered in this book.

It is worth mentioning that the application of interfacial thermal boundary conditions [Eq. (1.13)] is straightforward for Fourier heat conduction. However, mathematical implementations are required to correctly apply the interfacial and boundary conditions for non-Fourier models. For instance, the interfacial conditions should be applied in the Laplace domain for local non-Fourier models and/or we need to assume $\left.\frac{\partial q(x,t)}{\partial x}\right|_{x=S} = 0$, due to the continuity of heat flux in the interface, for nonlocal non-Fourier models.

1.3 Moisture Diffusion

Moisture, similar to temperature, could induce significant strains and stresses within a solid material. Moisture and temperature could also cause the reduction of strength and stiffness. For instance, the failure mode caused by moisture diffusion during the manufacturing and operation of fiber-reinforced composites and plastic encapsulated microcircuits, is a major reliability concern [1, 62, 63]. As a result, the determination of moisture distribution within solids could be as significant as temperature for design of composite structures.

Following the heat conduction study by Fourier [64], Fick [65] recognized that the diffusion of moisture in solids is similar to the heat conduction process. Therefore, the rate of moisture flux p is related to the moisture concentration gradient ∇m with the constitutive equation of moisture flux. The moisture flux, p [kg/m^{2}s], is the moisture transfer per unit time and per unit normal vector of the area, while moisture concentration, m [kg/m^3], is defined as the mass of moisture per unit volume of the dry solid [66, 67]. The moisture diffusion equation is obtained by employing the constitutive equation of moisture flux and the conservation equation of mass of moisture. In view of the similarity between the moisture diffusion and heat conduction, the Fickian and non-Fickian moisture diffusion equations are briefly reviewed in this Section.

1.3.1 *Fickian Moisture Diffusion*

The Fickian moisture diffusion equation, similar to Fourier heat conduction, assumes the proportionality between the moisture flux and moisture concentration as [1]:

$$p(x,t) = -k^m \nabla m(x,t) \tag{1.38}$$

where k^m [m^2/s] is the moisture diffusion coefficient, which is a positive scalar quantity or a second-order tensor for isotropic or anisotropic materials, respectively.

The negative sign in Eq. (1.38) implies that the moisture diffuses in the direction of the decreasing moisture concentration. Furthermore, the moisture coefficient has been shown to be dependent on temperature as [1]:

$$k^m = k_0^m \exp\left(-\frac{E_m}{R_g T}\right) \tag{1.39}$$

where k_0^m is a constant, E_m and R_g are, respectively, the activation energy and gas constant, and T is temperature in Kelvin. Some typical values of k_0^m and E_m have already been reported in [1, 68] for composite laminates. It has also been observed that moisture content has negligible effect on moisture diffusion coefficient k^m [69]; however, this observation may not hold for polymers, molding compounds, and organic substrates [63, 70, 71].

Furthermore, the conservation law for the mass of moisture, equivalent to energy conservation for temperature, is given by [66, 67]:

$$-\nabla.p(x,t) + \rho r_m(x,t) = \frac{\partial m(x,t)}{\partial t} \tag{1.40}$$

where r_m is the moisture source per unit time per unit mass of dry solid. The differential equation of transient Fickian moisture diffusion is derived by substituting Eq. (1.38) into Eq. (1.40) as follows:

$$\nabla.(k^m \nabla m(x,t)) + \rho r_m(x,t) = \frac{\partial m(x,t)}{\partial t} \tag{1.41}$$

For a homogenous isotropic material with moisture independent moisture diffusivity, Eq. (1.7) is further simplified as:

$$\nabla^2 m(x,t) + \frac{\rho r_m(x,t)}{k^m} = \frac{1}{k^m}\frac{\partial m(x,t)}{\partial t} \tag{1.42}$$

For steady-state moisture diffusion analysis, the right-hand side of Eqs. (1.41) and (1.42) are set zero and the moisture generation rate r_m is overlooked. Appropriate boundary and initial conditions for moisture or hygroscopic field should be specified for the moisture diffusion differential equation to a unique solution for the system of differential equations.

Similar to thermal boundary conditions, hygroscopic boundary conditions could be:

(a) Prescribed moisture concentration along the boundary surface (S):

$$m(x,t)|_{x=S} = \overline{m}(S,t) \tag{1.43}$$

where $\overline{m}$ is a prescribed function of position and time, or:

(b) Prescribed moisture flux across the boundary surface:

$$k^m \nabla m(x,t)|_{x=S} = \pm p_S(S,t) \tag{1.44}$$

where p_S is the prescribed moisture flux. Hygroscopic insulation boundary conditions could also be reached by substituting $p_S = 0$ in Eq. (1.44).

Due to the time derivatives of moisture concentration in the differential equation of moisture diffusion, initial conditions should also be specified. For the Fickian moisture diffusion, the initial condition for moisture concentration could be specified as:

$$m(x, t = 0) = m_0(x) \tag{1.45}$$

where $m_0(x)$ is a specified function of the spatial coordinate x.

When a multilayered medium is considered, hygroscopic interfacial conditions need to also be specified. The following hygroscopic interfacial conditions could be considered at the interface of two layers with a hygroscopic weak moisture diffusion [72]:

$$\left(m^{(1)}(x,t) - m^{(1)}(x,t)\right)\Big|_{x=S} = R^m p^{(1)}(S,t) \tag{1.46a}$$

$$p^{(1)}(S,t) = p^{(2)}(S,t) \tag{1.46b}$$

where R^m is the hygroscopic compliance constant or hygroscopic contact resistance for the imperfect interface. Moreover, $R^m = 0$, in Eq. (1.46), represents the perfectly bonded interfaces.

1.3.2 Non-Fickian Moisture Diffusion

Fickian moisture diffusion provides a good approximation for most of the engineering applications [73, 74]. However, it is well known that Fickian moisture diffusion exhibits an unrealistic infinite speed for propagation of mass of moisture due to the parabolic-type partial differential equation of Fickian moisture diffusion [Eq. (1.41)]. Moreover, the validity of traditional Fickian moisture diffusion equation breaks down for short-time inertial motion of mass and very high frequency of mass flux density [73]. Thus, non-Fickian moisture diffusion theories with a hyperbolic diffusion equation, which describe the moisture diffusion with a finite speed, have been developed. Non-Fickian moisture diffusion equations have recently found many practical application in superionic conductors, molten salts, laser drying, laser melting, and rapid solidification [75].

The non-Fickian moisture diffusion theories in the absence of a potential field are the same as non-Fourier heat conduction theories. Nonetheless, the potential field, which does not appear in non-Fourier heat conduction, plays an important role in non-Fickian moisture diffusion. One of the non-Fickian moisture diffusion theories was developed by Das [75] to accommodate the assumption of a local Maxwellian equilibrium. The constitutive non-Fourier moisture flux could be written as [73, 75]:

$$\left(1+\tau_p\frac{\partial}{\partial t}\right)p(x,t) = -k^m\nabla m(x,t) - \frac{\tau_p\nabla V(x)}{p_m}m(x,t) \tag{1.47}$$

where τ_p [s] is the relaxation time of the mass flux, p_m [kg] is the particle mass, and V is potential field. In the absence of $V(x)$, Eq. (1.47) is equivalent to the C-V model of non-Fourier heat conduction. The non-Fickian moisture diffusion equation is derived by omitting the moisture flux $p(x,t)$ between Eqs. (1.40) and (1.47). For a homogenous isotropic material with moisture independent moisture diffusivity, the non-Fickian moisture diffusion equation is written as:

$$\nabla^2 m(x,t) + \frac{\tau_p}{k^m p_m}\nabla.(\nabla V(x)m(x,t)) = \left(1+\tau_p\frac{\partial}{\partial t}\right)\left(\frac{1}{k^m}\frac{\partial m(x,t)}{\partial t} - \frac{\rho r_m(x,t)}{k^m}\right) \tag{1.48}$$

Equation (1.48) is a hyperbolic-type differential equation and reveals a wave-like behaviour for the propagation of moisture concentration. The finite hygroscopic wave speed $\left(C_D^m\right)$ predicted by Das model in Eq. (1.48) is obtained as:

$$C_D^m = \sqrt{\frac{k^m}{\tau_p}} \tag{1.49}$$

If the relaxation time of the mass flux τ_p and potential field $V(x)$ are neglected, Eq. (1.48) reduces to the Fickian moisture diffusion and C_D^m reaches infinity. Moreover, another effort for developing a non-Fickian moisture diffusion was made by Akbarzadeh [25] where the following dual-phase lag moisture diffusion equation was proposed analogous to the DPL heat conduction:

$$\left(1+\tau_p\frac{\partial}{\partial t}+\tau_p^2\frac{\partial^2}{\partial t^2}\right)p(x,t) = -k^m\left(1+\tau_m\frac{\partial}{\partial t}\right)\nabla m(x,t) - \frac{\tau_p\nabla V(x)}{p_m}m(x,t) \tag{1.50}$$

where τ_p and τ_m are the phase lags of moisture flux (the relaxation time of the mass flux) and moisture gradient, respectively. Although some of the non-Fickian moisture diffusion theories have been introduced in this section, the non-Fourier heat conduction has received more attention in the literature compared to non-Fickian moisture diffusion. As a result, we only focus in this book on the influence of Fourier/non-Fourier heat conduction and Fickian moisture diffusion on structural responses of smart materials.

References

1. Sih GC, Michopoulos J, Chou S-C (1986) Hygrothermoelasticity. Springer, Berlin
2. Brischetto S (2013) Hygrothermoelastic analysis of multilayered composite and sandwich shells. J Sandwich Struct Mater 15(2):168–202
3. Carslaw HS, Jaeger JC (1959) Conduction of heat in solids. Oxord at the Clarendon Press, Great Britain
4. Ozisik MN (1980) Heat conduction. Wiley, New York
5. Jiji LM (2009) Heat conduction. Springer, New York
6. Wang L, Zhou X, Wei X (2008) Heat conduction: mathematical models and analytical solutions. Springer, Berlin, Heidelberg
7. Lienhard JH (2008) A heat transfer text book. Phlogiston Press, Cambridge, MA
8. Wang L (1994) Generalized Fourier law. Int J Heat Mass Transf 37(17):2627–2634
9. Cengel YA, Boles MA (2006) Thermodynamics: an engineering approach, 5th edn. McGraw-Hill, Boston
10. Hetnarski RB, Eslami MR (2008) Thermal stresses-advanced theory and applications: advanced theory and applications, vol 158. Springer, Berlin
11. Parkus H (1976) Thermoelasticity. Springer, New York
12. Bejan A (1988) Advanced engineering thermodynamics. Wiley, New York
13. Clausius R (1865) On different forms of the fundamental equations of the mechanical theory of heat and their convenience for application. Physics (Leipzig) 125:313
14. Rankine WJM (1869) On the thermal energy of molecular vortices. Trans R Soc Edinburgh 25:557–566
15. Tian Z, Lee S, Chen G (2014) A comprehensive review of heat transfer in thermoelectric materials and devices. Ann Rev Heat Transf 17:425–483
16. Tzou DY (1996) Macro-to micro-scale heat transfer: the lagging behavior. CRC Press, Boca Raton
17. Eslami MR et al (2013) Theory of elasticity and thermal stresses. Springer, Berlin
18. Theodore L (2011) Heat transfer application for the practicing engineer. Wiley, Hoboken, NJ
19. Corral M (2008) Vector calculus. Michael Corral
20. Arpaci VS (1966) Conduction heat transfer. Addison-Wesley, Reading
21. Baher HD, Stephan K (2006) Heat and mass transfer, 2nd edn. Springer, Berlin
22. Akbarzadeh AH, Pasini D (2014) Phase-lag heat conduction in multilayered cellular media with imperfect bonds. Int J Heat Mass Transf
23. Chen T (2001) Thermal conduction of a circular inclusion with variable interface parameter. Int J Solids Struct 38(17):3081–3097
24. Chandrasekharaiah D (1998) Hyperbolic thermoelasticity: a review of recent literature. Appl Mech Rev 51(12):705–729
25. Akbarzadeh AH (2013) Multiphysical behaviour of functionally graded smart structures. Department of Mechanical Engineering, University of New Brunswick, Fredericton, NB, Canada
26. Antaki P (1995) Key features of analytical solutions for hyperbolic heat conduction
27. Babaei MH (2009) Multiphysics analysis of functionally graded piezoelectrics. University of New Brunswick, Canada
28. Cattaneo C (2011) Sulla conduzione del calore. In: Some aspects of diffusion theory. Springer, Berlin, pp 485–485
29. Vernotte P (1958) Les paradoxes de la theorie continue de l'equation de la chaleur. Comptes rendus de l'academie bulgare des sciences: sciences mathematiques et naturelles 246:3154–3155
30. Zhou L et al (2008) Electrical-thermal switching effect in high-density polyethylene/graphite nanosheets conducting composites. J Mater Sci 43(14):4886–4891
31. Chester M (1963) Second sound in solids. Phys Rev 131:2013–2015

32. Godoy S, García-Colín L (1997) Nonvalidity of the telegrapher's diffusion equation in two and three dimensions for crystalline solids. Phys Rev E 55(3):2127
33. Körner C, Bergmann H (1998) The physical defects of the hyperbolic heat conduction equation. Appl Phys A 67(4):397–401
34. Xu F, Seffen K, Lu T (2008) Non-Fourier analysis of skin biothermomechanics. Int J Heat Mass Transf 51(9):2237–2259
35. Tzou DY (1995) The generalized lagging response in small-scale and high-rate heating. Int J Heat Mass Transf 38(17):3231–3240
36. Tzou D (1995) A unified field approach for heat conduction from macro-to micro-scales. J Heat Transfer 117(1):8–16
37. Tzou DY (1995) Experimental support for the lagging behavior in heat propagation. J Thermophys Heat Transfer 9(4):686–693
38. Akbarzadeh A, Chen Z (2012) Heat conduction in one-dimensional functionally graded media based on the dual-phase-lag theory. Proc Inst Mech Eng, Part C: J Mech Eng Sci 227(4):744–759
39. Qiu T, Tien C (1992) Short-pulse laser heating on metals. Int J Heat Mass Transf 35(3):719–726
40. Qiu T, Tien C (1993) Heat transfer mechanisms during short-pulse laser heating of metals. J Heat Transfer 115(4):835–841
41. Akbarzadeh A, Pasini D (2014) Phase-lag heat conduction in multilayered cellular media with imperfect bonds. Int J Heat Mass Transf 75:656–667
42. Tzou DY, Dai W (2009) Thermal lagging in multi-carrier systems. Int J Heat Mass Transf 52 (5):1206–1213
43. Lord HW, Shulman Y (1967) A generalized dynamical theory of thermoelasticity. J Mech Phys Solids 15(5):299–309
44. Green A, Lindsay K (1972) Thermoelasticity. J Elast 2(1):1–7
45. Green A, Naghdi P (1993) Thermoelasticity without energy dissipation. J Elast 31(3):189–208
46. Choudhuri SR (2007) On a thermoelastic three-phase-lag model. J Therm Stresses 30(3):231–238
47. Ezzat MA, El-Bary AA, Fayik MA (2013) Fractional fourier law with three-phase lag of thermoelasticity. Mech Adv Mater Struct 20(8):593–602
48. Miranville A, Quintanilla R (2011) A phase-field model based on a three-phase-lag heat conduction. Appl Math Optim 63(1):133–150
49. Green A, Naghdi P (1885) A re-examination of the basic postulates of thermomechanics. Proc R Soc London. Ser A: Math Phys Sci 432:171–194
50. Akbarzadeh A, Fu J, Chen Z (2014) Three-phase-lag heat conduction in a functionally graded hollow cylinder. Trans Can Soc Mech. Eng 38(1):155–171
51. Dzieliński A, Sierociuk D, Sarwas G (2010) Some applications of fractional order calculus. Bull Polish Acad Sci: Tech Sci 58(4):583–592
52. Podlubny I (1998) Fractional differential equations: an introduction to fractional derivatives, fractional differential equations, to methods of their solution and some of their applications, vol 198. Academic press, New York
53. Stiassnie M (1979) On the application of fractional calculus for the formulation of viscoelastic models. Appl Math Model 3(4):300–302
54. Ezzat MA, El Karamany AS, Fayik MA (2012) Fractional order theory in thermoelastic solid with three-phase lag heat transfer. Arch Appl Mech 82(4):557–572
55. Jumarie G (2010) Derivation and solutions of some fractional Black–Scholes equations in coarse-grained space and time. Application to Merton's optimal portfolio. Comput Math Appl 59(3):1142–1164
56. Youssef HM (2010) Theory of fractional order generalized thermoelasticity. J Heat Transf 132 (6):061301
57. Kimmich R (2002) Strange kinetics, porous media, and NMR. Chem Phys 284(1):253–285

58. Povstenko YZ (2004) Fractional heat conduction equation and associated thermal stress. J Therm Stresses 28(1):83–102
59. Tzou D (2011) Nonlocal behavior in phonon transport. Int J Heat Mass Transf 54(1):475–481
60. Tzou D, Guo Z-Y (2010) Nonlocal behavior in thermal lagging. Int J Therm Sci 49(7): 1133–1137
61. Akbarzadeh AH, Cui Y, Chen ZT (2015) Non-local fractional phase-lag heat conduction: continuum to molecular dynamic simulation. Appl Phys Lett
62. Reddy JN (2004) Mechanics of laminated composite plates and shells: theory and analysis. CRC Press, New York
63. Shirangi M, Michel B (2010) Mechanism of moisture diffusion, hygroscopic swelling, and adhesion degradation in epoxy molding compounds, in moisture sensitivity of plastic packages of IC devices. Springer, Berlin, pp 29–69
64. Fourier J (1822) Theorie analytique de la chaleur, par M. Fourier. Chez Firmin Didot, père et fils
65. Fick A (1855) Ueber diffusion. Ann Phys 170(1):59–86
66. Smittakorn W (2001) A theoretical and experimental study of adaptive wood composites. Colorado State University
67. Heyliger WSPR (2000) A discrete-layer model of laminated hygrothermopiezoelectric plates. Mech Compos Mater Struct 7(1):79–104
68. Shen C-H, Springer GS (1976) Moisture absorption and desorption of composite materials. J Compos Mater 10(1):2–20
69. Augl J, Trabocco R (1975) Environmental degradation studies on carbon fiber reinforced epoxies. In: Workshop on durability of composite materials (held on September)
70. Chen X, Zhao S, Zhai L (2005) Moisture absorption and diffusion characterization of molding compound. J Electron Packag 127(4):460–465
71. Celik E, Guven I, Madenci E (2009) Experimental and numerical characterization of non-Fickian moisture diffusion in electronic packages. IEEE Trans Adv Packag 32(3): 666–674
72. Akbarzadeh A, Pasini D (2014) Multiphysics of multilayered and functionally graded cylinders under prescribed hygrothermomagnetoelectromechanical loading. J Appl Mech 81 (4):041018
73. Chen H-T, Liu K-C (2001) Numerical analysis of non-Fickian diffusion problems in a potential field. Numer Heat Transf: Part B: Fundam 40(3):265–282
74. Ding S, Petuskey WT (1998) Solutions to Fick's second law of diffusion with a sinusoidal excitation. Solid State Ionics 109(1):101–110
75. Das AK (1991) A non-Fickian diffusion equation. J Appl Phys 70(3):1355–1358

Chapter 2
Basic Problems of Non-Fourier Heat Conduction

2.1 Introduction

In this chapter, the non-Fourier heat conduction equations along with the boundary and initial conditions are solved for one-dimensional (1D) media with semi-infinite or finite dimensions in Cartesian, cylindrical, and spherical coordinate systems. In particular, semi-analytical solutions for C-V, DPL, TPL, and nonlocal C-V heat conduction models are provided. The influence of non-Fourier heat conduction theories on thermal responses of heterogeneous multilayered/functionally graded solid/cellular materials is also presented. Since applying Laplace transform to transient problems is a well-known methodology for dealing with the time-dependency of solutions, Laplace transform and numerical Laplace inversion techniques are first introduced in Sect. 2.2. Non-Fourier heat conduction in a homogeneous semi-infinite medium is considered in Sect. 2.3. Then non-Fourier heat conduction problem in finite homogenous and functionally graded media is studied in Sect. 2.4, while Sect. 2.5 studies the non-Fouirer heat conduction in multilayered media and disucss its application in porous or cellular media. Finally, a set of solutions is given in Sect. 2.6 for non-Fourier heat conduction problems in 1D media with finite dimensions.

2.2 Laplace Transform and Laplace Inversion

Laplace transform along with the method of separation of variables are frequently used to solve the time-dependent differential equations. Particularly, Laplace transform has been used in several studies on non-Fourier heat conduction. The Laplace transform of a function $f(x, t)$, denoted by $L\{f(x, t)\}$ or $f(x, s)$, is defined as:

$$f(x,t) = L^{-1}\{f(x,s)\} = \frac{1}{2\pi i}\int_{a-i\infty}^{a+i\infty} f(x,s)e^{st}\,ds \qquad (2.1)$$

Z. T. Chen and A. H. Akbarzadeh, *Advanced Thermal Stress Analysis of Smart Materials and Structures*, Structural Integrity 10,
https://doi.org/10.1007/978-3-030-25201-4_2

where s represents the Laplace variable. The time-dependent function $f(x,t)$, defined for $t \geq 0$, should be a piecewise continuous and of exponential order. A function $f(x,t)$ is said to be of exponential order if there exist constants c, $0 \leq M$, $0 \leq T$, such that:

$$|f(x,t)| \leq Me^{ct} \, for \, T \leq t \tag{2.2}$$

The inverse Laplace transform of a function $f(x,s)$, denoted by $L^{-1}\{f(x,t)\}$, is defined by the Bromwich integral formula as:

$$f(x,t) = L^{-1}\{f(x,s)\} = \frac{1}{2\pi i}\int_{a-i\infty}^{a+i\infty} f(x,s)e^{st}\, ds \tag{2.3}$$

where a is an arbitrary real number larger than all real parts of singularities of $f(x,s)$ and $i = \sqrt{-1}$ is the imaginary unit. Using Cauchy's residue theorem, the contour integration Eq. (2.3) could be analytically reduced to:

$$\frac{1}{2\pi i}\int_{a-i\infty}^{a+i\infty} f(x,s)e^{st}\, ds = \sum_{j=1} r_j \tag{2.4}$$

where r_j are the residues of $f(x,s)e^{st}$ at the singularities of $f(x,s)$. However, this analytical methodology could be cumbersome for complicated $f(x,s)$. As a result, numerical Laplace inversion techniques are commonly employed. Among various numerical approaches for inverse Laplace transform, the three major techniques using fast Fourier transform, Jacobi polynomial, and Reimann sum approximation are briefly reviewed here.

2.2.1 *Fast Laplace Inverse Transform*

Durbin used the fast Fourier transform (FFT) to speed up the computation time for Laplace inversion. According to the fast Laplace inverse transform (FLIT), proposed by Durbin, the Laplace inversion of $f(x,s)$ at time t_k is obtained by:

$$f(x,t_j) \approx C(j)\left[-\frac{1}{2}Re\{f(x,a)\} + Re\left\{\sum_{k=0}^{N-1}(A(x,k)+iB(x,k))W^{jk}\right\}\right] \tag{2.5}$$

where:

$$A(x,k) = \sum_{l=0}^{L} Re\left(f\left(x, a + i(k+lN)\frac{2\pi}{T_{period}}\right)\right),$$
$$B(x,k) = \sum_{l=0}^{L} Im\left(f\left(x, a + i(k+lN)\frac{2\pi}{T_{period}}\right)\right) \tag{2.6}$$
$$C(j) = \frac{2}{T_{period}} e^{aj\Delta t}, \Delta t = \frac{T_{period}}{N}, W = e^{i\frac{2\pi}{N}}, i = \sqrt{-1}$$

where T_{period} is the time period for performing the Laplace inversion and Δt stands for the time increment; *Re* and *Im* represent the real and imaginary parts of their arguments, respectively. Moreover, a is an arbitrary real number larger than all the real parts of the singularities present in the function $f(x,s)$ and L and N are two arbitrary parameters that affect the accuracy of the solutions. To minimize the numerical discretization and truncation errors, it is recommended to consider the following constraints for the arbitrary parameters:

$$5 \leq aT_{period} \leq 10, 50 \leq NL \leq 5000 \tag{2.7}$$

2.2.2 Reimann Sum Approximation

The Laplace inversion of $f(x,s)$ at time t could also be obtained by Reimann sum approximation of the Fourier integral transformed from Laplace inversion integral as:

$$f(x,t) = \frac{e^{\gamma t}}{t}\left[\frac{1}{2}F(x, s=\gamma) + Re\sum_{n=1}^{N} F\left(x, s = \gamma + \frac{in\pi}{t}\right)(-1)^n\right] \tag{2.8}$$

where γ and N are two real constants. The value of γ and truncation error determined by N dictate the accuracy of the Reimann-sum approximation. Furthermore, γ should satisfy the following relation to achieve a faster convergence:

$$\gamma t \cong 4.72 \tag{2.14}$$

As employed by Tzou for non-Fourier heat conduction analysis, the summation in Eq. (2.8) could be performed till the Cauchy norm is smaller than 10^{-15}.

2.2.3 Laplace Inversion by Jacobi Polynomial

Jacobi polynomial could also be employed to obtain the Laplace inversion of $f(x,s)$ as follows:

$$f(x,t) = \sum_{n=0}^{N} C_n P_n^{(0,\beta)}\left[2e^{-\delta t} - 1\right] \quad (2.10)$$

where δ is a real positive number and $\beta > -1$. Furthermore, $P_n^{(\alpha,\beta)}$ represents the Jacobi polynomial of degree n, defined as:

$$P_n^{(\alpha,\beta)}(z) = (1-z)^{-\alpha}(1+z)^{-\beta}\frac{(-1)^n}{2^n n!}\frac{d^n}{dz^n}\left[(1-z)^{n+\alpha}(1+z)^{n+\beta}\right] \quad (2.11a)$$

or:

$$P_n^{(\alpha,\beta)}(x) = \frac{\Gamma(\alpha+n+1)}{n!\Gamma(\alpha+\beta+n+1)}\sum_{k=0}^{n}\binom{n}{k}\frac{\Gamma(\alpha+\beta+n+k+1)}{\Gamma(\alpha+k+1)}\left(\frac{x-1}{2}\right)^k \quad (2.11b)$$

The unknown coefficients C_n in Eq. (2.10) are obtained by the following recurrence relation:

$$\delta f(x,(\beta+1+k)\delta) = \sum_{m=0}^{k}\frac{k(k-1)\cdots(k-(m-1))}{(k+\beta+1)(k+\beta+2)\cdots(k+\beta+1+m)}C_m \quad (2.12)$$

For an accurate approximation of Laplace inversion, it is recommended to choose the β and δ as:

$$-0.5 \le \beta \le 5 \quad (2.13a)$$

$$0.05 \le \delta \le 2 \quad (2.13b)$$

2.3 Non-Fourier Heat Conduction in a Semi-infinite Strip

To consider the effect of non-Fourier heat conduction on thermal responses, we consider a semi-infinite, isotropic, homogeneous 1D medium, as seen in Fig. 2.1. The solid medium is initially ($t = 0$) kept at a constant temperature T_0 and thermally disturbed from a stationary state, $\frac{\partial T(x,t)}{\partial t} = 0$ at $t = 0$. The surface temperature at $x = 0$ is abruptly raised to T_{WL} which leads to the propagation of thermal disturbance through the medium. However, temperature at a distance far from the heated zone keeps its initial value. The semi-infinite medium facilitates the examination of the way temperature distributes and thermal affected zone evolves, with no concern about the thermal wave separation/reflection from boundaries.

To accommodate comparing the heat conduction in a semi-infinite medium using Fourier, C-V, nonlocal C-V, and DPL theories, the heat conduction constitutive equation is written in the following nonlocal DPL form, in accordance with Eq. (1.35):

Fig. 2.1 A semi-infinite, homogeneous 1D medium with a suddenly raised surface temperature T_W

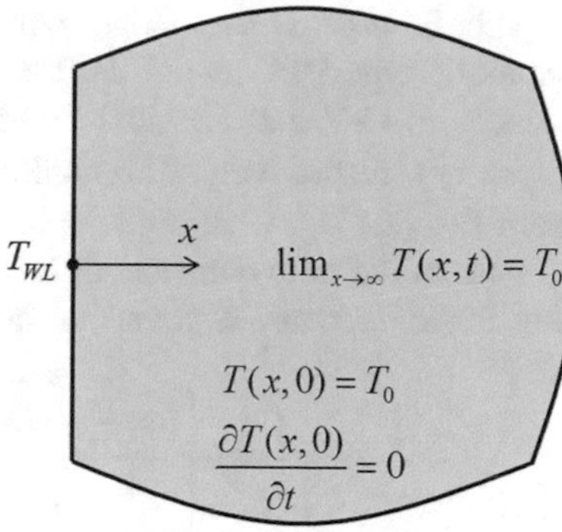

$$\left(1+\lambda_q\frac{\partial}{\partial x}+\tau_q\frac{\partial}{\partial t}+\frac{\tau_q^2}{2}\frac{\partial^2}{\partial t^2}\right)q(x,t) = -k^T\left(1+\tau_T\frac{\partial}{\partial t}\right)\frac{\partial T(x,t)}{\partial x} \tag{2.14}$$

where λ_q is the correlating length parameter for nonlocal analysis. The correlating length is equivalent, by two times, to the length parameters in the thermomass model of heat transfer in dielectric lattices. Specifically, λ_q and τ_q could be correlated to the mean free time and the mean free path in microscale heat transport [1]. Eliminating the heat flux $q(x,t)$ between Eq. (2.14) and 1D form of the energy Eq. (1.6) leads to:

$$\left(1+\tau_T\frac{\partial}{\partial t}\right)\frac{\partial^2 T(x,t)}{\partial x^2} = \left(1+\lambda_q\frac{\partial}{\partial x}+\tau_q\frac{\partial}{\partial t}+\frac{\tau_q^2}{2}\frac{\partial^2}{\partial t^2}\right)\frac{1}{\alpha}\frac{\partial T(x,t)}{\partial t} \tag{2.15}$$

The following non-dimensional parameters are also introduced for analysis:

$$\theta = \frac{T-T_0}{T_w-T_0}, \beta = \frac{t}{\tau_q}, \xi = \frac{x}{\sqrt{\alpha\tau_q}}, Z = \frac{\tau_T}{\tau_q}, L = \frac{\lambda_q}{\sqrt{\alpha\tau_q}} \tag{2.16}$$

The heat conduction Eq. (2.15) and the initial and boundary conditions could be written in terms of the non-dimensional parameters. To identify different heat conduction models, two artificial coefficients A and B are included in the non-dimensional heat conduction equation as follows:

$$\frac{\partial^2\theta(\xi,\beta)}{\partial\xi^2}+Z\frac{\partial^3\theta(\xi,\beta)}{\partial\xi^2\partial\beta} = \frac{\partial\theta(\xi,\beta)}{\partial\beta}+L\frac{\partial^2\theta(\xi,\beta)}{\partial\xi\partial\beta}+A\frac{\partial^2\theta(\xi,\beta)}{\partial\beta^2}+B\frac{\partial^3\theta(\xi,\beta)}{\partial\beta^3} \tag{2.17a}$$

$$\theta(\xi,0) = \frac{\partial\theta(\xi,0)}{\partial\beta} = 0 \qquad (\textit{Initial conditions}) \tag{2.17b}$$

$$\theta(0,\beta) = 1,\ \lim_{\xi\to\infty}\theta(\xi,\beta) = 0 \qquad (\textit{Boundary conditions}) \tag{2.17c}$$

As seen in Eq. (2.17), the thermal responses are characterized by Z, L, A, and B parameters. For the case of $L=0$, $A=1$, $B=\frac{1}{2}$ and $Z\neq 0$, Eq. (2.17) reduces to hyperbolic-type DPL model, while $L = 0$, $A = 1$, $B = 0$, and $Z \neq 0$ lead to parabolic-type DPL model. Furthermore, for $L = 0$, $A = 1$, and $B = 0$, Eq. (2.17) reduces to C-V and classical Fourier models when Z is set to 0 $(\tau_T = 0)$ and 1 $(\tau_q = \tau_T)$, respectively. The nonlocal C-V (NL C-V) model could also be derived from Eq. (2.17), by setting $L \neq 0$, $A = 1$, and $B = Z = 0$.

Due to the time dependency of transient thermal responses in Eq. (2.17), solution for temperature is found in the Laplace transform as:

$$\tilde{\theta}(\xi, s) = \frac{1}{s}\exp\left[\frac{Ls - \sqrt{(Ls)^2 + 4s(1 + As + Bs^2)(1 + Zs)}}{2(1 + Zs)}\xi\right] \tag{2.18}$$

shown by Tzou [1–3], using the partial expansion technique and the limiting theorem in Laplace transform, Eq. (2.18) presents the thermal wave behavior for nonlocal C-V $\left(C^T_{NL\,C\text{-}V}\right)$ model with the following thermal wave speed:

$$C^T_{NL\,C\text{-}V} = \sqrt{\frac{\alpha}{\tau_q} + \left(\frac{\lambda_q}{2\tau_q}\right)^2} + \frac{\lambda_q}{2\tau_q} = \sqrt{\left(C^T_{C\text{-}V}\right)^2 + \left(\frac{\lambda_q}{2\tau_q}\right)^2} + \frac{\lambda_q}{2\tau_q} \tag{2.19}$$

which reveals that the $C^T_{NL\,C\text{-}V} > C^T_{C\text{-}V}$. Reimann sum approximation is then employed to numerically transform temperature in Laplace domain, given in Eq. (2.18), to time domain.

A code in MATLAB could be developed to numerically conduct the Laplace inversion of Eq. (2.18). Reimann sum approximation has been used here for the Laplace inversion. Figure 2.2 compares temperature distribution at non-dimensional time $\beta = 1$ for classical Fourier, C-V, diffusive-like DPL, wave-like DPL, and NL C-V heat conduction models. As shown in Fig. 2.2, C-V, wave-like DPL, and NL C-V models result in a wave-like behavior for temperature and reveals a sharp wavefront in thermal wave which divides the thermal response domain into the heat affected and unaffected zones. In accordance with Eq. (2.19), NL C-V model predicts a higher thermal wave speed compared to the C-V model. Furthermore, as derived in Eq. (1.22), thermal wave speed of the wave-like DPL model is related to the C-V model as: $C^T_{DPL} = C^T_{C\text{-}V}\sqrt{2Z}$. As a result:

$$C^T_{DPL} < C^T_{C\text{-}V} \; for \; Z < \frac{1}{2} \tag{2.20a}$$

$$C^T_{DPL} \geq C^T_{C\text{-}V} \; for \; Z \geq \frac{1}{2} \tag{2.20b}$$

For the assumed $Z = 10$ in Fig. 2.2, the wave front of the wave-like DPL model is ahead of the C-V and NL C-V models. As observed in Figs. 2.2 and 2.3, the mixed derivative term $Z\frac{\partial^3\theta(\xi,\beta)}{\partial\xi^2\partial\beta}$ in Eq. (2.17a) removes the singularity at the thermal wavefront, compared to the wave-like DPL model. As Z increases in the diffusive-like DPL model the thermal wavefront is completely destroyed and

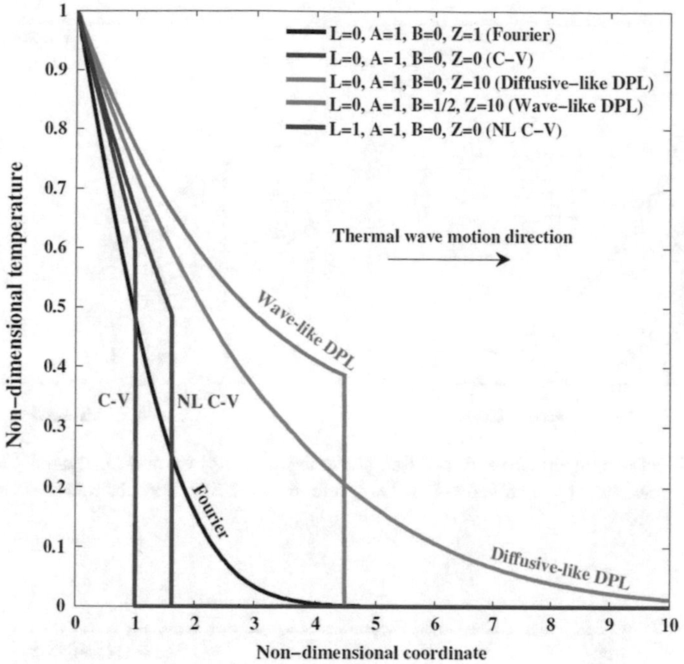

Fig. 2.2 Non-dimensional temperature distribution predicted by Fourier, C-V, diffusive-like DPL, hyperbolic-type DPL, and NL C-V heat conduction models at non-dimensional time $\beta = 1$

temperature responses show a diffusive behavior same as classical Fourier heat conduction, Fig. 2.3a. Although both of the diffusive-like DPL and classical Fourier heat conduction models do not show a finite thermal wave speed, the thermally affected zones are not the same for these two models. While for $Z > 1$, the diffusive-like DPL model reveals a wider affected zone compared to the Fourier heat conduction; the affected zone is narrower for $Z < 1$. For diffusive-like DPL model, thermal wave with discontinue temperature distribution around the wave-front is detected in Fig. 2.3b. Thermal wave speed increases by increasing Z. As opposed to diffusive-like DPL, decreasing Z could result in a diverged and noisy temperature distribution for very low values of $(Z \sim 0)$.

The effect of non-dimensional correlation length L on temperature distribution is illustrated in Fig. 2.4, at non-dimensional time $\beta = 1$ using the NL C-V heat conduction model. As seen in this figure, the NL C-V model reduces to the C-V model for = 0 $\left(\lambda_q = 0\right)$. As L increases, the wavefront of the NL C-V thermal wave advances and temperature in the heat affected zone is raised. The NL model for $L < 2$ has been shown by Tzou to be identical to the thermomass heat transfer model in phonon gas.

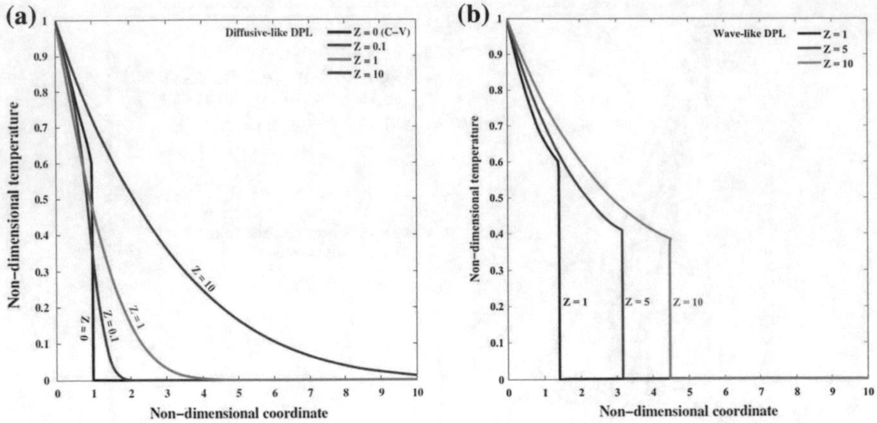

Fig. 2.3 Effect of temperature and heat flux phase lag ratio (Z) on non-dimensional temperature distribution predicted by **a** diffusive-like DPL and **b** wave-like DPL at non-dimensional time $\beta = 1$

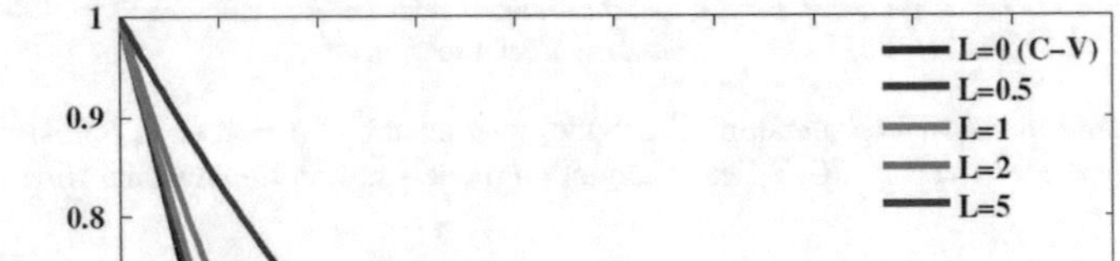

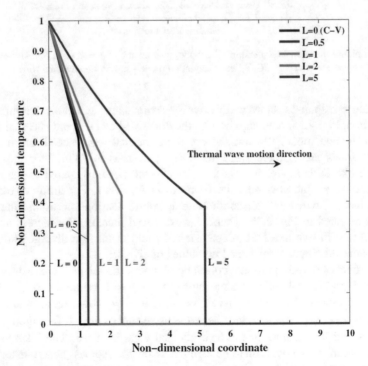

Fig. 2.4 Effect of non-dimensional correlation length (L) on non-dimensional temperature distribution at non-dimensional time $\beta = 1$ using NL C-V heat conduction model. [Reproduced from [1] with permission from Elsevier Masson SAS]

2.4 Nonlocal Phase-Lag Heat Conduction in a Finite Strip

As shown in Fig. 2.2, C-V, NL C-V, and wave-like DPL heat conduction models result in a sharp wavefront of temperature with an infinite temperature gradient across the thermal wavefront. To remove the singularity of thermal wavefront in C-V and NL-CV model, the introduction of τ_T has been shown to be effective [1]. However, this approach can not remove the singularity of thermal wave observed in the wave-like DPL/TPL models unless a fractional-order is used for the heat conduction. As a result, we use in this section the wave-like NL FTPL, introduced in Eq. (1.36), for a heat conduction analysis in a finite, isotropic, homogeneous 1D strip. As shown in Fig. 2.5, the strip is initially at T_0 and a stationary state $\frac{\partial T(x,t)}{\partial t} = 0$. The surface temperature of the left side $(x = 0)$ is suddenly raised to T_{WL}, while the surface temperature of the right side of the strip is kept at initial temperature.

According to Eq. (1.36), the heat conduction constitutive equation is written in the wave-like NL FTPL form in the one-dimensional Cartesian coordinate as:

$$\begin{aligned} &\left(1 + \lambda_q \frac{\partial}{\partial x} + \tau_q \frac{\partial}{\partial t} + \frac{\tau_q^2}{2}\frac{\partial^2}{\partial t^2}\right) q(x,t) \\ &= -k^T I^{\alpha_F - 1}\left(1 + \tau_T \frac{\partial}{\partial t}\right)\frac{\partial T(x,t)}{\partial x} - k^{*T} I^{\alpha_F - 1}\left(1 + \tau_\upsilon \frac{\partial}{\partial t}\right)\frac{\partial \upsilon(x,t)}{\partial x} \end{aligned} \tag{2.21}$$

Equation (2.21) is one of the forms of the wave-like NL FTPL heat conduction derived by the first-order Taylor series expansion of λ_q in space and the second-order Taylor series expansion of τ_q and the first-order Taylor series expansion of τ_T and τ_υ in time. The other types of wave-like NL FTPL heat conduction could be achieved with a similar approach given in Ref. [4] for the TPL heat conduction. The current NL FTPL is not only able to remove the singularity of thermal wavefront, but also is capable to take into account size-dependency, sub-diffusion or superdiffusion, and phonon-electron interaction in heat transport. In Eq. (2.21), α_F is the order of Riemann-Liouville fractional integral [5, 6]. The heat conduction equation of NL FTPL could be obtained by eliminating the heat flux $q(x,t)$ between Eq. (2.21) and energy Eq. (1.6). Taking the divergence and time-derivative of Eq. (2.21) and eliminating $\nabla . q$ via the time-derivation of energy Eq. (2.21) leads to the following heat conduction in 1D Cartesian coordinate:

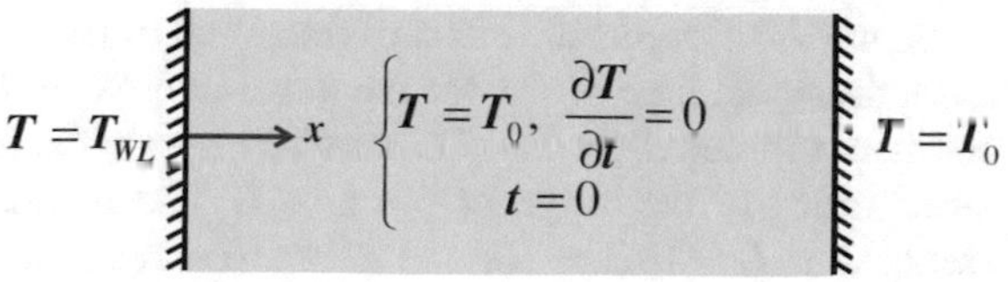

Fig. 2.5 Thermal boundary and initial conditions of a 1D finite strip

$$I^{\alpha_F-1}\left(1+\tau_T\frac{\partial}{\partial t}\right)\frac{\partial^3 T(x,t)}{\partial x^2\partial t}+\frac{k^{*T}}{k^T}I^{\alpha_F-1}\left(1+\tau_\upsilon\frac{\partial}{\partial t}\right)\frac{\partial^2 T(x,t)}{\partial x^2}$$
$$=\left(1+\lambda_q\frac{\partial}{\partial x}+\tau_q\frac{\partial}{\partial t}+\frac{\tau_q^2}{2}\frac{\partial^2}{\partial t^2}\right)\frac{1}{\alpha}\frac{\partial^2 T(x,t)}{\partial t^2} \tag{2.22}$$

Following the non-dimensional parameters defined in Eq. (2.16), Eq. (2.22) and initial and boundary conditions can be rewritten as:

$$\frac{\partial^2\theta(\zeta,\beta)}{\partial\beta^2}+L\frac{\partial^3\theta(\zeta,\beta)}{\partial\beta^2\partial\zeta}+\frac{\partial^3\theta(\zeta,\beta)}{\partial\beta^3}+\frac{1}{2}\frac{\partial^4\theta(\zeta,\beta)}{\partial\beta^4}$$
$$=I^{\alpha_F-1}C_K\frac{\partial^2\theta(\zeta,\beta)}{\partial\zeta^2}+I^{\alpha_F-1}(1+C_KZ^*)\frac{\partial^3\theta(\zeta,\beta)}{\partial\zeta^2\partial\beta}+I^{\alpha_F-1}Z\frac{\partial^4\theta(\zeta,\beta)}{\partial\zeta^2\beta^2} \tag{2.23a}$$

$$\theta(\xi,0)=\frac{\partial\theta(\xi,0)}{\partial\beta}=0\,(\textit{Initial conditions}) \tag{2.23b}$$

$$\theta(0,\beta)=1,\theta(\zeta_R,\beta)=0\ (\textit{Boundary conditions}) \tag{2.23c}$$

where $C_K=\frac{\tau_q k^{*T}}{k^T}$ and $Z^*=\frac{\tau_\upsilon}{\tau_q}$; ζ_R represents the non-dimensional length of the medium, which is assumed $\zeta_R=10$ for the numerical results. Solution of Eq. (2.23a) in Laplace domain could be found as:

$$\tilde{\theta}(\zeta,s)=\frac{e^{r_2\zeta_R}}{e^{r_2\zeta_R}-e^{r_1\zeta_R}}e^{r_1\zeta_R}-\frac{e^{r_1\zeta_R}}{e^{r_2\zeta_R}-e^{r_1\zeta_R}}e^{r_2\zeta_R} \tag{2.24}$$

in which r_1 and r_2 are characteristic roots of Eq. (2.23a) in the Laplace domain:

$$r_{1,2}=\frac{Ls^2\pm\sqrt{L^2s^4+4s^{-\alpha_F+3}(C_K+(1+C_KZ^*)s+Zs^2)\left(1+s+\frac{s^2}{2}\right)}}{2s^{-\alpha_F+1}(C_K+(1+C_KZ^*)s+Zs^2)} \tag{2.25}$$

It is worth reminding the following identity for the Laplace transform of Riemann-Liouville fractional integral as [5]:

$$L\{I^{\alpha_F}f(t)\}=\frac{1}{s^{\alpha_F}}L\{f(t)\}\quad \alpha_F>0 \tag{2.26}$$

To retrieve the temperature in the time domain, the Reimann sum approximation has been used here.

Figure 2.6 presents the effect of fractional order α_F on temperature distribution at no-dimensional time $\beta=1$ for the wave-like NL FTPL heat conduction. The temperature contours of the 1D slab are also given in the inset of Fig. 2.6. The fractional order varies in the range of $0<\alpha_F\leq 1$. The nonlocal and phase-lag rations are assumed as $L=1$, $Z=10$, and $Z^*=10$, respectively. For $\alpha_F=1$, the wave-like

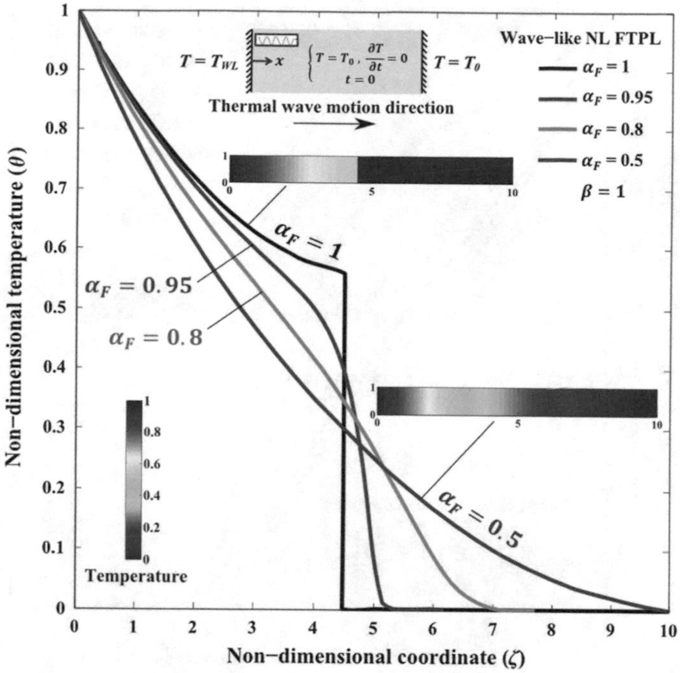

Fig. 2.6 Effect of fractional order (α_F) on non-dimensional temperature distribution at non-dimensional time $\beta = 1$ using NL FTPL heat conduction model ($L = 1$, $Z = 10$, $Z^* = 10$). Reproduced from [7] with permission from the Royal Society of Chemistry

NL FTPL reduces to a wave-like NL TPL with a sharp wavefront with a transition from a thermally affected zone ($\zeta \leq 4.48$) to an unaffected one ($\zeta > 4.48$), which leads to an infinite temperature gradient across the thermal wavefront. As seen in Fig. 2.6, fractional order α_F is an alternative for removing the singularity of thermal wave in NL TPL heat conduction. A slight variation of α_F from 1 to 0.95 effectively smoothens the thermal wavefront and the transition between the heat affected and unaffected zones. The effectiveness of α_F on smoothening the thermal wave front is more evident for lower values of α_F, e.g. $\alpha_F = 0.85$. Further decrease of α_F could completely destroyed the thermal wave front, as seen for $\alpha_F = 0.5$. Temperature level of the heat affected zone decreases while the temperature level of thermally unaffected zone increases by a decrease in the value of α_F. The effect of α_F on thermal wave propagation in wave-like NL FTPL heat conduction is similar to the effect of Z on diffusive-like Nonlocal DPL (diffusive-like NL DPL) heat conduction. As found by Zou and Guo [1], an increase in the value of Z in the NL DPL heat conduction could effectively remove the discontinuity of thermal wavefront while results in an increase in temperature level.

Figure 2.7 illustrates temperature distribution and temperature contours in the 1D medium at non-dimensional time $\beta = 1$ for a wave-like NL FTPL with $\alpha_F = 0.8$, $Z = 10$, and $Z^* = 10$. Effect of non-dimensional TPL parameter (C_K)

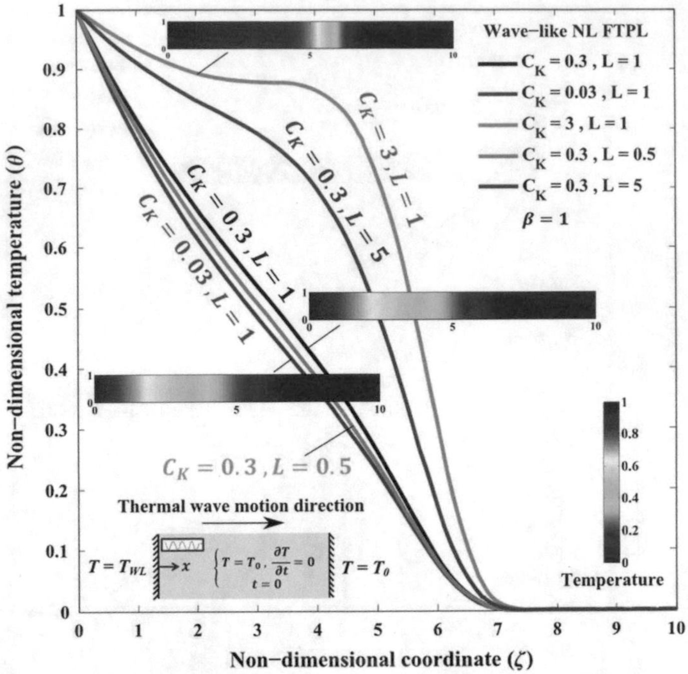

Fig. 2.7 Effect of non-dimensional TPL parameter (C_K) and non-dimensional correlation length (L) on non-dimensional temperature distribution at non-dimensional time $\beta = 1$ using NL FTPL heat conduction model ($\alpha_F = 0.8$, $Z = 10$, and $Z^* = 10$) [7]

and non-dimensional correlation length (L) on the thermal behavior is specifically investigated. Smooth thermal wavefront is detected in the thermal responses due to the fractional heat conduction of the order $\alpha_F = 0.8$. As seen in Fig. 2.7, C_K and L have similar effect on thermal responses; temperature is increased by an increase in either C_K or L. Moreover, an increase in C_K or L leads to an enhance in temperature gradient around the thermal wavefront. While the correlation length L alters the thermal wave speed in NL C-V heat conduction, the effect of L on the thermal wave speed of wave-like NL FTPL is not considerable.

To reveal the effect of phase lag ratios on thermal responses of wave-like NL FTPL heat conduction, Fig. 2.8 shows temperature distribution at non-dimensional time $\beta = 1$ for alternative values of Z and Z^*. The fractional order of heat conduction and TPL parameter are assumed as: $C_K = 0.3$ and $\alpha_F = 0.8$. As seen in Fig. 2.8, increasing the phase-lag ratio Z, which means either an increase in the phase-lag of temperature gradient τ_T or a decrease in the phase-lag of heat flux τ_q, increases the temperature level throughout the heat affected zone of the medium and increases the thermal wave speed. Enhancing the phase-lag ratio Z^*, equivalently an increase in the phase-lag of thermal displacement τ_υ or a decrease in the phase-lag of heat flux τ_q, leads to an increase in the temperature level. Thermal wave speed, however, does not change by τ_υ in the wave-like NL FTPL model.

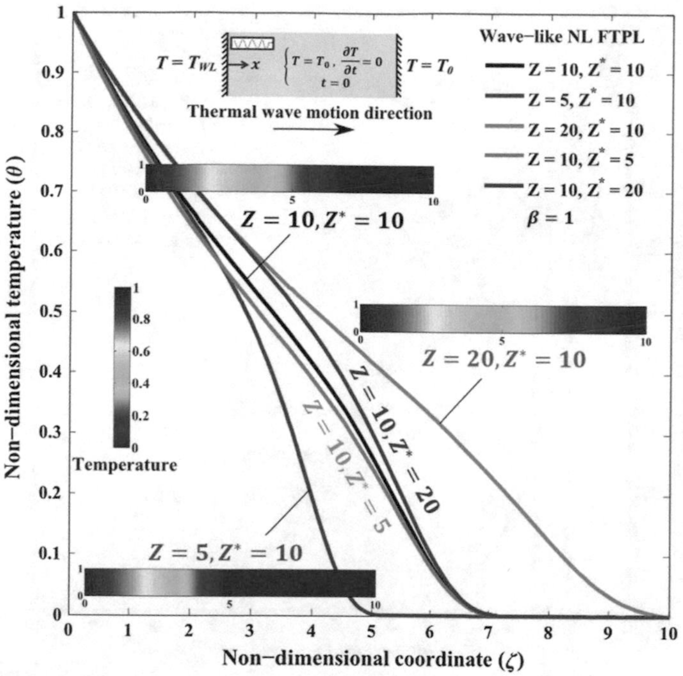

Fig. 2.8 Effect of non-dimensional phase-lag ratios Z and Z^* on non-dimensional temperature distribution at non-dimensional time $\beta = 1$ using NL FTPL heat conduction model: ($C_K = 0.3$, $\alpha_F = 0.8$) [7]

Comparing the current observations for the thermal wave speed of wave-like NL FTPL heat conduction with those found for wave-like TPL model in Eq. (1.27) and Refs. [4, 8] shows that all phase-lags have the same effect on the thermal wave speed of both aforementioned heat conduction models.

To summarize the characteristics of alternative heat conduction in a continuum scale, temperature distribution developed within a semi-infinite slab subjected to an increased temperature on the left side is illustrated in Fig. 2.9. The thermal boundary conditions are the same as those assumed in Fig. 2.9 for a semi-infinite one-dimensional strip. Deduced from Eq. (2.23), temperature distribution at non-dimensional time $\beta = 1$ is compared in Fig. 2.9 for the Fourier, C-V, nonlocal C-V (NL C-V) ($L = 1$), diffusive-like and wave-like DPL ($Z = 1$), wave-like fractional DPL (FDPL) ($Z = 10$, $\alpha_F = 0.95$), wave-like TPL ($Z = Z^* = 10$, $C_K = 0.3$), wave-like FTPL ($\alpha_F = 0.95$, $Z = Z^* = 10$, $C_K = 0.3$), wave-like NL TPL ($L = 1$, $Z = Z^* = 10$, $C_K = 0.3$), and wave-like NL FTPL ($L = 1$, $\alpha_F = 0.95$, $Z = Z^* = 10$, $C_K = 0.3$) continuum heat conduction models. The propagation of thermal disturbance in the form of thermal wave can be observed in all the aforementioned heat conduction models, except the Fourier and diffusive-like DPL models. Temperature profile of C-V, NL C-V, wave-like DPL, and wave-like TPL, wave-like NL TPL heat conduction models reveals unrealistic sharp thermal

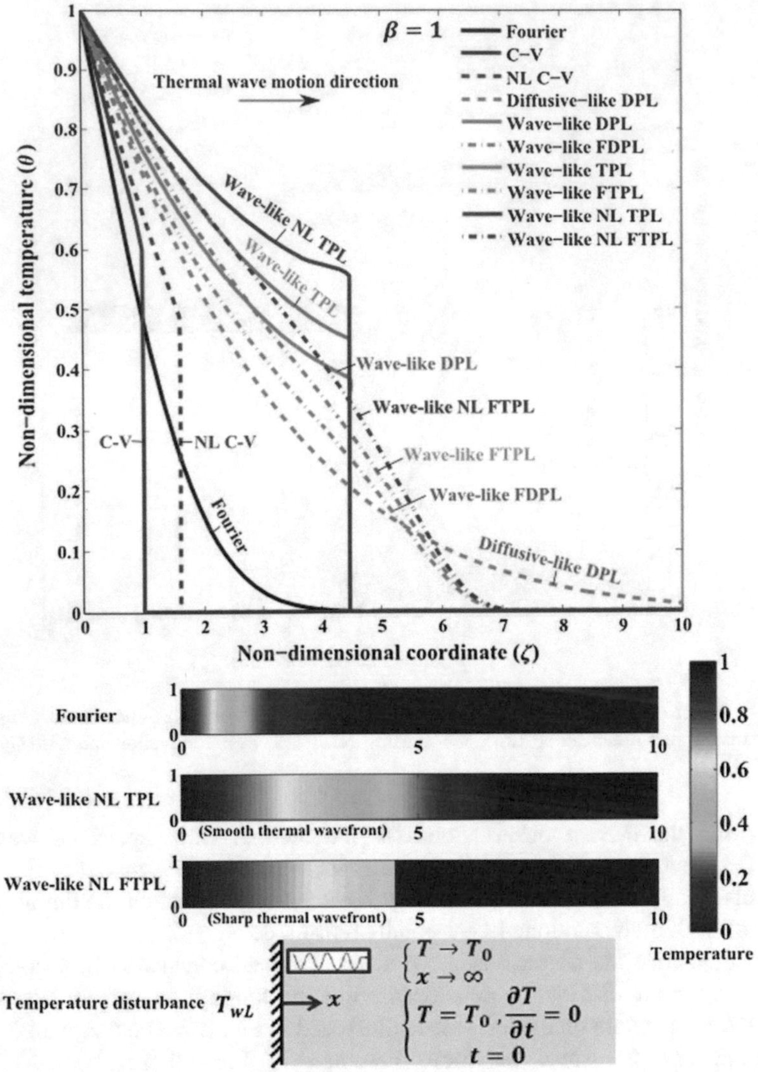

Fig. 2.9 Temperature profile of alternative continuum heat conduction models at non-dimensional time $\beta = 1$ in an infinite homogenous medium subjected to an abrupt temperature increase on its left side [7]

wavefront which divides the thermal domain intro heat affected and heat unaffected zones causing the unrealistic discontinuity of temperature and resulting thermal strain [7]. Reported by Tzou [1], thermal wave speed of NL C-V model $(C_{NL\,C\text{-}V})$ is higher than the C-V model $(C_{C\text{-}V})$: $(C_{NL\,C\text{-}V}) = \frac{\sqrt{4+L^2}+L}{2} C_{C\text{-}V}$ where the dimensional thermal wave speed of C-V is: $C_{C\text{-}V} = \sqrt{\frac{\alpha}{\tau_q}} \left(x = C_{C\text{-}V} t = \sqrt{\frac{\alpha}{\tau_q}} t \right)$. Thermal wave speed of wave-like DPL $\left(C_{wave\text{-}like\,DPL}\right)$, wave-like TPL $\left(C_{wave\text{-}like\,TPL}\right)$, and wave-like NL TPL $\left(C_{wave\text{-}like\,NL\,TPL}\right)$ are equal and their thermal wave speed can be mathematically expressed as: $C_{wave\text{-}like\,DPL} = C_{wave\text{-}like\,TPL} = C_{wave\text{-}like\,NL\,TPL} = \sqrt{2Z}C_{C\text{-}V}$. Thermal wave speed of different continuum non-Fourier heat conduction models depends on the phase-lags of temperature gradient and heat flux, thermal diffusivity, and correlating length and can be compared as: $C_{NL\,C\text{-}V} > C_{C\text{-}V} > C_{wave\text{-}like\,DPL} = C_{wave\text{-}like\,TPL} = C_{wave\text{-}like\,NL\,TPL}$ for $Z < \frac{2+L^2+L\sqrt{4+L^2}}{4}$.

$$C_{wave\text{-}like\,DPL} = C_{wave\text{-}like\,TPL} = C_{wave\text{-}like\,NL\,TPL} \geq C_{NL\,C\text{-}V} \geq C_{C\text{-}V}$$
$$for\, Z \geq \frac{2+L^2+L\sqrt{4+L^2}}{4} \tag{2.27}$$

As shown in Fig. 2.9, while both classical Fourier heat conduction and diffusive-like DPL models do not show finite thermal wave speed, the thermal affected zones are not the same for these models. For $Z > 1$, the diffusive-like DPL model leads to a wider affected zone compared to the classical Fourier heat conduction, while the thermal affected zone is narrower for $Z<1$. Introducing time-fractional derivatives in non-Fourier heat conduction models adds $I^{\alpha_F-1}Z\frac{\partial^4\theta(\zeta,\beta)}{\partial\zeta^2\partial\beta^2}$ term in the heat conduction Eq. (2.23a) and effectively destroys the singularity of temperature field around the thermal wavefront. Smooth variation of temperature is observed in the temperature profile of all fractional non-Fourier heat conduction. In addition, thermal wave speed is the same for all wave-like FDPL, FTPL, and NL FTPL models; however, NL FTPL model leads to a higher temperature range for the thermal affected zone. The NL FTPL heat conduction, as a recently introduced non-Fourier nonlocal continuum heat conduction model [7], simultaneously detects thermal wave propagation, removes the discontinuity of temperature at thermal wavefront, and enables taking into account the effect of length scale and microstructural heat transport on the conductive heat transport.

2.4.1 *Molecular Dynamics to Determine Correlating Nonlocal Length*

Determining the value of correlating nonlocal length is one of the most intricate challenges for application of recently developed nonlocal non-Fourier heat conduction models to nanoscale materials. Since the correlating nonlocal length is an intrinsic property of material, it is required to be determined for each material. The experimental testing is a cumbersome task for measuring the thermal nonlocal length. While the experimental testing is inevitable for accurate determination of nonlocal length, molecular dynamics (MD) and atomistic simulation are feasible methods for determining nonlocal length by comparing the characteristics of thermal wave in nonlocal non-Fourier heat conduction and MD thermal results. Herein, we introduce the MD approach for measuring thermal nonlocal length of

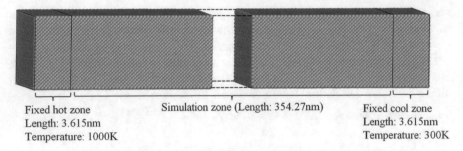

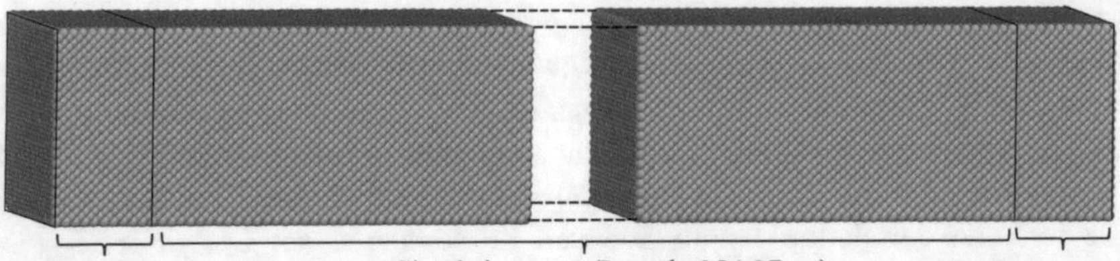

Fig. 2.10 A nano-slab considered for MD simulation of thermal wave propagation [7]

copper. In specific, we present the results of MD simulation for a relatively-long nano-slab subjected to thermal excitation on its left side, an example which resembles thermal wave propagation in one-dimensional heat transport.

The MD simulation is conducted by *LAMMPS* software [9] for a copper single-crystalline nano-slab of 361.5 nm length, 7.23 nm width, and 7.23 nm height as shown schematically in Fig. 2.10. The steps required to be taken for the MD simulation are [7]:

(1) Creating the copper nano-slab by face-centred-cubic lattices.
(2) Initializing the atoms with random velocities.
(3) Equilibrating the nano-slab at room temperature 300 K for 20 picosecond (ps) under Noose-Hoover thermostat (NVT) ensembles [10]. We fix the temperature of both hot and cool zones of the equilibrated nano-slab by rescaling their atoms at each time step.
(4) Increasing the temperature of the fixed hot zone of the nano-slab to 1000 K temperature, a condition that replicates the thermal boundary condition of one-dimensional continuum NL FTPL heat conduction.

We apply MD simulation for time steps of 1 femtosecond (fs) and for a total time period of 10 ps before thermal wavefront reaches the right end of nano-slab. Heat transport in solids is carried out by electrons and phonons. The atomistic interactions between atoms and electrons are introduced into the MD simulation through the embedded atom method (EAM) potential defined as [11]:

$$E = \sum_i F^i\left(\sum_{i \neq j} \rho^i(r_{ij})\right) + \frac{1}{2}\sum_{ij, i \neq j} \phi^{ij}(r_{ij}) \tag{2.28}$$

where r_{ij} represents the distance between atoms i and j, ρ^i is the contribution of the electron charge density, F^i is the summation of individual embedding function of atom i, and ϕ^{ij} represents a pairwise potential function between atoms. To determine temperature distribution along the slab length, we divide the slab to finite numbers of segments (here 100 segments) and obtain the average temperature of each segment as:

$$T_{seg} = \frac{\sum_i m v_i^2}{2 N_{seg} k_B} \tag{2.29}$$

In this equation, m, v_i, N_{seg}, and k_B are, respectively, atomic mass, velocity of atom i, number of atoms in each segment, and Boltzmann constant and T_{seg} present the average value at each segment. As seen in Eq. (2.29), average thermodynamic temperature is related to the mean square velocity of atoms.

Figure 2.11 shows temperature profile in the copper nano-slab at different time. Similar to the temperature profile observed in a nano argon film [12], temperature evolves in the nano-slab in the form of thermal wave. Thermal wave is observed to travel from the hot surface on the left side of the nano-slab towards the cold surface on the right with estimated thermal wave speed of $C_{MD} = 23 - 25 \times 10^3$ m/s. If we correlate the thermal wave speed estimated by MD simulation with the speed of sharp thermal wavefront in NL C-V model, the correlating length λ_q defined in Eq. (2.14) can be estimated as:

$$\lambda_q = \tau_q \left(C_{NL\,C\text{-}V} - \frac{C_{C\text{-}V}^2}{C_{NL\,C\text{-}V}} \right) \approx \tau_q \left(C_{MD} - \frac{C_{C\text{-}V}^2}{C_{MD}} \right) = 5.79 \sim 7.11\,\text{nm} \tag{2.30}$$

where $C_{C\text{-}V}$ and $C_{NL\,C\text{-}V}$ represent the thermal wave speed predicted by the C-V model and NL C-V models, respectively. While MD results in Fig. 2.11 show the propagation of temperature disturbance in the form of thermal wave, slight temperature rise in atoms (locations) ahead of thermal wavefront is observed in temperature distribution. This observation is compatible with the characteristics of thermal wave predicted by the NL FTPL heat conduction and those reported in Ref. [8] for phase-lag heat conduction in homogenous and heterogeneous porous materials.

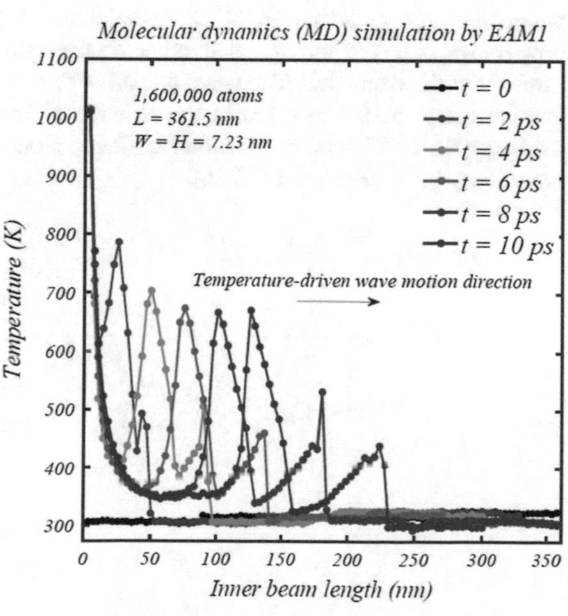

Fig. 2.11 Temperature distribution in a nano-slab at different time obtained by MD simulation [7]

2.4.1.1 Nonlocal Heat Conduction in Functionally Graded Materials

Most of biological materials with extreme mechanical and thermochemical properties, e.g. Moso culm bamboo [13], dento-enamel-junction of natural teeth [14], and the Humboldt squid beak [15], reveal a multi-scale hierarchical and functionally graded (FG) microstructure. Examples of the extreme properties of functionally graded materials (FGMs) are resistant to contact damage, cracking, deformation, thermal stresses, and heat flow due to the gradation of microstructural morphology, porosity, and chemical/material ingredients in FGMs [16–18]. FGMs enable the engineering of advanced materials with tuned multiphysics properties to satisfy mechanical, hygrothermal, electrical, and biological requirements for structural design in a wide range of applications as thermal barriers, bone tissues and implants, thermoelectric generators, and energy harvesters. Advances in powder metallurgy [19], laser cutting [20], and additive manufacturing/3D printing [21] have also facilitated fabrication and the arbitrary variation of material composition and micro-architecture of FGMs.

Due to the importance of FGMs, we present here the temperature evolution and thermal wave propagation in an FGM nano-slab. The material properties of FGMs can be arbitrarily tailored within FGMs through the variation of volume fraction of constituent solid components (Two-phase solid FGMs) or relative density of porous materials (Single-phase porous FGMs). To be able to obtain closed-form solutions for transient temperature in the Laplace domain, we adopt here an exponential function for variation of thermal conductivity (k), material constant of the TPL theory (k^*), and specific heat per volume (ρc) through the length of FGM nano-slab:

$$k = k_0 e^{n_G x}, k^* = k_0^* e^{n_G x}, (\rho c_p) = C_0 e^{n_G x} \tag{2.31}$$

where n_G represents the FGM exponential index for the variation of material properties [22]. We assume that other thermosphysical properties are constant throughout the nano-slab. The wave-like NL FTPL heat conduction equation for the exponentially graded medium in the absence of heat source can be obtained by using the NL FTPL heat conduction Eq. (2.21) along with the energy Eq. (1.6) and non-dimensional parameters (2.16):

$$\begin{aligned}
&(1+n'L)\frac{\partial^2\theta(\zeta,\beta)}{\partial\beta^2} + L\frac{\partial^3\theta(\zeta,\beta)}{\partial\beta^2\partial\zeta} + \frac{\partial^3\theta(\zeta,\beta)}{\partial\beta^3} + \frac{1}{2}\frac{\partial^4\theta(\zeta,\beta)}{\partial\beta^4} \\
&= I^{\alpha_F-1}C_K\frac{\partial^2\theta(\zeta,\beta)}{\partial\zeta^2} + n' I^{\alpha_F-1}\frac{\partial^2\theta(\zeta,\beta)}{\partial\beta\partial\zeta} \\
&\quad + I^{\alpha_F-1}(1+C_K Z^*)\frac{\partial^3\theta(\zeta,\beta)}{\partial\zeta^2\partial\beta} + I^{\alpha_F-1}Z\frac{\partial^4\theta(\zeta,\beta)}{\partial\zeta^2\partial\beta^2}
\end{aligned} \tag{2.32}$$

where $n' = n_G\sqrt{\alpha\tau_q}$. The closed-form solution of Eq. (2.32) in the Laplace domain is the same as the one provided in Eq. (2.24) with the characteristics roots modified as follows:

$$r_{1,2} = \frac{(Ls^2 - n'S^{-\alpha_F+2}) \pm \sqrt{\begin{array}{c}(Ls^2 - n'S^{-\alpha_F+2}) \\ +4s^{-\alpha_F+3}(C_K + (1 + C_K Z^*)s + Zs^2) \\ \left(1 + n'L + s + \frac{s^2}{2}\right)\end{array}}}{2s^{-\alpha_F+1}(C_K + (1 + C_K Z^*)s + Zs^2)} \tag{2.33}$$

Temperature can then be retrieved in the time domain by using a numerical Laplace inversion technique.

Figure 2.12 shows the effect of FGM exponential index n_G [presented in Eq. (2.31)] on the characteristics of NL FTPL thermal wave at the non-dimensional time $\beta = 1$. As seen in this figure, the NL FTPL thermal wave speed is constant and independent of non-homogeneity index n_G for exponential type of FGM materials; a phenomenon caused by the absence of non-homogeneity index parameter $n' = n_G\sqrt{\alpha\tau_q}$ in terms of the highest order of temperature derivatives in NL FTPL

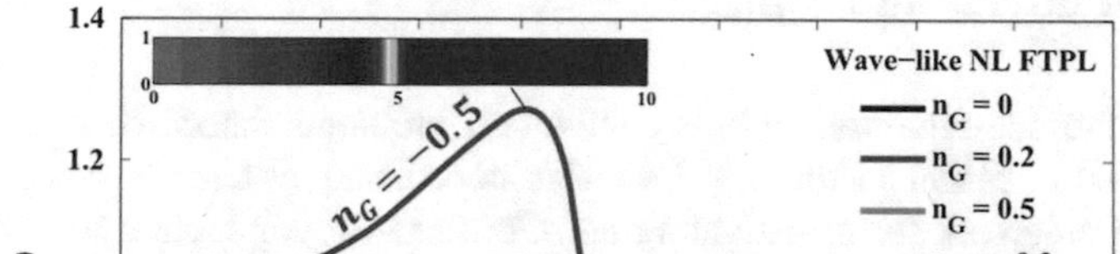

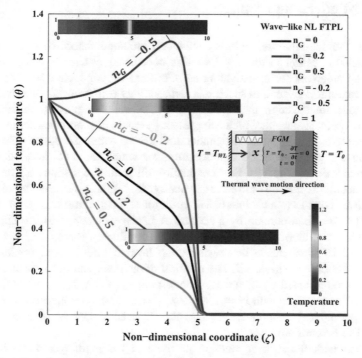

Fig. 2.12 Effect of material non-homogeneity index of FGM nano-slab on non-dimensional temperature distribution at non-dimensional time ($\beta = 1$) using NL FTPL heat conduction model ($L = 1, C_K = 0.3, \alpha_F = 0.95, Z = 10, Z^* = 10$) [7]

differential equation of heat conduction [Eq. (2.32)]. It is important mentioning that while the exponential material gradation does not alter the thermal wave speed based on the NL FTPL model for this specific FGM medium, thermal wave speed can vary within an FGM medium with the FGM non-homogeneity index for FGM materials with an arbitrary variation of material properties [8, 23, 24]. Figure 2.12 shows that material gradation can effectively tailor temperature within the thermal affected zone of FGM medium. In addition, increasing the value of FGM exponential index n_G from −0.5 to 0.5 can remarkably reduce temperature within the thermal affected zone of the FGM medium. Interestingly, decreasing the FGM exponential index n_G can also magnify temperature at the thermal wavefront causing that the temperature within the FGM medium exceeds the temperature at the boundaries; for example, the maximum temperature within an FGM medium with $n_G = -0.5$ is about 30% higher than the maximum temperature occurred within a homogenous medium $n_G = 0$. Consequently, material gradation can potentially improve the performance of advanced materials used in extreme environmental conditions if the material gradation index is optimized.

2.5 Three-Phase-Lag Heat Conduction in 1D Strips, Cylinders, and Spheres

Till now, the one-dimensional heat conduction problems discussed in this chapter were limited to planar medium in Cartesian coordinate system. In this section, we present a framework for investigating non-Fourier heat conduction in 1D media in general coordinate system. In specific, a methodology is introduced for solving TPL heat conduction equation in a multilayered 1D (solid or porous) medium in a general coordinate system, which can present thermal wave propagation in 1D rod, 1D infinitely-long axisymmetric cylinder, and 1D axisymmetric sphere. Each layer of medium is assumed to be homogenous, for which the TPL heat conduction equation is written. The TPL heat conduction differential equations for the multilayered medium can be solved analytically in Laplace domain by applying appropriate boundary and interfacial equations. Temperature is then can be retrieved in the time domain by a numerical Laplace Inversion to investigate the characteristics of thermal wave in general 1D coordinate systems.

Figure 2.13 illustrates a heterogeneous *N*-layered multilayered medium in a general 1D coordinate system $\vec{x}$. The position of the inner and outer surfaces of the medium is represented by x_i and x_o. In addition, $x_n\,(n = 1, 2, \ldots, N)$ is the inner surface of the n*th* layer with $x_1 = x_i$ and $x_{N+1} = x_o$. The heterogeneous medium is initially at ambient temperature T_0 and the material properties in each layer is assumed to be constant.

To deal with the heat conduction problem in the multilayered medium, we introduce the TPL heat conduction equation in the following form [25]:

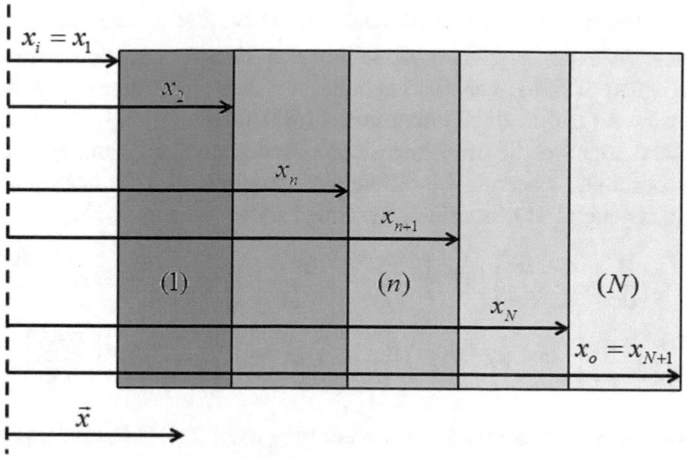

Fig. 2.13 A Multilayered medium in a general 1D coordinate system. [Reproduced from [8] with permission from Elsevier]

$$\vec{q}^{(n)}\left(\vec{x}, t + \tau_q^{(n)}\right) = -\left[K^{(n)}\vec{\nabla}T\left(\vec{x}, t + \tau_T^{(n)}\right) + K^{*(n)}\vec{\nabla}\upsilon\left(\vec{x}, t + \tau_\upsilon^{(n)}\right)\right] \tag{2.34}$$

In this equation, superscript 'n' represents the layer number of the multilayered medium and associated material properties in that layer. As mentioned earlier in this chapter, $\vec{q}^{(n)}$, $T^{(n)}$, $\upsilon^{(n)}$, $K^{(n)}$, and $K^{*(n)}$ $(n = 1, 2, \ldots, N)$ are, respectively, heat flux vector, absolute temperature, thermal displacement $\left(\dot{\upsilon}^{(n)} = T^{(n)}\right)$, thermal conductivity, and material constant characteristics of the TPL theory; $\tau_q^{(n)}$, $\tau_T^{(n)}$, and $\tau_\upsilon^{(n)}$ also represent, respectively, the phase-lag of heat flux, temperature gradient, and thermal displacement gradient in each layer. To develop the wave-like and diffusive-like TPL heat conduction models, we expand the heat conduction equation of (2.34) by using the second-order Taylor series expansion for $\tau_q^{(n)}$ and the first-order for $\tau_T^{(n)}$ and $\tau_\upsilon^{(n)}$ as follows [8]:

$$\begin{aligned} &\left(1 + \tau_q^{(n)}\frac{\partial}{\partial t} + \frac{\tau_q^{(n)^2}}{2}\frac{\partial^2}{\partial t^2}\right)\vec{q}^{(n)} \\ &= -\left(\left(K^{(n)} + K^{*(n)}\tau_\upsilon^{(n)}\right)\vec{\nabla}T^{(n)} + K^{(n)}\tau_T^{(n)}\frac{\partial}{\partial t}\vec{\nabla}T^{(n)} + K^{*(n)}\vec{\nabla}\upsilon^{(n)}\right) \end{aligned} \tag{2.35}$$

where $\frac{\partial}{\partial t}\upsilon^{(n)} = T^{(n)}$. The energy conservation equation can also be written as:

$$-\vec{\nabla}.\vec{q}^{(n)} + R^{(n)} = \rho^{(n)}c_p^{(n)}\frac{\partial T}{\partial t} \tag{2.36}$$

where $R^{(n)}$, $\rho^{(n)}$, and $c_p^{(n)}$ are internal heat generation density, material density, and specific heat in layer n. Heat conduction and energy equations introduced in Eqs. (2.35) and (2.36) can be applied to any coordinate systems from one-dimensional (1D) to three-dimensional (3D) media. If Eqs. (2.35) and (2.36) are combined together in one-dimensional media in Cartesian, cylindrical, and spherical coordinate systems, the differential equation of TPL heat conduction is obtained in a general 1D coordinate system $\vec{x}$ shown in Fig. 2.13:

$$\left(1+\tau_q^{(n)}\frac{\partial}{\partial t}+\frac{\tau_q^{(n)^2}}{2}\frac{\partial^2}{\partial t^2}\right)\left[\rho^{(n)}c_p^{(n)}\frac{\partial^2 T^{(n)}}{\partial t^2}-\frac{\partial R^{(n)}}{\partial t}\right]$$
$$=\left(\frac{\partial}{\partial x}+\frac{m}{x}\right)\left[\left(K^{(n)}+K^{*(n)}\tau_v^{(n)}\right)\frac{\partial^2 T^{(n)}}{\partial x\partial t}+K^{(n)}\tau_T^{(n)}\frac{\partial^3 T^{(n)}}{\partial x\partial t^2}+K^{*(n)}\frac{\partial T^{(n)}}{\partial x}\right] \tag{2.37}$$

where $m=0$, $m=1$, and $m=2$ are respectively used for 1D Cartesian, cylindrical, and spherical coordinate systems. Similar to the non-dimensional parameters introduced in Eq. (2.16), we use the following non-dimensional parameters for the heat conduction analysis in multilayered media:

$$\zeta=\frac{K'^{(1)}t}{x_o^2},\eta=\frac{x}{x_o},\theta^{(n)}=\frac{T^{(n)}-T_0}{T_0},Q_x^{(n)}=\frac{x_o q_x^{(n)}}{K^{(1)}T_0},\varepsilon_0^{(n)}=\frac{K'^{(1)}\tau_q^{(n)}}{x_o^2}$$
$$\delta_0^{(n)}=\frac{K'^{(1)}\tau_T^{(n)}}{x_o^2},\alpha_0^{(n)}=\frac{K'^{(1)}\tau_v^{(n)}}{x_o^2},K'^{(1)}=\frac{K^{(1)}}{\rho^{(1)}c_p^{(1)}},K'^{(n)}=\frac{K^{(n)}}{\rho^{(n)}c_p^{(n)}} \tag{2.38}$$
$$\eta_\gamma=\frac{x_i}{x_o},\theta_{wi}=\frac{T_{wi}-T_0}{T_0},\theta_{wo}=\frac{T_{wo}-T_0}{T_0}$$

where T_{wi} and T_{wo} are the temperature on the inner and outer surfaces. In the absence of internal heat generation, Eq. (2.37) can be rewritten in the following non-dimensional form:

$$\left(1+\varepsilon_0^{(n)}\frac{\partial}{\partial\zeta}+\frac{\varepsilon_0^{(n)^2}}{2}\frac{\partial^2}{\partial\zeta^2}\right)\frac{\partial^2\theta^{(n)}}{\partial\zeta^2}$$
$$=\left[C_T^{(n)^2}+\left(\frac{K'^{(n)}}{K'^{(1)}}+C_T^{(n)^2}\alpha_0^{(n)}\right)\frac{\partial}{\partial\zeta}+\frac{K'^{(n)}}{K'^{(1)}}\delta_0^{(n)}\frac{\partial^2}{\partial\zeta^2}\right]\left[\frac{m}{\eta}\frac{\partial\theta^{(n)}}{\partial\eta}+\frac{\partial^2\theta^{(n)}}{\partial\eta^2}\right] \tag{2.39}$$

where $C_T^{(n)^2}=\frac{K^{*(n)}x_0^2}{\rho^{(n)}c_p^{(n)}K'^{(1)^2}}$. The thermal wave speed in each layer for C-V, wave-like DPL, and wave-like TPL models can deduced from Eq. (2.39) as follows:

$$C_{C\text{-}V}^{(n)}=\sqrt{\frac{K'^{(n)}}{K'^{(1)}\varepsilon_0^{(n)}}}\qquad (C\text{-}V) \tag{2.40a}$$

$$C_{DPL}^{(n)} = C_{TPL}^{(n)} = \frac{1}{\varepsilon_o^{(n)}} \sqrt{\frac{2K'^{(n)} \delta_0^{(n)}}{K'^{(1)}}} \qquad (\textit{Wave-like DPL and TPL}) \tag{2.40b}$$

It is found that the non-dimensional thermal wave speed in 1D media depends on the phase-lag of the heat flux, phase-lag of temperature gradient, and the material properties of each layer of a multilayered composite. The difference of material properties and interfacial imperfection between neighboring layers of multilayered media can cause the separation of the thermal wave into transmitted and reflected parts [8, 26].

Applying Laplace transform to Eq. (2.39) and considering zero initial conditions lead to:

$$\eta^2 \frac{\partial \tilde{\theta}^{(n)}}{\partial \eta^2} + m\eta \frac{\partial \tilde{\theta}^{(n)}}{\partial \eta} - D_{TPL}^{(n)} \eta^2 \tilde{\theta}^{(n)} = 0 \tag{2.41}$$

where

$$D_{TPL}^{(n)} = \frac{\left(1 + \varepsilon_0^{(n)} s + \frac{\varepsilon_0^{(n)^2}}{2} s^2\right) s^2}{C_T^{(n)^2} + \left(\frac{K'^{(n)}}{K'^{(1)}} + C_T^{(n)^2} \alpha_0^{(n)^2}\right) s + \frac{K'^{(n)}}{K'^{(1)}} \delta_0^{(n)} s^2} \tag{2.42}$$

The differential Eq. (2.41) can be solved in terms of Bessel functions as:

$$\tilde{\theta}^{(n)}(\eta, s) = \eta^{\frac{1-m}{2}} \left(A_1^{(n)} J_{G^{(n)}}\left(I^{(n)}\eta\right) + A_2^{(n)} Y_{G^{(n)}}\left(I^{(n)}\eta\right)\right) \tag{2.43}$$

where

$$G^{(n)} = \left|\frac{m-1}{2}\right|, I^{(n)} = \sqrt{-D_{TPL}^{(n)}} \tag{2.44}$$

and $A_1^{(n)}$ and $A_2^{(n)}$ $(n = 1, 2, \ldots, N)$ are integration constants, and $J_{G^{(n)}}$ and $Y_{G^{(n)}}$ are $G^{(n)}$th-order Bessel functions of the first and second kind. The heat flux can also be written in the Laplace domain using Eqs. (2.35) and (2.43):

$$\tilde{Q}_x^{(n)}(\eta, s) = \frac{P_{TPL}^{(n)}}{2} \eta^{\frac{-m-1}{2}} \begin{pmatrix} A_1^{(n)}\left(M^{(n)} J_{G^{(n)}}\left(I^{(n)}\eta\right) - 2I^{(n)}\eta J_{G^{(n)}+1}\left(I^{(n)}\eta\right)\right) \\ +A_2^{(n)}\left(M^{(n)} Y_{G^{(n)}}\left(I^{(n)}\eta\right) - 2I^{(n)}\eta Y_{G^{(n)}+1}\left(I^{(n)}\eta\right)\right) \end{pmatrix} \tag{2.45}$$

where

$$M^{(n)} = 2G^{(n)} - (m-1), P_{TPL}^{(n)} = -\frac{C_T^{(n)^2} + \left(\frac{K^{(n)}}{K^{(1)}} + C_T^{(n)^2} \alpha_0^{(n)}\right) s + \frac{K^{(n)}}{K^{(1)}} \delta_0^{(n)} s^2}{s\left(1 + \varepsilon_0^{(n)} s + \frac{\varepsilon_0^{(n)^2}}{2} s^2\right)} \tag{2.46}$$

The integration constants in Eqs. (2.43) and (2.45) are obtained by satisfying the thermal boundary and interfacial conditions:

$$\theta^{(1)}(\eta,\zeta)|_{\eta=\eta_\gamma} = \theta_{wi} f_i(\zeta) \tag{2.47a}$$

$$\frac{\chi_T^{(j)} K^{(1)}}{x_o} Q_x^{(j)}(\eta,\zeta)|_{\eta=\eta_{j+1}} = \left(\theta^{(j)}(\eta,\zeta) - \theta^{(j+1)}(\eta,\zeta)\right)\Big|_{\eta=\eta_{j+1}} \tag{2.47b}$$

$$Q_x^{(j)}(\eta,\zeta)|_{\eta=\eta_{j+1}} = Q_x^{(j+1)}(\eta,\zeta)|_{\eta=\eta_{j+1}} \tag{2.47c}$$

$$\theta^{(N)}(\eta,\zeta)|_{\eta=1} = \theta_{wo} f_o(\zeta) \qquad (j = 1, 2, \ldots, N-1) \tag{2.47d}$$

where $\chi_T^{(j)}$ stands for the thermal compliance constant or thermal contact resistance for the imperfect interface between layers j and $j+1$; f_i and f_o are temporal functions for the applied transient thermal boundary conditions on the inner and/or outer surfaces of the multilayered medium. The bonding imperfection with thermally weak conduction has taken into account in Eq. (2.47) to conduct a reliable thermal analysis for multilayered composites [27, 28]. The perfectly bonded interfaces are associated with $\chi_T^{(j)} = 0$. Although a Heaviside step function, $f_i(\zeta) = f_o(\zeta) = H(\zeta)$, is considered for the time-dependent functions in this case study, similar procedure can be followed for other types of transient thermal disturbances [23]. Using Eqs. (2.43) and (2.45) and the thermal boundary and interfacial conditions (2.47) in the Laplace domain results in the following algebraic equation that allows obtaining the integration constants:

$$[K_{TPL}]_{2N\times 2N}\{X_{TPL}\}_{2N\times 1} = \{F_{TPL}\}_{2N\times 1} \tag{2.48}$$

where $[K_{TPL}]$ is a $2N \times 2N$ matrix, $\{X_{TPL}\}$ is a $2N \times 1$ vector of integration constants $\{X_{TPL}\}^T = \left\{ A_1^{(1)} \quad A_2^{(1)} \quad \cdots \quad A_1^{(N)} \quad A_2^{(N)} \right\}$, and $\{F_{TPL}\}$ is a $2N \times 1$ vector. By solving Eq. (2.48), the transient temperature change and heat flux in the Laplace domain are obtained in the Laplace domain. Finally, temperature change and heat flux can be retrieved in the time domain by implementing a numerical Laplace inversion technique introduced in Sect. 2.2. If this methodology is applied, we can investigate the effect of bonding interface, material heterogeneity, and continuous variation of material properties on thermal responses of advanced materials.

2.5.1 *Effect of Bonding Imperfection on Thermal Wave Propagation*

To examine the effect of bonding interface on transient thermal responses, Fig. 2.14 illustrates the effect of imperfectly bonded interface on the temperature and heat flux distribution of a bilayered cylinder ($m = 1$) with inner radius $x_i = 0.6$ and

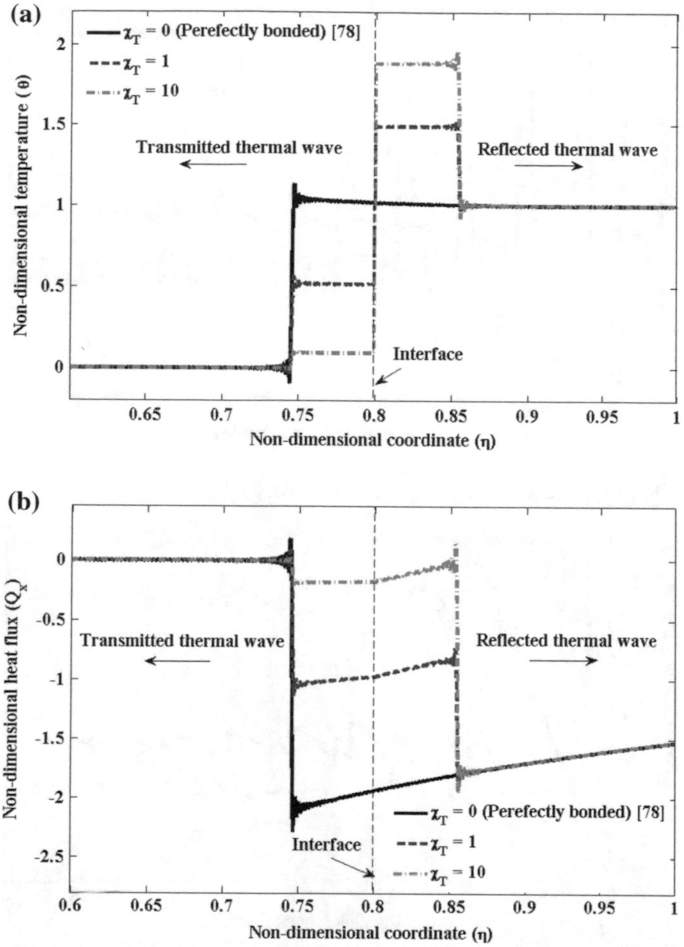

Fig. 2.14 Effect of thermal compliance on the distribution of **a** temperature and **b** heat-flux at dimensionless time $\zeta = 0.126$ ($m = 1$, $\varepsilon_0 = 0.35, \delta_0 = 0.25$, *Hyperbolic DPL*) [8]

outer radius $x_i = 1$. Both bonded layers are made of the same materials, i.e. copper (Cu). The outer surface of the bilayered cylinder is subjected to a sudden temperature rise $T_{wo} = 600$ K and the hyperbolic DPL heat conduction model is used for the thermal analysis. Figure 2.14a, b depict temperature and heat flux at the non-dimensional time $\zeta = 0.126$. Thermal excitation causes the thermal wave to propagate towards the inner surface of the cylinder. The bonding imperfection causes the separation of the initial thermal wave into the transmitted and a reflected parts, each travelling, at a given thermal wave speed of C_{DPL}, towards the inner and outer surfaces of cylinder, respectively. While temperature is discontinuous at the interface of the bilayered cylinder, the radial heat flux is continuous at the interface which is compatible with the thermal boundary conditions considered in Eq. (2.47).

Figure 2.15 reveals that the absolute value of the transient and steady-state temperature and heat flux in the middle of the inner layer ($\eta = 0.7$) decreases by

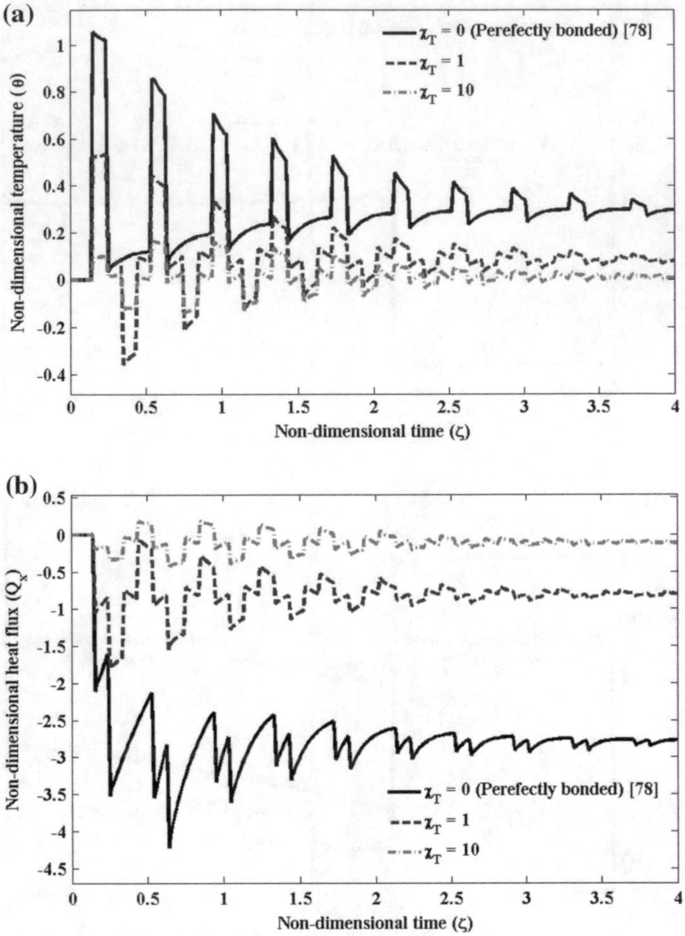

Fig. 2.15 Effect of thermal compliance on: **a** temperature and **b** heat flux time-histories at $\eta = 0.7$ $(m = 1, \varepsilon_0 = 0.35, \delta_0 = 0.25, Hyperbolic\ DPL)$ [8]

stiffening the thermal compliance of the bonding interface. Figures 2.14 and 2.15 show that thermal compliance of imperfectly bonded interface amplifies temperature difference at the interface of cylinder and reduces the heat flux transmitted through the interface.

2.5.2 *Effect of Material Heterogeneity on Thermal Wave Propagation*

Although bonding imperfections often trigger thermal wave separation in a multilayered system, material heterogeneity is generally the main culprit for the thermal wave separation. Figure 2.16 shows the influence of heterogeneity of the middle

(a)
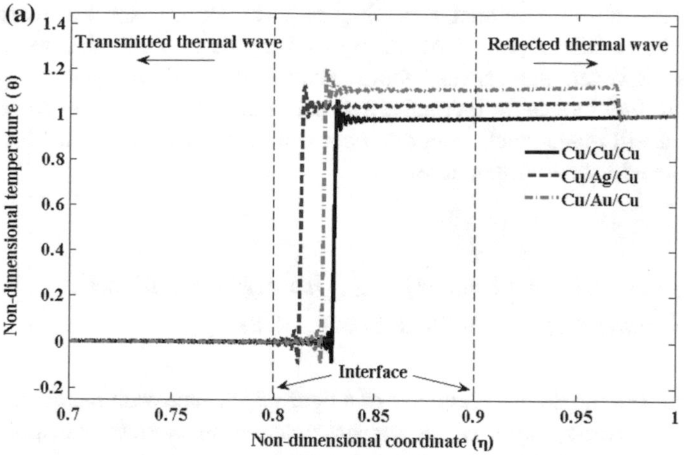

(b)
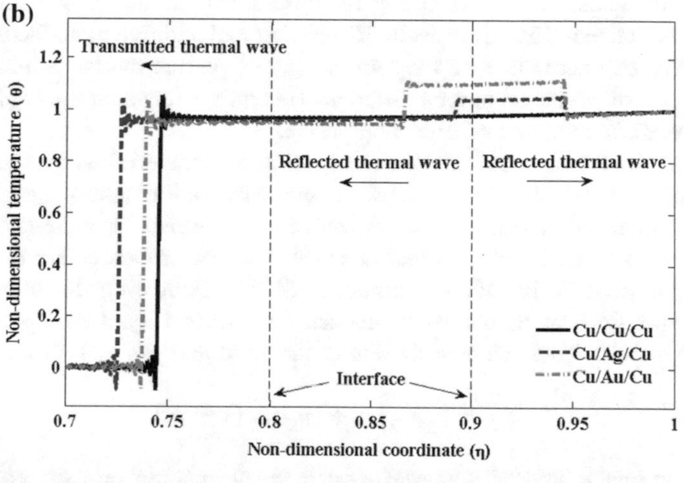

Fig. 2.16 Effect of material heterogeneity in the middle layer of a sandwich slab on temperature distribution at **a** $\zeta = 0.084$ and **b** $\zeta = 0.126$ non-dimensional time. ($m = 0, n = 3, \chi_T = 0,\ \varepsilon_0 = 0.35, \delta_0 = 0.25, \alpha_0 = 0.15, C_T^2 = 2$, *Hyperbolic TPL*) [8]

layer in a sandwich slab ($m = 0$) on thermal responses predicted by the hyperbolic TPL model. The position of the inner and outer surfaces are assumed to be $x_i = 0.7$ and $x_o = 1$. The sandwich slab is perfectly bonded ($\chi_T = 0$) and is made by the inner and outer layers of Cu while the middle layer of the slab is of Cu, Ag, or Au. We track the thermal waves propagation towards the inner layer of the slab for three different layer arrangements: Cu/Cu/Cu, Cu/Ag/Cu, and Cu/Au/Cu. Thermal wave separation occurs only for the heterogeneous material arrangements of Cu/Ag/Cu, and Cu/Au/Cu. This is caused by the meeting of thermal wave front at the interface between layers with dissimilar material properties, thereby generating waves with reflected and transmitted portions. This phenomenon is depicted in Fig. 2.16a, b at two different non-dimensional time. Figure 2.16a shows that the thermal wave speed at the middle layer of Ag is higher than that of Au due to its thermal properties; however, the transient temperature at the middle layer of Ag is lower than Au. If time increases, thermal wave will transmit and reflects from all interfaces and it will reflect back from the inner and outer surfaces of the slab to finally reach the steady-state temperature.

2.5.3 Thermal Response of a Lightweight Sandwich Circular Panel with a Porous Core

We present here the thermal response of a lightweight sandwich panel with a porous core, which is commonly used for aerospace, automotive, electronics, biomedicine applications. These lightweight porous materials are recently of great significance due to their capabilities for satisfying multiple functionalities if their microarchitecture is optimized. The advances in 3D printing and additive manufacturing have also enabled engineers to tune the morphology of porous materials and fabricate this new type of advanced porous materials. Herein, we focus on the application of porous materials as a core of sandwich panels.

Since a fully detailed microscale analysis of porous materials is computationally expensive, a multiscale model based on homogenization theory and micromechanics is commonly used. In homogenization, a representative volume element of the porous materials is selected and effective material properties are obtained by applying the periodic boundary conditions [29–32]. Following this approach, the effective specific heat of porous foams can be obtained by the classical rule of mixture which includes the contribution of the solid and gas as follows [33]:

$$\overline{\rho c_p} = (\rho_s c_{p_s})\rho_r + (\rho_g c_{p_g})(1 - \rho_r) \tag{2.49}$$

where subscripts "*s*" and "*g*" represent respectively the material properties of solid and gas, and the overbar parameters specify the effective material properties of foams; ρ_r is the relative density defined as: $\rho_r = \frac{\overline{\rho}}{\rho_s}$. If the heat convection of the gas is neglected due to the small size of pores, we can approximate its thermal conductivity [33]:

$$\overline{K} = \frac{1}{3}\left(\rho_r + 2\rho_r^{\frac{3}{2}}\right)K_s + (1 - \rho_r)K_g \tag{2.50}$$

We can assume that the specific heat and thermal conductivity of the gas to be equal to those of dry air: $\rho_g c_{p_g} = 1.006 \times 10^3$ J/kgK and $K_g = 0.025$ W/mK. We use the effective material properties presented in Eqs. (2.49) and (2.50) to investigate heat conduction in sandwich panels with a porous core.

When thermal disturbance hits a sandwich panel, the relative density of the cellular foam core controls the thermal response of the lightweight sandwich structure. We examine a three-layer perfectly bonded sandwich cylinder ($m = 1$) with inner and outer solid layers of Cu and a porous middle layer of Cu with the relative density of ρ_r. The inner and outer radius of the sandwich cylinder are $x_i = 0.7$ and $x_o = 1$, respectively. The thermal compliance constants for all interfaces are expressed by the index: $\chi_T = \frac{\chi_T^{(j)} K^{(1)}}{x_o}$ $(j = 1, 2, \ldots, N-1)$.

Temperature and heat flux distribution obtained by the hyperbolic DPL model is shown in Figs. 2.17 and 2.18. The figures confirm the impact of relative density; in particular in porous middle layer, the thermal wave speed decreases with the relative density. The reduction of relative density in the middle layer increases the transient temperature in the middle and outer layers of the cylinder, while it

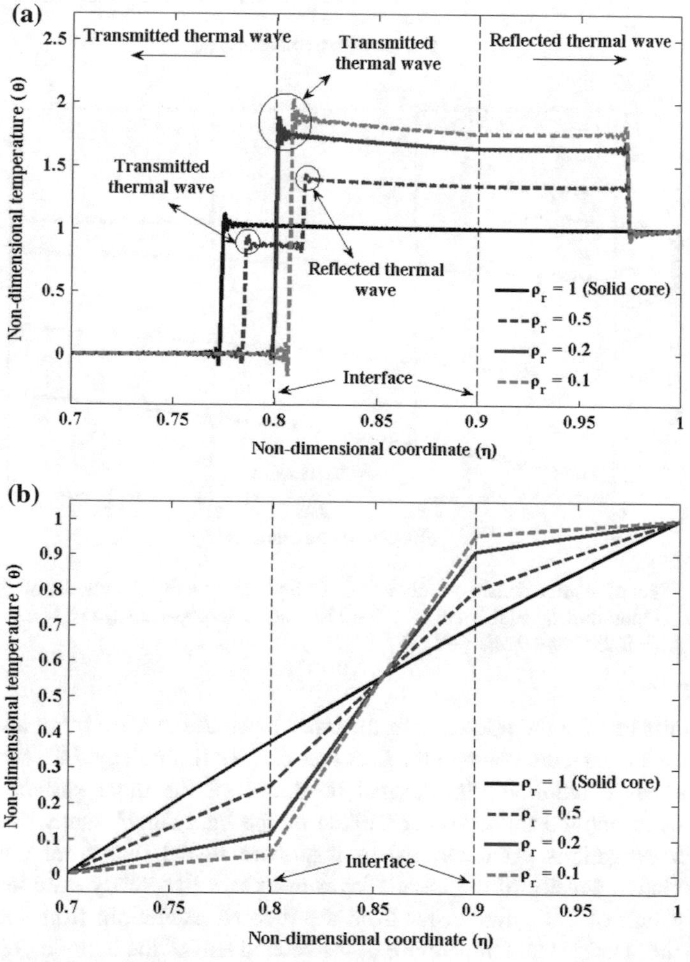

Fig. 2.17 Effect of relative density of the porous middle layer in a sandwich cylinder on the temperature distribution at non-dimensional time: **a** $\zeta = 0.112$ and **b** steady-state ($m = 1$, $n = 3$, $\chi_T = 0$, $\varepsilon_0 = 0.35$, $\delta_0 = 0.25$, *Hyperbolic DPL*) [8]

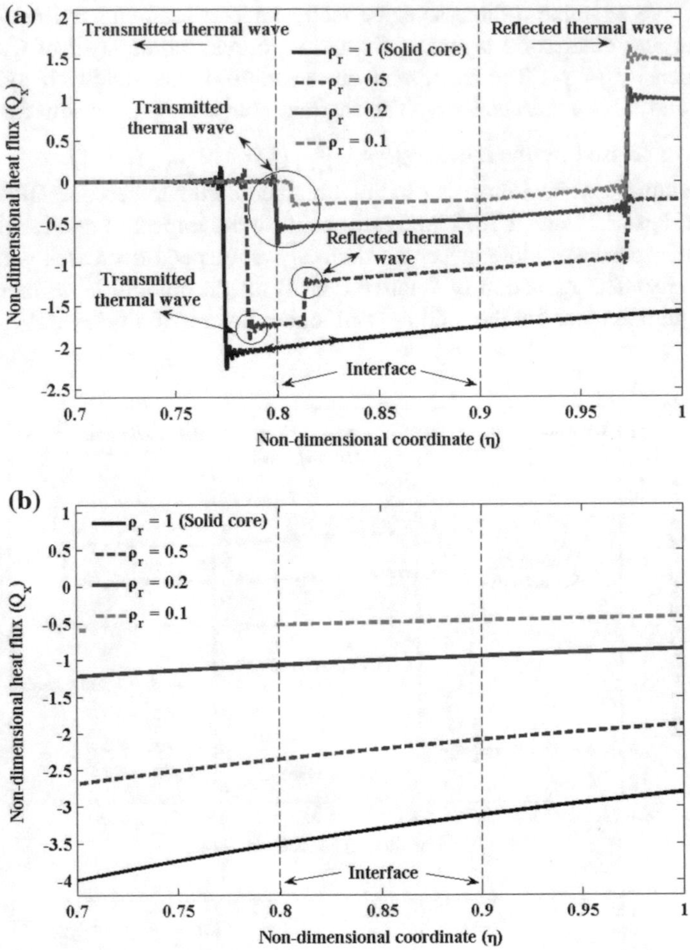

Fig. 2.18 Effect of relative density of porous middle layer in a sandwich cylinder on the heat flux distribution at non-dimensional time: **a** $\zeta = 0.112$ and **b** steady-state ($m = 1$, $n = 3$, $\chi_T = 0$, $\varepsilon_0 = 0.35$, $\delta_0 = 0.25$, *Hyperbolic DPL*) [8]

decreases the transient temperature in the inner layer. A decrease in relative density also lessens the absolute value of the heat flux passing through the inner and middle layer, whereas it amplifies the thermal insulation of the inner surface from the thermal shock applied on the outer surface of the lightweight sandwich cylinder. Further investigations on the transient responses reveal that from one hand a reduced relative density of the middle layer decreases the steady-state temperature of the first half of the cylinder, far from the thermal excitation; from the other, it increases the steady-state temperature of the second half of the cylinder, close to the thermal disturbance. The heat flux within the cylinder, however, decreases with the relative density of the middle layer [8].

It is worth mentioning that the multilayered methodology presented in this section can also be used for investigating the thermal response of FG solid and porous materials. In this case, FG materials are divided into a finite number of homogenous layers in which the material properties of each layer are obtained according to the function associated with the variation of material properties in FG materials. This method is called "piecewise homogenous layer" for the simulation of FGMs, which is consistent with the methods which are used by 3D printing or additive manufacturing techniques for fabrication of FGMs.

2.6 Dual-Phase-Lag Heat Conduction in Multi-dimensional Media

All heat conduction problems yet discussed in this chapter have been limited to 1D media. However, the heat conduction differential equations introduced earlier in this chapter can be applied to 2D and 3D problems in all coordinate systems. The solution procedure for 2D and 3D problems are usually more complex than 1D problems. For general types of thermal boundary conditions, numerical method, e.g. finite element method or boundary element method, are efficient for solving the heat conduction problems in spatial coordinate systems. Closed-form solutions for 2D and 3D problems may also be developed for specific thermal boundary conditions. In this section, we introduce a semi-analytical methodology for the heat conduction analysis of 2D and 3D problems in the form of cylindrical/spherical panel based on the DPL model.

2.6.1 DPL Heat Conduction in Multi-dimensional Cylindrical Panels

We consider a radially graded FG cylindrical panel of the inner and outer radii r_i and r_o, azimuthal angle φ_0, and length L. As shown in Fig. 2.19, we consider the cylindrical coordinate system (r, φ, z). The initial temperature of the FG cylindrical panel is assumed to be ambient temperature T_∞.

The DPL heat conduction equation in heterogeneous materials can be written as [34]:

$$\left(1+\tau_q\frac{\partial}{\partial t}+\frac{\tau_q^2}{2}\frac{\partial^2}{\partial t^2}\right)\left(\rho c_p\frac{\partial T}{\partial t}-R\right)=\vec{\nabla}.\left(K\left(1+\tau_T\frac{\partial}{\partial t}\right)\vec{\nabla}T\right) \tag{2.51}$$

To simplify the solution procedure, we assume that phase lags to be constant. Quintanilla [35] proved the stability of Eq. (2.51) for $\frac{\tau_\theta}{\tau_q}>\frac{1}{2}$. To investigate the

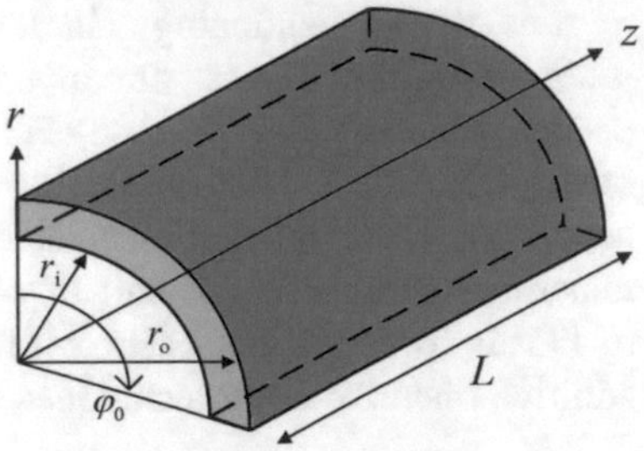

Fig. 2.19 Radially graded FG cylindrical panel. [Reproduced from [34] with permission from Springer]

temperature field in the heterogeneous cylindrical structures, heat conduction Eq. (2.51) can be written in 3D cylindrical coordinate system (r, φ, z):

$$\begin{aligned}\left(1+\tau_q\frac{\partial}{\partial t}+\frac{\tau_q^2}{2}\frac{\partial^2}{\partial t^2}\right)\rho c_p\frac{\partial T}{\partial t}&=\frac{1}{r}\frac{\partial}{\partial r}\left[rK\left\{\frac{\partial T}{\partial r}+\tau_T\frac{\partial^2 T}{\partial t\partial r}\right\}\right]\\&+\frac{1}{r^2}\frac{\partial}{\partial\varphi}\left[K\left\{\frac{\partial T}{\partial\varphi}+\tau_T\frac{\partial^2 T}{\partial t\partial\varphi}\right\}\right]+\frac{\partial}{\partial z}\left[K\left\{\frac{\partial T}{\partial z}+\tau_T\frac{\partial^2 T}{\partial t\partial z}\right\}\right]\end{aligned}\tag{2.52}$$

Considering the thermal boundary conditions and the geometry of the cylindrical panel, Eq. (2.52) can be used to solve the DPL heat conduction problem in an axisymmetric cylinder with an infinite length (r space), a cylinder with finite length [(r, z) space], an infinitely long cylindrical panel [(r, φ) space], and a cylindrical panel with finite length [(r, φ, z) space]. The FG cylindrical panels are assumed to have a continuous transition of material properties, except for the phase lags, according to a power law formulation:

$$K(\eta)=K_0\eta^{n_1},\ \rho(\eta)=\rho_0\eta^{n_2},\ c_p(\eta)=c_{p0}\eta^{n_3}\tag{2.53}$$

where K_0, ρ_0, and c_{p0} are constants; n_j $(j = 1, 2, 3)$ are non-homogeneity indices and $\eta=\frac{r}{r_0}$. To simplify our analysis, the following non-dimensional parameters are used:

$$\begin{aligned}&\zeta=\frac{K_0' t}{r_o^2},\ \varepsilon_0=\frac{K_0'\tau_q}{r_o^2},\ \delta_0=\frac{K_0'\tau_T}{r_o^2},\ \eta=\frac{r}{r_o},\ \overline{\varphi}=\frac{\varphi}{\varphi_o},\ \overline{z}=\frac{z}{L},\ r_\gamma=\frac{r_i}{r_o}\\&\theta=\frac{T-T_\infty}{T_{wo}-T_\infty},\ T_\gamma=\frac{T_{wi}-T_\infty}{T_{wo}-T_\infty},\ Q_r=\frac{r_o q_r}{K_0 T_\infty},\ Q_\varphi=\frac{r_o\varphi_0 q_\varphi}{K_0 T_\infty},\ Q_z=\frac{L q_z}{K_0 T_\infty}\end{aligned}\tag{2.54}$$

where T_{wi} and T_{wo} are temperature on the inner and outer surfaces of the cylindrical panel and $K_0'=\frac{K_0}{\rho_0 c_0}$; q_r, q_φ, and q_z are the radial, azimuthal, and longitudinal heat fluxes, respectively.

We consider an FG cylindrical panel with a length L and azimuthal angle φ_0. The non-dimensional form of the DPL heat conduction equation for the FG panel is written as:

$$\eta^{n_2+n_3}\left(1+\varepsilon_0\frac{\partial}{\partial\zeta}+\frac{\varepsilon_0^2}{2}\frac{\partial^2}{\partial\zeta^2}\right)\frac{\partial\theta}{\partial\zeta}$$
$$=\left[\delta_0\frac{\partial}{\partial\zeta}+1\right]\left[\eta^{n_1}\frac{\partial^2\theta}{\partial\eta^2}+(n_1+1)\eta^{n_1-1}\frac{\partial\theta}{\partial\eta}+\frac{1}{\varphi_0^2}\eta^{n_1-2}\frac{\partial^2\theta}{\partial\overline{\varphi}^2}+\left(\frac{r_o}{L}\right)^2\eta^{n_1}\frac{\partial^2\theta}{\partial\overline{z}^2}\right] \tag{2.55}$$

To develop a semi-analytical solution for the heat conduction Eq. (2.55), we assume the thermal boundary and initial conditions in the following form:

$$\begin{aligned}\theta(\eta,\overline{\varphi},\overline{z},\zeta)|_{\eta=r_\gamma}&=T_\gamma\sin(m\pi\overline{\varphi})\sin(p\pi\overline{z})\\ \theta(\eta,\overline{\varphi},\overline{z},\zeta)|_{\eta=1}&=\sin(m\pi\overline{\varphi})\sin(p\pi\overline{z})\end{aligned} \tag{2.56a}$$

$$\theta(\eta,\overline{\varphi},\overline{z},\zeta)|_{\zeta=0}=0,\quad \frac{\partial}{\partial\zeta}\theta(\eta,\overline{\varphi},\overline{z},\zeta)|_{\zeta=0}=0 \tag{2.56b}$$

To satisfy the thermal boundary conditions of Eq. (2.56a), temperature change is assumed as follows:

$$\theta(\eta,\overline{\varphi},\overline{z},\zeta)=\theta_{mp}(\eta,\zeta)\sin(m\pi\overline{\varphi})\sin(p\pi\overline{z}) \tag{2.57}$$

where m and p stand for the mode number and $\theta_{mp}(\eta,\zeta)$ is an unknown function for temperature. Using Eqs. (2.55) and (2.57) and implementing the Laplace transform lead to:

$$\eta^2\frac{\partial^2\tilde{\theta}_{mp}}{\partial\eta^2}+(n_1+1)\eta\frac{\partial\tilde{\theta}_{mp}}{\partial\eta}-\left(E\eta^{n_2+n_3-n_1+2}+\left(\frac{n\pi r_o}{L}\right)^2\eta^2+\left(\frac{p\pi}{\varphi_0}\right)^2\right)\tilde{\theta}_{mp}=0 \tag{2.58}$$

where E is defined as: $E=\frac{s\left(\frac{\varepsilon_0^2}{2}s^2+\varepsilon_0 s+1\right)}{\delta_0 s+1}$. For $n_1=n_2+n_3$, the solution of Eq. (2.58) can be found as:

$$\tilde{\theta}_{mp}(\eta,s)=\eta^{-\frac{n_1}{2}}(A_1J_G(I\eta)+A_2Y_G(I\eta)) \tag{2.59}$$

where

$$G=\sqrt{\left(\frac{n_1}{2}\right)^2+\left(\frac{m\pi}{\varphi_0}\right)^2},I=i\sqrt{E+\left(\frac{p\pi r_o}{L}\right)^2} \tag{2.60}$$

and the integration constants A_1 and A_2 are obtained as:

$$\begin{aligned} A_1 &= -\frac{Y_G(Ir_\gamma) + r_\gamma^{\frac{n_1}{2}} T_\gamma Y_G(I)}{s\left(-Y_G(Ir_\gamma)J_G(I) + J_G(Ir_\gamma)Y_G(I)\right)} \\ A_2 &= \frac{J_G(Ir_\gamma) - r_\gamma^{\frac{n_1}{2}} T_\gamma J_G(I)}{s\left(-Y_G(Ir_\gamma)J_G(I) + J_G(Ir_\gamma)Y_G(I)\right)} \end{aligned} \tag{2.61}$$

The radial, azimuthal, and longitudinal heat fluxes in the Laplace domain can be obtained as:

$$\begin{aligned} \tilde{Q}_r(\eta, \overline{\varphi}, \overline{z}, s) &= \frac{P}{2}\eta^{\frac{n_1}{2}-1}\left[A_1\left(MJ_G(I\eta) - 2HI\eta^H J_{G+1}(I\eta)\right)\right. \\ &\quad \left. + A_2\left(MY_G(I\eta) - 2HI\eta^H Y_{G+1}(I\eta)\right)\right]\sin(m\pi\overline{\varphi})\sin(p\pi\overline{z}) \\ \tilde{Q}_\varphi(\eta, \overline{\varphi}, \overline{z}, s) &= (m\pi P)\eta^{\frac{n_1}{2}-1}\left[A_1 J_G(I\eta) + A_2 Y_G(I\eta)\right]\cos(m\pi\overline{\varphi})\sin(p\pi\overline{z}) \\ \tilde{Q}_z(\eta, \overline{\varphi}, \overline{z}, s) &= (p\pi P)\eta^{\frac{n_1}{2}}\left[A_1 J_G(I\eta) + A_2 Y_G(I\eta)\right]\sin(m\pi\overline{\varphi})\cos(p\pi\overline{z}) \end{aligned} \tag{2.62}$$

where P and M are defined as:

$$P = -\left(\frac{T_{wo} - T_\infty}{T_\infty}\right)\left(\frac{\delta_0 s + 1}{\frac{\varepsilon_0^2}{2}s^3 + \varepsilon_0 s^2 + s}\right), M = 2G - n_1 \tag{2.63}$$

The steady-state temperature change can also be defined as:

$$\theta_s(\eta, \overline{\varphi}, \overline{z}) = \theta_{mps}(\eta)\sin(m\pi\overline{\varphi})\sin(p\pi\overline{z}) \tag{2.64}$$

For $n_1 = n_2 + n_3$, Eq. (2.64) is substituted into the steady-state form of Eq. (2.55):

$$\eta^2\frac{\partial^2\theta_{mps}}{\partial\eta^2} + (n_1 + 1)\eta\frac{\partial\theta_{mps}}{\partial\eta} - \left(\left(\frac{m\pi}{\varphi_0}\right)^2 + \left(\frac{p\pi r_o}{L}\right)^2\right)\theta_{mps} = 0 \tag{2.65}$$

The solution of Eq. (2.65) can be obtained as follows

$$\theta_{mps}(\eta) = \eta^{-\frac{n_1}{2}}\left(B_1 J_{G_s}(I_s\eta) + B_2 Y_{G_s}(I_s\eta)\right) \tag{2.66}$$

where $G_s = \sqrt{\left(\frac{n_1}{2}\right)^2 + \left(\frac{m\pi}{\varphi_0}\right)^2}$ and $I_s = \left(\frac{p\pi r_o}{L}\right)i$. The integration constants B_1 and B_2 in Eq. (2.66) are obtained as:

$$B_1 = \frac{-Y_{G_s}(I_s r_\gamma) + r_\gamma^{\frac{n_1}{2}} T_\gamma Y_{G_s}(I_s)}{-Y_{G_s}(I_s r_\gamma) J_{G_s}(I_s) + J_{G_s}(I_s r_\gamma) Y_{G_s}(I_s)}$$
$$B_2 = \frac{J_{G_s}(I_s r_\gamma) - r_\gamma^{\frac{n_1}{2}} T_\gamma J_{G_s}(I_s)}{-Y_{G_s}(I_s r_\gamma) J_{G_s}(I_s) + J_{G_s}(I_s r_\gamma) Y_{G_s}(I_s)} \tag{2.67}$$

Finally, temperature and heat flux are retrieved in the time domain by implementing a numerical Laplace inversion technique. It is worth mentioning that Eq. (2.55) shows the thermal wave speed in the radial direction based on the DPL model $[C_{DPL}(\eta)]$ within axisymmetric infinitely-long hollow FG cylinder depends on the location of thermal wavefront since $[C_{DPL}(\eta)]$ can be expressed as:

$$C_{DPL}(\eta) = \sqrt{\frac{2\delta_0}{\varepsilon_0}\eta^{\frac{n_1 - n_2 - n_3}{2}}} \tag{2.68}$$

According to Eq. (2.68), when the non-homogeneity indices follow an specific relation, i.e. $n_1 = n_2 + n_3$, the radial thermal wave speed is independent of both the radial coordinate and the non-homogeneity indices. To numerically confirm that the mathematical conclusion is also valid for cylindrical panels with a finite length, the temperature distribution along the radial direction is depicted in Fig. 2.20 for an FG cylindrical panel with the azimuthal angle $\phi_0 = \frac{\pi}{2}$ and length $L = 1$ for different combinations of non-homogeneity indices, in which $n_1 = n_2 + n_3$. The radial temperature distribution has been shown at the mid-section of the panel with $\overline{\phi} = 0.5$ and $\overline{z} = 0.5$. As shown in Fig. 2.20, all the wavefronts are in the same position at the non-dimensional time $\zeta = 0.126$ independent of the value of non-homogeneity

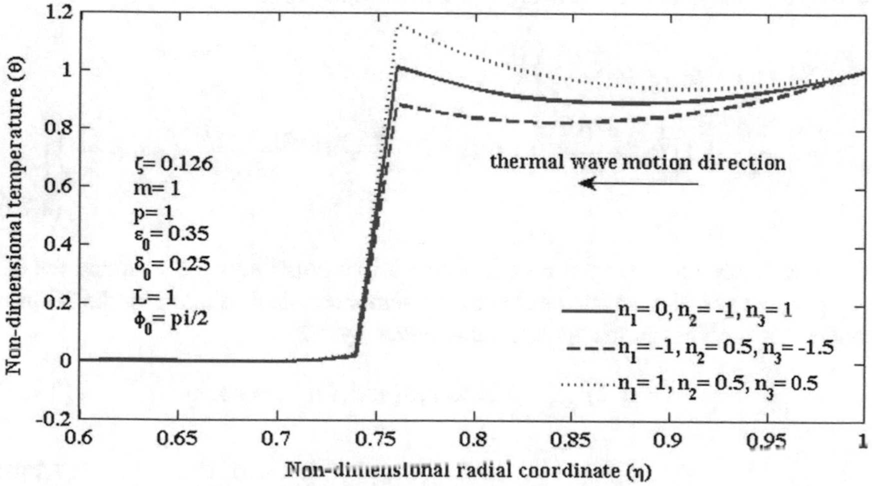

Fig. 2.20 In an FG cylindrical panel with a finite length ($\phi_0 = \frac{\pi}{2}$ and $L = 1$), thermal wavefront location is independent of the non-homogeneity indices when $n_1 = n_2 + n_3$ [34]

indices. This observation corroborate the DPL thermal wave speed is the same at different locations of FG cylindrical panels when $n_1 = n_2 + n_3$.

2.6.2 DPL Heat Conduction in Multi-dimensional Spherical Vessels

To obtain temperature field in heterogeneous spherical vessels, we should write DPL heat conduction Eq. (2.51) in the spherical coordinate system (r, φ, ψ) presented in Table 1.1. The thermal boundary conditions for the spherical vessel can be assumed to be: (1) Spherically symmetric one-dimensional or (2) Axisymmetric two-dimensional. As a result, the heat conduction equation is simplified in the form of the following partial differential equations in the spherical coordinate system:

$$\begin{aligned}&\left(1+\tau_q\frac{\partial}{\partial t}+\frac{\tau_q^2}{2}\frac{\partial^2}{\partial t^2}\right)\rho c_p\frac{\partial T}{\partial t}\\&=\frac{1}{r^2}\frac{\partial}{\partial r}\left[Kr^2\left(1+\tau_T\frac{\partial}{\partial t}\right)\frac{\partial T}{\partial r}\right]+\frac{1}{r\sin\varphi}\frac{\partial}{\partial\varphi}\left[\frac{\sin\varphi}{r}K\left(1+\tau_T\frac{\partial}{\partial t}\right)\frac{\partial T}{\partial\varphi}\right]\end{aligned}\tag{2.69}$$

In this section, we focus on the heat conduction in spherically axisymmetric two-dimensional problem. The material properties of the spherical vessel is assumed to vary radially according to the power law formulation, similar to those introduced in Sect. 2.6.1. Except for phase lags, which are assumed constant, all other thermal properties varies according to Eq. (2.53). To simplify the solution procedure, we employ the non-dimensional parameters of Eq. (2.54).

Using the assumed material properties for the FG spherical vessel, we can rewrite Eq. (2.69) in the following non-dimensional form:

$$\begin{aligned}&\eta^{n_2+n_3}\left(1+\varepsilon_0\frac{\partial}{\partial\zeta}+\frac{\varepsilon_0^2}{2}\frac{\partial^2}{\partial\zeta^2}\right)\frac{\partial\theta}{\partial\zeta}\\&=\left[\delta_0\frac{\partial}{\partial\zeta}+1\right]\left[\eta^{n_1}\frac{\partial^2\theta}{\partial\eta^2}+(n_1+2)\eta^{n_1-1}\frac{\partial\theta}{\partial\eta}+\eta^{n_1-2}\frac{1}{\sin\varphi}\frac{\partial}{\partial\varphi}\left(\sin\varphi\frac{\partial\theta}{\partial\varphi}\right)\right]\end{aligned}\tag{2.70}$$

The axisymmetric thermal boundary and initial conditions are assumed for the FG spherical vessel to enable us obtaining a semi-analytical solution for the 2D heat conduction problem in the spherical coordinate system [36]:

$$\theta(\eta,\varphi,\zeta)|_{\eta=r_\gamma}=T_\gamma\cos\varphi,\theta(\eta,\varphi,\zeta)|_{\eta=1}=\cos\varphi\tag{2.71a}$$

$$\theta(\eta,\varphi,\zeta)|_{\zeta=0}=0,\frac{\partial}{\partial\zeta}\theta(\eta,\varphi,\zeta)|_{\zeta=0}=0\tag{2.71b}$$

Considering the above-mentioned thermal boundary conditions, temperature can be written as:

$$\theta(\eta, \varphi, \zeta) = \theta_1(\eta, \zeta)\cos\varphi \tag{2.72}$$

where $\theta_1(\eta, \zeta)$ is an unknown temperature that is needed to be determined using the initial and boundary thermal conditions. Substituting Eq. (2.72) into Eq. (2.70) and performing the Laplace transform with regard to the initial conditions (2.71b), lead to:

$$\eta^2 \frac{\partial^2 \tilde{\theta}_1}{\partial \eta^2} + (n_1 + 2)\eta \frac{\partial \tilde{\theta}_1}{\partial \eta} - \left(E\eta^{n_2+n_3-n_1+2} + 2\right)\tilde{\theta}_1 = 0 \tag{2.73}$$

where E is defined as: $E = \dfrac{s\left(\frac{\varepsilon_0^2}{2}s^2 + \varepsilon_0 s + 1\right)}{\delta_0 s + 1}$. The differential Eq. (2.73) can be solved as:

$$\begin{aligned} &\tilde{\theta}(\eta, s) = \eta^{-\frac{n_1+1}{2}}\left(A_1 J_G(I\eta^H) + A_2 Y_G(I\eta^H)\right) \qquad (for\, n_1 - n_2 - n_3 \neq 2) \\ &\tilde{\theta}(\eta, s) = A_1\eta^{\lambda_1} + A_2\eta^{\lambda_2} \qquad (for\, n_1 - n_2 - n_3 = 2) \end{aligned} \tag{2.74}$$

where

$$\begin{aligned} &G = \frac{2\sqrt{\left(\frac{n_1+1}{2}\right)^2 + 2}}{n_2 + n_3 - n_1 + 2}, \lambda_{1,2} = \frac{-(n_1+1) \pm \sqrt{(n_1+1)^2 + 4(E+2)}}{2} \\ &H = 1 + \frac{n_2 + n_3 - n_1}{2}, \; I = \frac{2\sqrt{-E}}{n_2 + n_3 - n_1 + 2} \end{aligned} \tag{2.75}$$

The integration constants A_1 and A_2 can be also obtained in the Laplace domain as:

For $n_1 - n_2 - n_3 \neq 2$:

$$\begin{aligned} A_1 &= \frac{-Y_G\left(Ir_\gamma^H\right) + r_\gamma^{\frac{n_1+1}{2}} T_\gamma Y_G(I)}{s\left(-Y_G\left(Ir_\gamma^H\right)J_G(I) + J_G\left(Ir_\gamma^H\right)Y_G(I)\right)} \\ A_2 &= \frac{J_G\left(Ir_\gamma^H\right) - r_\gamma^{\frac{n_1+1}{2}} T_\gamma J_G(I)}{s\left(-Y_G\left(Ir_\gamma^H\right)J_G(I) + J_G\left(Ir_\gamma^H\right)Y_G(I)\right)} \end{aligned} \tag{2.76a}$$

For $n_1 - n_2 - n_3 = 2$:

$$A_1 = \frac{T_\gamma - r_\gamma^{\lambda_2}}{s\left(r_\gamma^{\lambda_1} - r_\gamma^{\lambda_2}\right)}$$
$$A_2 = \frac{r_\gamma^{\lambda_1} - T_\gamma}{s\left(r_\gamma^{\lambda_1} - r_\gamma^{\lambda_2}\right)} \tag{2.76b}$$

Non-dimensional radial and polar heat fluxes in the Laplace domain are also obtained as:

For $n_1 - n_2 - n_3 \neq 2$:

$$\begin{aligned} \tilde{Q}_r(\eta, \varphi, s) = {} & \frac{P}{2}\eta^{\frac{n_1-3}{2}}\left[A_1\left(MJ_G\left(I\eta^H\right) - 2IH\eta^H J_{G+1}\left(I\eta^H\right)\right)\right. \\ & \left. + A_2\left(MY_G\left(I\eta^H\right) - 2IH\eta^H Y_{G+1}\left(I\eta^H\right)\right)\right]\cos\varphi \\ \tilde{Q}_\varphi(\eta, \varphi, s) = {} & -P\eta^{\frac{n_1-3}{2}}\left[A_1 J_G\left(I\eta^H\right) + A_2 Y_G\left(I\eta^H\right)\right]\sin\varphi \end{aligned} \tag{2.77a}$$

and for $n_1 - n_2 - n_3 = 2$:

$$\begin{aligned} \tilde{Q}_r\left(\eta, \overline{\phi}, s\right) &= P\left(A_1\lambda_1\eta^{n_1+\lambda_1-1} + A_2\lambda_2\eta^{n_1+\lambda_2-1}\right)\cos\varphi \\ \tilde{Q}_\phi\left(\eta, \overline{\phi}, s\right) &= -P\left(A_1\eta^{n_1+\lambda_1-1} + A_2\eta^{n_1+\lambda_2-1}\right)\sin\varphi \end{aligned} \tag{2.77b}$$

where M and P are defined as:

$$M = 2GH - (n_1 + 1),\ P = -\left(\frac{T_{wo} - T_\infty}{T_\infty}\right)\left(\frac{\delta_0 s + 1}{\frac{\varepsilon_0^2}{2}s^3 + \varepsilon_0 s^2 + s}\right) \tag{2.78}$$

The temperature in time domain can then be obtained by implementing a numerical Laplace inversion technique. It is worth mentioning that while thermal wave can propagate in multiple directions in 2D or 3D spherical vessels, the radial thermal wave speed in all 1D, 2D, and 3D structures are the same.

To clarify the effects of each non-homogeneity indices on the thermal responses of an axisymmetric hollow sphere (2D) based on the DPL heat conduction theory, among the three different non-homogeneity indices n_1 (thermal conductivity index), n_2 (density index), and n_3 (specific heat index), two of them are kept constant and only one varies. Figure 2.21a–c show the temperature history of the mid-plane in three different cases: (a) $n_2 = n_3 = 1$; (b) $n_1 = n_3 = 1$; and (c) $n_1 = n_2 = 1$. Figure 2.21a reveals that increasing the non-homogeneity index of thermal conductivity n_1 leads to higher transient and steady-state temperature. Although increasing the non-homogeneity indices of density n_2 and specific heat n_3 increases the amplitudes of transient temperature, the steady-state temperature does not change by tailoring n_2 and n_3 as shown in Fig. 2.21b, c.

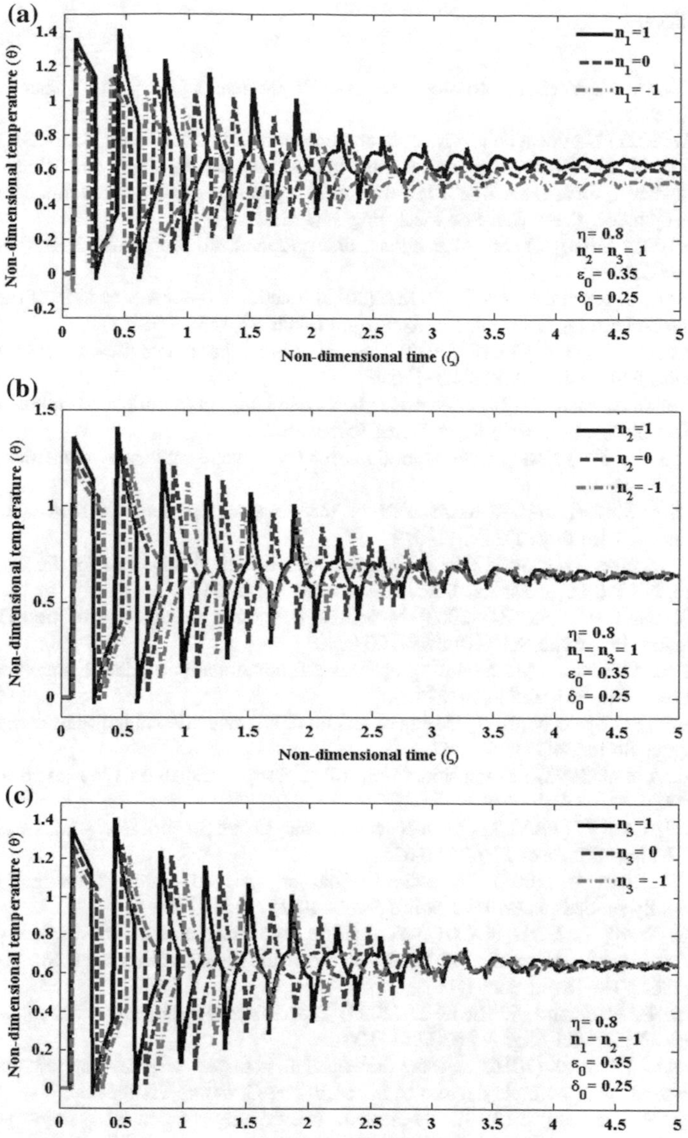

Fig. 2.21 Effect of non-homogeneity indices on temperature time-history of the mid-plane of the axisymmetric (2D) hollow sphere: **a** Effect of thermal conductivity, **b** Effect of density, and **c** Effect of specific heat. [Reproduced from [36] with permission from World Scientific Publishing Co., Inc]

References

1. Tzou D, Guo Z-Y (2010) Nonlocal behavior in thermal lagging. Int J Therm Sci 49(7): 1133–1137
2. Tzou D (2011) Nonlocal behavior in phonon transport. Int J Heat Mass Transf 54(1):475–481
3. Tzou DY (2014) Macro-to microscale heat transfer: the lagging behavior. Wiley
4. Akbarzadeh A, Fu J, Chen Z (2014) Three-phase-lag heat conduction in a functionally graded hollow cylinder. Trans Can Soc Mech Eng 38(1):155
5. Youssef HM (2010) Theory of fractional order generalized thermoelasticity. J Heat Transf 132(6):061301
6. Ezzat MA, El Karamany AS, Fayik MA (2012) Fractional order theory in thermoelastic solid with three-phase lag heat transfer. Arch Appl Mech 82(4):557–572
7. Akbarzadeh A, Cui Y, Chen Z (2017) Thermal wave: from nonlocal continuum to molecular dynamics. RSC Adv 7(22):13623–13636
8. Akbarzadeh A, Pasini D (2014) Phase-lag heat conduction in multilayered cellular media with imperfect bonds. Int J Heat Mass Transf 75:656–667
9. Plimpton S (1995) Fast parallel algorithms for short-range molecular dynamics. J Comput Phys 117(1):1–19
10. Nosé S (1984) A unified formulation of the constant temperature molecular dynamics methods. J Chem Phys 81(1):511–519
11. Foiles S, Baskes M, Daw M (1986) Embedded-atom-method functions for the fcc metals Cu, Ag, Au, Ni, Pd, Pt, and their alloys. Phys Rev B 33(12):7983
12. Liu Q, Jiang P, Xiang H (2008) Molecular dynamics simulations of non-Fourier heat conduction. Progr Nat Sci 18(8):999–1007
13. Tan T et al (2011) Mechanical properties of functionally graded hierarchical bamboo structures. Acta Biomater 7(10):3796–3803
14. Rahbar N, Soboyejo W (2011) Design of functionally graded dental multilayers. Fatigue Fract Eng Mater Struct 34(11):887–897
15. Miserez A et al (2008) The transition from stiff to compliant materials in squid beaks. Science 319(5871):1816–1819
16. Reddy J, Chin C (1998) Thermomechanical analysis of functionally graded cylinders and plates. J Therm Stresses 21(6):593–626
17. Vel SS, Batra R (2003) Three-dimensional analysis of transient thermal stresses in functionally graded plates. Int J Solids Struct 40(25):7181–7196
18. Guo L, Wang Z, Noda N (2012) A fracture mechanics model for a crack problem of functionally graded materials with stochastic mechanical properties. Proc R Soc A 468(2146), 2939–2961. The Royal Society
19. Kieback B, Neubrand A, Riedel H (2003) Processing techniques for functionally graded materials. Mater Sci Eng, A 362(1):81–106
20. Overvelde JT et al (2016) A three-dimensional actuated origami-inspired transformable metamaterial with multiple degrees of freedom. Nat Commun 7:10929
21. Bartlett NW et al (2015) A 3D-printed, functionally graded soft robot powered by combustion. Science 349(6244):161–165
22. Akbarzadeh A, Abedini A, Chen Z (2015) Effect of micromechanical models on structural responses of functionally graded plates. Compos Struct 119:598–609
23. Akbarzadeh A, Chen Z (2013) Heat conduction in one-dimensional functionally graded media based on the dual-phase-lag theory. Proc Inst Mech Eng Part C J Mech Eng Sci 227(4): 744–759
24. Babaei M, Chen Z (2008) Hyperbolic heat conduction in a functionally graded hollow sphere. Int J Thermophys 29(4):1457–1469
25. Choudhuri SR (2007) On a thermoelastic three-phase-lag model. J Therm Stresses 30(3): 231–238

26. Ramadan K (2009) Semi-analytical solutions for the dual phase lag heat conduction in multilayered media. Int J Therm Sci 48(1):14–25
27. Akbarzadeh A, Chen Z (2013) On the harmonic magnetoelastic behavior of a composite cylinder with an embedded polynomial eigenstrain. Compos Struct 106:296–305
28. Chen T (2001) Thermal conduction of a circular inclusion with variable interface parameter. Int J Solids Struct 38(17):3081–3097
29. Wang M, Pan N (2008) Modeling and prediction of the effective thermal conductivity of random open-cell porous foams. Int J Heat Mass Transf 51(5):1325–1331
30. Arabnejad S, Pasini D (2013) Mechanical properties of lattice materials via asymptotic homogenization and comparison with alternative homogenization methods. Int J Mech Sci 77:249–262
31. Akbarzadeh A et al (2016) Electrically conducting sandwich cylinder with a planar lattice core under prescribed eigenstrain and magnetic field. Compos Struct 153:632–644
32. Mirabolghasemi A, Akbarzadeh AH, Rodrigue D, Therriault D (2019) Thermal conductivity of architected cellular metamaterials. Acta Materialia 174:61–80
33. Ashby MF, Cebon D (1993) Materials selection in mechanical design. Le Journal de Physique IV 3(C7):C7-1–C7-9
34. Akbarzadeh A, Chen Z (2012) Transient heat conduction in a functionally graded cylindrical panel based on the dual phase lag theory. Int J Therm 33(6):1100–1125
35. Quintanilla R, Racke R (2006) A note on stability in dual-phase-lag heat conduction. Int J Heat Mass Transf 49(7):1209–1213
36. Akbarzadeh AH, Chen ZT (2014) Dual phase lag heat conduction in functionally graded hollow spheres. Int J Appl Mech 6(1):1450002

Chapter 3
Multiphysics of Smart Materials and Structures

3.1 Smart Materials

In this section, the definition of smart materials are presented. The concept of multiphysics is introduced and different types of coupled multiphysical fields are elucidated. Moreover, the piezoelectric and piezomagnetic materials as the two commonly used smart materials are introduced. Finally, some potential applications of these advanced smart materials are mentioned.

Multiphysics involves the investigation of the interaction among different physical fields in multiple simultaneous physical phenomena; multiphysical simulation typically leads to a set of coupled systems of partial differential equations [1, 2]. As an example, the coupled physical fields can be displacement, electric potential, magnetic potential, temperature, and moisture concentration in a *hygrothermomagnetoelectroelastic* medium. The interaction of multiple physical fields may be observed in natural (wood, bone, and liquid crystals) or synthetic (piezoelectric, piezomagnetic, magnetoelectroelastic, magnetostrictive, and polyelectrolyte gel) smart materials [3]. Piezoelectric materials exhibit interesting phenomena; as seen in Fig. 3.1, an electric field is generated when piezoelectric materials are mechanically deformed and vice versa. The intrinsic property makes the mechanical displacement and electric potential coupled. Common piezoelectric materials are made of ceramics subjected to a strong DC electric field so that a permanent dipole moment is aligned during the poling process. The process induces the piezoelectricity and anisotropy property in piezoelectric materials [4, 5].

Piezomagnetic materials possess a magnetoelastic coupling similar to the electroelastic coupling in piezoelectric materials; they are mechanically strained when subjected to a magnetic field and vice versa. In the same way, magnetoelectric coupling is identified as the effect of magnetic (or electric) field on the dielectric polarization (or magnetization) of smart materials. This phenomenon can be observed directly in single-phase multiferroics, or indirectly by stress or strain in magnetoelectroelastic (MEE) composites. As seen in Fig. 3.2, the in-plane

Z. T. Chen and A. H. Akbarzadeh, *Advanced Thermal Stress Analysis of Smart Materials and Structures*, Structural Integrity 10,
https://doi.org/10.1007/978-3-030-25201-4_3

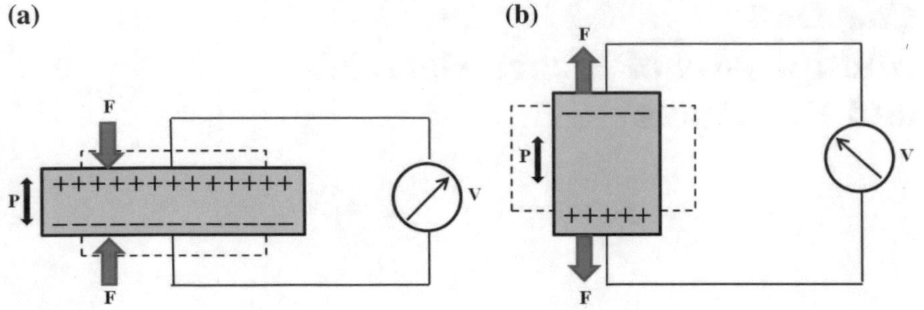

Fig. 3.1 Effect of mechanical stresses on the generation of electric potential in a vertically polarized piezoelectric material

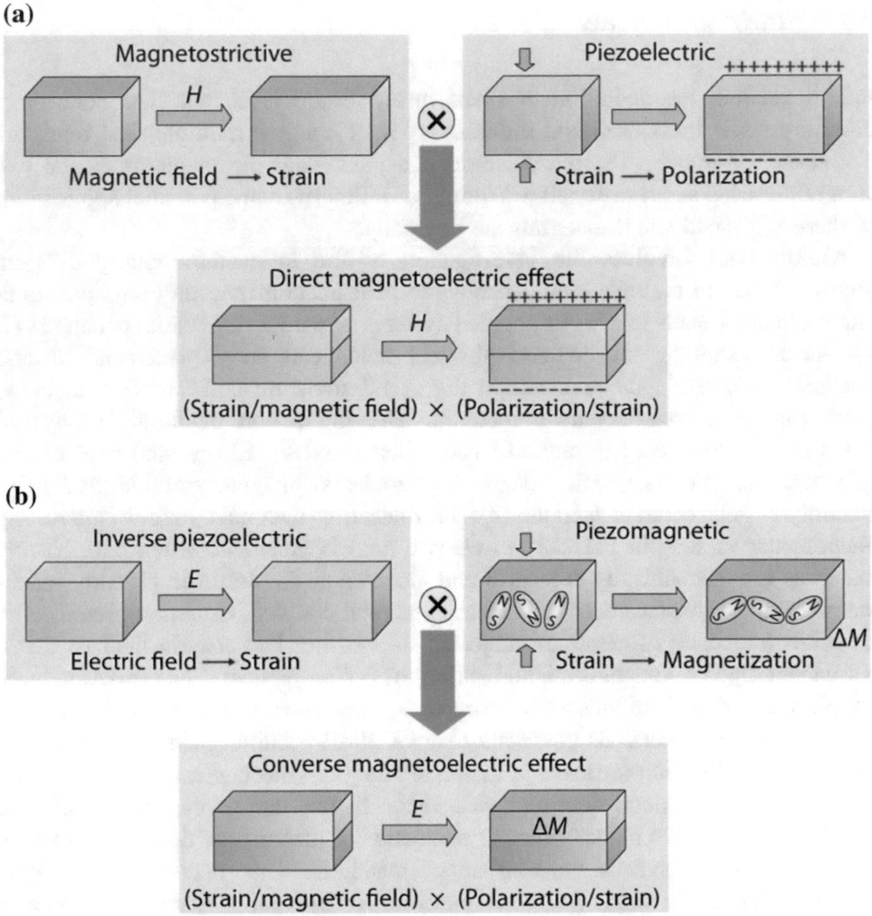

Fig. 3.2 Magnetoelectric coupling in magnetoelectroelastic (MEE) composites: **a** Direct effect; **b** Converse effect. [Reproduced from [7] with permission from Springer Nature]

magnetic field H leads to strain in the magnetic component due to the magnetostrictive effect, which is mechanically transferred to the ferroelectric component inducing a dielectric polarization because of the piezoelectric effect. Conversely, the magnetoelectric coupling is observed when the MEE composite is subjected to an electric field E which results in the magnetization change ΔM [6, 7].

Multiphysical materials are frequently called smart, intelligent, active, or adaptive materials. Due to their multifunctional capabilities, these materials may be found in the following physical fields: electromagnetic, hygrothermal, thermoelastic, magnetoelectroelastic, optothermoelectromagnetoelastic, and hygrothermopiezoelectric fields. For instance, piezoelectric, piezomagnetic, electrostrictive, magnetostrictive, magnetoelectroelastic, and photovoltaic materials as well as electro/magnetorheological fluids are some of the common synthetic smart materials that are being used in different applications in science and technology [8].

Some of the synthetic multiphysical or smart materials exhibit a full coupling among different physical fields; however, natural ones rarely do so. Smart materials have been used in vibration and damping suppression, noise reduction, controlled active deformation, health monitoring, and improved fatigue and corrosion resistance. They are reportedly being used in transportation and aerospace industries [3, 9]. Smart structures with piezoelectric and/or piezomagnetic patches to control the structural vibration are of great interest. These types of active control have been employed in axisymmetric shells [10, 11] and laminated beams [12]. Smart materials could also be employed in active noise control. A numerical approach for the design of smart lightweight structures was presented in reference [13] for active reduction of noise and vibration. Furthermore, smart materials can be used for active shape control, health monitoring, and damage detection of structural elements [14]. A new development is a wear detection system for train wheels by detecting the vibration behaviour of the entire wheel caused by surface change on the rolling contact area. Another application of smart materials is in energy harvesting, which is the process of changing parasitic mechanical energy into electrical energy. This energy can be used for driving electrical circuits or storage in a battery [15, 16]. In the following sections the piezoelectric and magnetoelectroelastic fields are formulated. Then, the equations of motion for analyzing smart hollow cylinders, subjected to different physical fields, are derived. Sample results are then presented to describe the behavior of various types of smart materials.

3.1.1 Piezoelectric Materials

The linear constitutive equations for a multiphysical piezo electric medium are written as [17, 18]:

$$\begin{aligned} \sigma_{ij} &= C_{ijkl}\varepsilon_{kl} - e_{kij}E_k - \beta_{ij}\vartheta - \xi_{ij}m, \quad (i,j,k,l = 1,2,3) \\ D_i &= e_{ijk}\varepsilon_{jk} + \varepsilon_{ij}E_j + \gamma_i\vartheta + \chi_i m, \quad (i,j,k,l = 1,2,3) \end{aligned} \tag{3.1}$$

in which σ_{ij} and D_i are, respectively, stress and electric displacement; ε_{ij}, E_k, ϑ and m are strain, electric field, temperature change, and moisture concentration change with respect to the reference state; C_{ijkl}, e_{kij} and ϵ_{ij} are elastic, piezoelectric, dielectric, and coefficients, respectively; β_{ij}, ξ_{ij}, γ_i and χ_i are thermal stress, hygroscopic stress, pyroelectric and hygroelectric, coefficients. Furthermore, $\vartheta = T - T_0$ and $m = M - M_0$, in which T and M are the absolute temperature and moisture concentration while T_0 and M_0 represent the stress-free temperature and moisture concentration.

The stress and strain tensors for the considered problem are symmetric which result in the following symmetrical properties:

$$\begin{aligned} &\sigma_{ij} = \sigma_{ji}, \ \varepsilon_{ij} = \varepsilon_{ji}, \ C_{ijkl} = C_{jikl} = C_{ijlk} = C_{klij}, \ e_{kij} = e_{kji}, \\ &\epsilon_{ij} = \epsilon_{ji}, \ \beta_{ij} = \beta_{ji}, \ \zeta_{ij} = \zeta_{ji} \end{aligned} \tag{3.2}$$

The symmetric properties lower the total number of independent coefficients in Eq. (3.1). For a general case of triclinic system with the least symmetry, there exist 21 elastic, 18 piezoelectric, 6 dielectric 6 thermal stress, 6 hygroscopic stress, 3 pyroelectric and 3 hygroelectric constants. Nonetheless, the number of constants depends on the symmetry of crystal structure of multiphysical materials. The nonzero matrix elements for some piezoelectric materials such as quartz, lithium niobate $LiNbO_3$, cadmium sulphide C_dS, polarized ceramic, and gallium arsenide GaAs could be found in [19]. It is more convenient to utilize a system with abbreviated subscripts for material properties to simplify the problems in elasticity. Since stress and strain tensors are symmetric, each component can be specified by one subscript rather than two as follows [19]:

$$\begin{aligned} &\sigma_{11} = \sigma_1, \sigma_{22} = \sigma_2, \sigma_{33} = \sigma_3, \sigma_{23} = \sigma_4, \sigma_{13} = \sigma_5, \sigma_{12} = \sigma_6 \\ &\varepsilon_{11} = \varepsilon_1, \varepsilon_{22} = \varepsilon_2, \varepsilon_{33} = \varepsilon_3, \varepsilon_{23} = \frac{1}{2}\varepsilon_4, \varepsilon_{13} = \frac{1}{2}\varepsilon_5, \varepsilon_{12} = \frac{1}{2}\varepsilon_6 \end{aligned} \tag{3.3}$$

The same logic is used for the following material properties:

$$c_{\alpha\delta} = C_{ijkl}, e_{k\alpha} = e_{kij}, \beta_\alpha = \beta_{ij}, \xi_\alpha = \xi_{ij} \quad (i,j,k,l = 1,2,3; \ \alpha,\delta = 1,2,\ldots,6) \tag{3.4}$$

It should be mentioned that the contracted notations for stresses and strains are not applied in this book.

3.1.1.1 Potential Field Equations

The strain and electric fields are related to their potentials by the following gradient equations. The relation between strain and displacement components for small strain is:

$$\varepsilon_{ij} = \frac{1}{2}\left(u_{i,j} + u_{j,i}\right) \tag{3.5}$$

where u_i is the displacement component and a comma denotes partial differentiation with respect to the space variables. The quasi-stationary electric field equations in the absence of free conducting electromagnetic current are expressed as:

$$E_i = -\phi_{,i} \tag{3.6}$$

where ϕ is the scalar electric potential.

3.1.2 Magnetoelectroelastic Materials

In order to consider the effect of magnetic field in the constitutive equations, Eq. (3.1) must be modified to:

$$\begin{aligned} \sigma_{ij} &= C_{ijkl}\varepsilon_{kl} - e_{kij}E_k - d_{kij}H_k - \beta_{ij}\vartheta - \xi_{ij}m \\ D_i &= e_{ijk}\varepsilon_{jk} + \epsilon_{ij}E_j + g_{ij}H_j + \gamma_i\vartheta + \chi_i m \qquad (i,j,k,l = 1,2,3) \\ B_i &= d_{ijk}\varepsilon_{jk} + g_{ij}E_j + \mu_{ij}H_j + \tau_i\vartheta + \upsilon_i m \end{aligned} \tag{3.7}$$

in which B_i and H_k are respectively magnetic induction and magnetic field. d_{kij}, g_{kij}, and μ_{ij} are piezomagnetic, magnetoelectric, and magnetic permeability coefficients, respectively; τ_i and υ_i are pyromagnetic, and hygromagnetic coefficients. Similar to piezoelectric materials, the symmetry of stress and strain tensors leads to the following symmetrical properties:

$$d_{kij} = d_{kji}, \quad g_{ij} = g_{ji}, \quad \mu_{ij} = \mu_{ji} \tag{3.8}$$

These symmetric properties result in 18 piezomagnetic, 6 magnetoelectric and 6 magnetic permeability constants for the most general case of triclinic system. As an example, the constitutive equations (3.7) for orthotropic and radially polarized and magnetized materials in a cylindrical coordinate system (r, θ, z) can be written as:

$$\begin{Bmatrix} \sigma_{rr} \\ \sigma_{\theta\theta} \\ \sigma_{zz} \\ \sigma_{z\theta} \\ \sigma_{rz} \\ \sigma_{r\theta} \end{Bmatrix} = \begin{bmatrix} c_{33} & c_{13} & c_{23} & 0 & 0 & 0 \\ c_{13} & c_{11} & c_{12} & 0 & 0 & 0 \\ c_{23} & c_{12} & c_{22} & 0 & 0 & 0 \\ 0 & 0 & 0 & 2c_{66} & 0 & 0 \\ 0 & 0 & 0 & 0 & 2c_{44} & 0 \\ 0 & 0 & 0 & 0 & 0 & 2c_{55} \end{bmatrix} \begin{Bmatrix} \varepsilon_{rr} \\ \varepsilon_{\theta\theta} \\ \varepsilon_{zz} \\ \varepsilon_{z\theta} \\ \varepsilon_{rz} \\ \varepsilon_{r\theta} \end{Bmatrix} - \begin{bmatrix} e_{33} & 0 & 0 \\ e_{31} & 0 & 0 \\ e_{32} & 0 & 0 \\ 0 & 0 & 0 \\ 0 & 0 & e_{24} \\ 0 & e_{15} & 0 \end{bmatrix} \begin{Bmatrix} E_r \\ E_\theta \\ E_z \end{Bmatrix}$$

$$- \begin{bmatrix} d_{33} & 0 & 0 \\ d_{31} & 0 & 0 \\ d_{32} & 0 & 0 \\ 0 & 0 & 0 \\ 0 & 0 & d_{24} \\ 0 & d_{15} & 0 \end{bmatrix} \begin{Bmatrix} H_r \\ H_\theta \\ H_z \end{Bmatrix} - \begin{Bmatrix} \beta_1 \\ \beta_2 \\ \beta_3 \\ 0 \\ 0 \\ 0 \end{Bmatrix} \vartheta - \begin{Bmatrix} \xi_1 \\ \xi_2 \\ \xi_3 \\ 0 \\ 0 \\ 0 \end{Bmatrix} \vartheta$$

$$\begin{Bmatrix} D_r \\ D_\theta \\ D_z \end{Bmatrix} = \begin{bmatrix} e_{33} & e_{31} & e_{32} & 0 & 0 & 0 \\ 0 & 0 & 0 & 0 & 0 & 2e_{15} \\ 0 & 0 & 0 & 0 & 2e_{24} & 0 \end{bmatrix} \begin{Bmatrix} \varepsilon_{rr} \\ \varepsilon_{\theta\theta} \\ \varepsilon_{zz} \\ \varepsilon_{z\theta} \\ \varepsilon_{rz} \\ \varepsilon_{r\theta} \end{Bmatrix} + \begin{bmatrix} \epsilon_{33} & 0 & 0 \\ 0 & \epsilon_{11} & 0 \\ 0 & 0 & \epsilon_{22} \end{bmatrix} \begin{Bmatrix} E_r \\ E_\theta \\ E_z \end{Bmatrix}$$

$$+ \begin{bmatrix} g_{33} & 0 & 0 \\ 0 & g_{11} & 0 \\ 0 & 0 & g_{22} \end{bmatrix} \begin{Bmatrix} H_r \\ H_\theta \\ H_z \end{Bmatrix} + \begin{Bmatrix} \gamma_1 \\ \gamma_2 \\ \gamma_3 \end{Bmatrix} \vartheta + \begin{Bmatrix} \chi_1 \\ \chi_2 \\ \chi_3 \end{Bmatrix} m$$

$$\begin{Bmatrix} B_r \\ B_\theta \\ B_z \end{Bmatrix} = \begin{bmatrix} d_{33} & d_{31} & d_{32} & 0 & 0 & 0 \\ 0 & 0 & 0 & 0 & 0 & 2d_{15} \\ 0 & 0 & 0 & 0 & 2d_{24} & 0 \end{bmatrix} \begin{Bmatrix} \varepsilon_{rr} \\ \varepsilon_{\theta\theta} \\ \varepsilon_{zz} \\ \varepsilon_{z\theta} \\ \varepsilon_{rz} \\ \varepsilon_{r\theta} \end{Bmatrix} + \begin{bmatrix} g_{33} & 0 & 0 \\ 0 & g_{11} & 0 \\ 0 & 0 & g_{22} \end{bmatrix} \begin{Bmatrix} E_r \\ E_\theta \\ E_z \end{Bmatrix}$$

$$+ \begin{bmatrix} \mu_{33} & 0 & 0 \\ 0 & \mu_{11} & 0 \\ 0 & 0 & \mu_{22} \end{bmatrix} \begin{Bmatrix} H_r \\ H_\theta \\ H_z \end{Bmatrix} + \begin{Bmatrix} \tau_1 \\ \tau_2 \\ \tau_3 \end{Bmatrix} \vartheta + \begin{Bmatrix} \upsilon_1 \\ \upsilon_2 \\ \upsilon_3 \end{Bmatrix} m \tag{3.9}$$

3.1.2.1 Potential Field Equations

The quasi-stationary magnetic field equations in the absence of free conducting electromagnetic current are expressed as:

$$H_i = -\varphi_{,i} \tag{3.10}$$

where φ is the scalar magnetic potential.

3.1.2.2 Conservation Equations

The conservation or divergence equations for a hygrothermomagnetoelectroelastic medium are provided in this section. The equation of motion is:

$$\sigma_{ij,j} + f_i = \rho u_{i,tt} \tag{3.11}$$

where f_i, ρ, and t, respectively, stand for body force, density, and time. In magneto-hygrothermoelectroelastic analysis, an electrically conducting elastic solid subjected to an external magnetic field experiences the Lorentz force via the electromagnetic-elastic interaction which works as a body force in Eq. (3.11) as follows [20]:

$$\begin{aligned} &\vec{J} = \nabla \times \vec{h},\ \nabla \times \vec{e} = -\mu \vec{h}_{,t},\ \nabla . \vec{h} = 0,\ \vec{e} = -\mu(\vec{u}_{,t} \times \vec{H}) \\ &\vec{h} = \nabla \times (\vec{u} \times \vec{H}),\ \vec{f} = \mu(\vec{J} \times \vec{H}) \end{aligned} \tag{3.12}$$

in which, $\vec{J}$, $\vec{h}$, $\vec{e}$, $\vec{u}$, $\vec{H}$, and $\vec{f}$ are, respectively, the electric current density, perturbation of magnetic field, perturbation of electric field, displacement, magnetic intensity, and the Lorenz force vectors; μ represents the magnetic permeability. Maxwell's electromagnetic equations or equations of charge and current conservation are written as [21]:

$$D_{i,i} = \rho_e, \quad B_{i,i} = 0 \tag{3.13}$$

in which ρ_e is the charge density.

Furthermore, the classical energy conservation equation is [22]:

$$q_{i,i} + \rho(c_v \vartheta_{,t} - R) = 0 \tag{3.14}$$

where q_i, c_v, and R are heat flux component, specific heat at constant volume, and internal heat source per unit mass, respectively. However, Biot [23] introduced the effects of elastic term in the energy equation to obtain more accurate results for thermoelastic analysis. The energy equation (3.14) was modified for thc classical, coupled thermoelasticity as follows:

$$q_{i,i} + \rho(S_{,t} T_0 - R) = 0 \tag{3.15}$$

where S denotes the entropy per unit mass and is defined as:

$$\rho S = \beta_{ij}\varepsilon_{ij} + \frac{\rho c_v}{T_0}\vartheta \tag{3.16}$$

Through Eqs. (3.15) and (3.16), the energy equation is coupled with the strain rate. Considering the advent of smart materials with coupled multiphysical interactions, the classical coupled thermoelasticity equations could be modified to also consider the coupling effects of electric, magnetic, and hygroscopic fields on the energy equations [24]. Accordingly, Eq. (3.16) could be written in the following form for classical, coupled hygrothermomagnetoelectroelasticity:

$$\rho S = \beta_{ij}\varepsilon_{ij} + \gamma_i E_i + \tau_i H_i + \frac{\rho c_v}{T_0}\vartheta + d_t m \tag{3.17}$$

where d_t is the specific heat-moisture coefficient. On the other hand, the conservation law for the mass of moisture in the absence of a moisture source is given by [25]:

$$p_{i,i} + m_{,t} = 0 \tag{3.18}$$

in which, p_i represents the moisture flux component that is the rate of moisture transfer per unit area.

3.1.2.3 Fourier Heat Conduction and Fickian Moisture Diffusion

The following Fourier heat conduction theory which relates the heat flux q_i to the temperature gradient is the most widely used theory in the literature:

$$q_i = -k_{ij}^T \vartheta_{,j} \tag{3.19}$$

Furthermore, the diffusion of moisture in a solid is basically the same as that of temperature. As a result, the Fickian moisture diffusion equation for moisture flux p_i can be defined similar to Fourier heat conduction equation as follows:

$$p_i = -\zeta_{ij}^H m_{,j} \tag{3.20}$$

In the above equations, k_{ij}^T and ζ_{ij}^H are the thermal conductivity and moisture diffusivity coefficients, respectively. Substituting Eq. (3.19) into (3.14) and Eq. (3.20) into (3.18) lead to a diffusion-like equations with parabolic-type governing differential equations for temperature and moisture concentration. To consider the possible effect of other physical fields on the heat and mass flux, Eqs. (3.19) and (3.20) could be modified as [3, 18]:

$$\begin{aligned} q_i &= k_{ijkl}^M \varepsilon_{kl,j} - k_{ijk}^E E_{k,j} - k_{ijk}^B H_{k,j} - k_{ij}^T \vartheta_{,j} - k_{ij}^H m_{,j} \\ p_i &= \zeta_{ijkl}^M \varepsilon_{kl,j} - \zeta_{ijk}^E E_{k,j} - \zeta_{ijk}^B H_{k,j} - \zeta_{ij}^T \vartheta_{,j} - \zeta_{ij}^H m_{,j} \end{aligned} \tag{3.21}$$

where k_{ijkl}^{M}, k_{ijk}^{E}, k_{ijk}^{B}, k_{ij}^{H}, ζ_{ijkl}^{M}, ζ_{ijk}^{E}, ζ_{ijk}^{B}, ζ_{ij}^{T} $(i,j,k,l = 1,2,3)$ are, respectively, strain-thermal conductivity, electric-thermal conductivity, magnetic-thermal conductivity, moisture-thermal conductivity (Dufour effect), strain-moisture diffusivity, electric-moisture diffusivity, magnetic-moisture diffusivity, heat-moisture diffusivity (Soret effect) coefficients. These coefficients represent the degree of thermal and mechanical, thermal and electrical, thermal and magnetic, thermal and hygroscopic, hygroscopic and mechanical, hygroscopic and electrical, hygroscopic and magnetic, and hygroscopic and thermal field interactions.

The conventional heat conduction and moisture diffusion theories based on the classical Fourier and Fickian laws lead to an infinite speed of thermal and moisture wave propagation due to the parabolic-type heat and mass transport equations. Fourier and Fickian laws assume instantaneous hygrothermal responses and a quasi-equilibrium thermodynamic condition. The classical diffusion theories have been widely used in heat and mass transfer problems; however, the heat and mass transmission is observed to be a non-equilibrium phenomenon, and they propagate with a finite speed for applications involving very low temperature, high temperature gradients, short-pulse heating, laser drying, laser melting and welding, rapid solidification, very high frequencies of heat and mass flux densities, and micro temporal and spatial scales [26]. Consequently, different non-Fourier and non-Fickian heat and mass transfer theories have been developed to remove these drawbacks.

3.1.3 Advanced Smart Materials

Functionally graded materials (FGMs) have become considerably important in extremely high temperature environments such as rocket nozzles and chemical plants. In 1984, the concept of FGMs was proposed in Japan as thermal barrier materials [27]. As shown in Table 3.1, FGMs are composite materials, microscopically non-homogeneous, in which material properties vary continuously with respect to spatial coordinates. FGMs are typically made from a mixture of ceramic and metal or a combination of different metals. The advantage of using FGMs can be expressed as their sustainability in high temperature environments while maintaining their structural integrity. The ceramic constituents of FGMs provide the high temperature resistance due to their low thermal conductivity. On the other hand, the ductile metal constituent of FGMs impedes fracture due to high temperature gradient in a very short period of time as seen in laser impulse applications [28]. The smooth and continuous changes of material properties and thermomechanical stresses in FGMs distinguish them from the conventional laminated composites with a mismatch of material properties across the laminate interfaces. The laminated composites are prone to debonding, crack initiation, and the presence of residual stresses due to the difference in thermal expansion coefficients of different layers. The continuous transition of volume fraction in FGMs eliminates the deficiency.

Table 3.1 Characteristics of FGMs [27]

	FGM	Non-FGM
Property: 1. Mechanical strength 2. Thermal conductivity	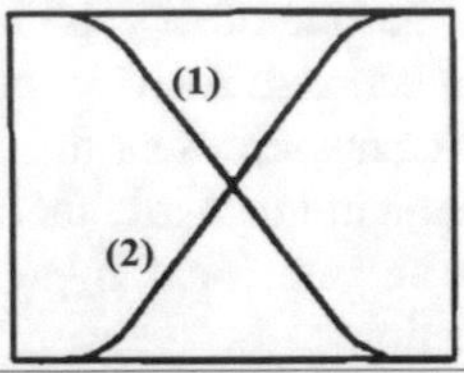	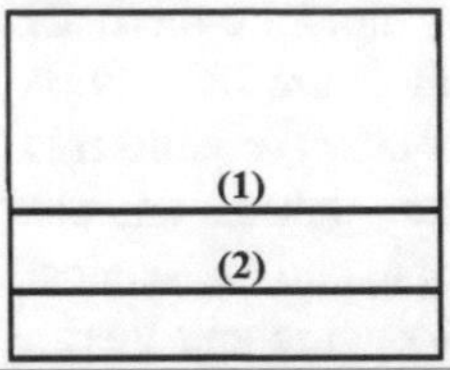
Constituent elements: 1. Ceramic ● 2. Metal ○ 3. Fiber □	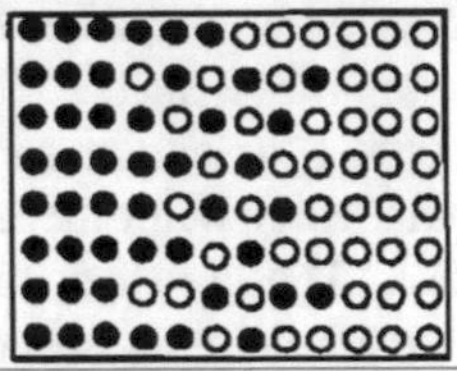	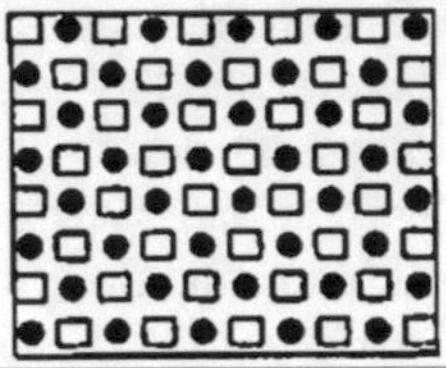

This gradual variation in material properties reduces the likelihood of delamination caused by stress concentration, in-plane and transverse thermal stresses, and the stress intensity factors [2, 28].

FGMs were first introduced as thermal barriers to withstand high temperature changes; nonetheless, they have lots of applications in modern industry. For size reduction and enhancement of the reliability of electric power equipment, FGMs with spatial distribution of dielectric permittivity have been used recently [29]. Some of the applications of FGMs in biomedical engineering such as implants for bone and knee joint replacement are mentioned in [30]. The normal and shear stresses in a double-layered pressure vessel due to internal pressure and thermal loadings were reduced by using FGM materials [31]. Moreover, there are various applications of FGMs in aerospace structures, fusion reactors, turbine rotors, fly-wheels, gears, wear resistant linings, thermoelectric generators, prostheses, etc.

3.2 Thermal Stress Analysis in Homogenous Smart Materials

In this section, the constitutive relations and governing equations for solving the two thermomagnetoelastic and thermo-magnetoelectroelastic problems are presented. Consider an infinitely long, hollow cylinder rotating at a constant angular velocity ω as shown in Fig. 3.3. The cylinder is magnetized and polarized in the radial direction. The inner and outer radii of the cylinder are a and b, respectively. The cylinder experiences the magnetic scalar potential, φ, electric scalar potential, ϕ, and pressure, P, at the inner and outer surfaces. The inner surface of the cylinder is subjected to the temperature change, $\vartheta_a = T_a - T_\infty$, where T_a is the absolute temperature at the inner surface and T_∞ is the ambient temperature. The outer

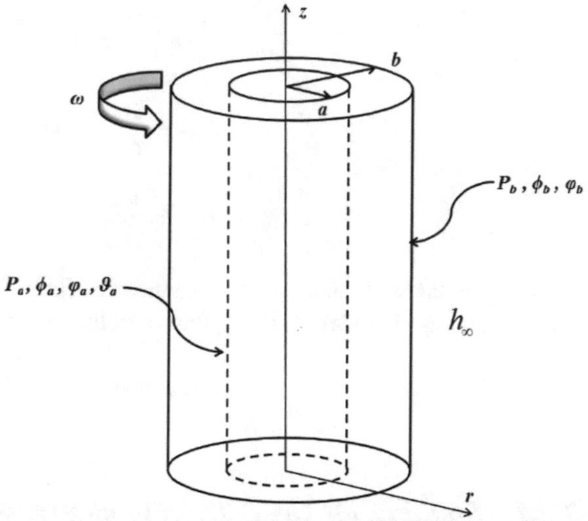

Fig. 3.3 Rotating hollow cylinder and its boundary conditions [Reproduced from [35] with permission from Taylor & Francis Ltd.]

surface is under convection boundary condition with the heat convectivity coefficient, h_∞. Subscripts "a" and "b" are employed to indicate the load on the inner and outer surfaces, respectively.

The non-zero components of strain, electric, and magnetic fields for the axisymmetric, plane strain problem are written as:

$$\bar{\varepsilon}_{rr} = u_r, \quad \bar{\varepsilon}_{\theta\theta} = \frac{u}{r}, \quad E_r = -\phi_{,r}, \quad H_r = -\varphi_{,r} \tag{3.22}$$

where $u = u_r$ is the radial displacement; r and θ are the radial and circumferential coordinates. Using constitutive Eqs. (3.9) and (3.22), one can obtain:

$$\sigma_{rr} = c_{33}u_r + c_{13}\frac{u}{r} + e_{33}\phi_r + d_{33}\varphi_{,r} - \beta_1\vartheta \tag{3.23a}$$

$$\sigma_{\theta\theta} = c_{13}u_r + c_{11}\frac{u}{r} + e_{31}\phi_r + d_{31}\varphi_{,r} + d_{31}\varphi_{,r} - \beta_3\vartheta \tag{3.23b}$$

$$D_r = e_{33}u_r + e_{31}\frac{u}{r} - \varepsilon_{33}\phi_{,r} - g_{33}\varphi_{,r} + \gamma_1\vartheta \tag{3.23c}$$

$$B_r = d_{33}u_r + d_{31}\frac{u}{r} - g_{33}\phi_{,r} - \mu_{33}\varphi_{,r} + \tau_1\vartheta \tag{3.23d}$$

in which, $c_{mn} = C_{ijkl}$, $e_{mk} = e_{pij}$, $d_{mk} = d_{pij}$ $(i, j, k, l, p = 1, 2, 3;\ m, n = 1, 2, \ldots, 6)$, $\beta_1 = \beta_{11}$, and $\beta_3 = \beta_{33}$. The governing equations for a rotating magnetoelectroelastic cylinder under axisymmetric loading, when the body force, free charge density, and current density are absent, are expressed as:

$$\sigma_{rr,r} + \frac{1}{r}(\sigma_{rr} - \sigma_{\theta\theta}) = \rho_d u_{,tt} \tag{3.24a}$$

$$D_{r,r} + \frac{1}{r}D_r = 0 \tag{3.24b}$$

$$B_{r,r} + \frac{1}{r}B_r = 0 \tag{3.24c}$$

where ρ_d is the mass density, and t stands for time. Furthermore, the inertial effect for the rotating cylinder with angular velocity ω can be written as:

$$u_{,tt} = -r\omega^2 \tag{3.25}$$

3.2.1 Solution for the a Thermomagnetoelastic FGM Cylinder

The solution for a thermomagnetoelastic FGM rotating hollow cylinder is obtained in this section. It is assumed that the material properties of the FGM cylinder vary according to a power law along the radial direction as follows:

$$\chi(r) = \chi_0\left(\frac{r}{b}\right)^{2N} \tag{3.26}$$

where $\chi(r)$, χ_0, and N represent, respectively, the general material properties of the cylinder, their values at the outer surface, and the non-homogeneity parameter. Substituting Eqs. (3.23), (3.25), and (3.26) into Eq. (3.24) leads to the following coupled governing differential equations in terms of displacement and magnetic potential:

$$\begin{aligned} &r^2 c_{330} u_{,rr} + r(2N+1)c_{330}u_{,r} + (2Nc_{130} - c_{110})u + r^2 d_{330}\phi_{,rr} \\ &\quad + r((2N+1)d_{330} - d_{310})\phi_{,r} - r^2\beta_{10}\vartheta_{,r} - r((2N+1)\beta_{10} - \beta_{30})\vartheta + \rho_{d_0}\omega^2 = 0 \end{aligned} \tag{3.27a}$$

$$\begin{aligned} &r^2 d_{330}u_{,rr} + r((2N+1)d_{330} + d_{310})u_{,r} + 2Nd_{310}u - r^2 u_{330}\phi_{,rr} \\ &-r(2N+1)\mu_{330}\phi_{,r} + r^2\tau_{10}\vartheta_{,r} + r(2N+1)\tau_{10}\vartheta = 0 \end{aligned} \tag{3.27b}$$

Using the following non-dimensional parameters:

$$\alpha = \frac{c_{110}}{c_{330}},\ \beta = \frac{e_{310}}{e_{330}},\ \delta = \frac{c_{130}}{c_{330}},\ v = \frac{d_{310}}{d_{330}},\ \eta = \frac{\beta_{330}}{\beta_{110}},\ \gamma = \frac{\varepsilon_{330}c_{330}}{e_{330}^2}$$
$$\zeta = \frac{g_{330}c_{330}}{d_{330}e_{330}},\ \lambda = \frac{\mu_{330}c_{330}}{d_{330}^2},\ X = \frac{\gamma_{10}c_{330}}{e_{330}\beta_{10}},\ Y = \frac{\tau_{10}c_{330}}{d_{330}\beta_{10}} \tag{3.28}$$

as well as a new electric potential, a new magnetic potential, and a new temperature change:

$$\Phi = \frac{e_{330}}{c_{330}}\phi, \quad \Psi = \frac{d_{330}}{c_{330}}\varphi, \quad \Theta = \frac{\beta_{10}}{c_{330}}\vartheta \tag{3.29}$$

Equation (3.27) can be rewritten in the following form:

$$r^2u_{,rr} + r(2N+1)u_{,r} + (2N\delta - \alpha)u + r^2\Psi_{,rr}$$
$$+ r(2N+1-v)\Psi_{,r} - r^2\Theta_{,r} - r(2N+1-\eta)\Theta + \frac{\rho_{d_0}\omega^2r^3}{c_{330}} = 0 \tag{3.30a}$$

$$r^2u_{,rr} + r(2N+1+v)u_{,r} + 2Nvu - r^2\lambda\Psi_{,rr}$$
$$-r(2N+1)\lambda\Psi_{,r} + r^2Y\Theta_{,r} + r(2N+1)Y\Theta = 0 \tag{3.30b}$$

Employing the normalized radial coordinate, $\rho = \frac{r}{a}$, Eq. (3.30) is rearranged as follows:

$$\rho^2u_{,\rho\rho} + \rho(2N+1)u_{,\rho} + (2N\delta - \alpha)u + \rho^2\Psi_{,\rho\rho}$$
$$+ \rho(2N+1-v)\Psi_{,\rho} - \rho^2a\Theta_{,\rho} - \rho(2N+1-\eta)a\Theta + \frac{\rho_{d0}\omega^2\rho^3a^3}{c_{330}} = 0 \tag{3.31a}$$

$$\rho^2u_{,\rho\rho} + \rho(2N+1+v)u_{,\rho} + 2Nvu - \lambda\rho^2\Psi_{,\rho\rho}$$
$$-\rho(2N+1)\lambda\Psi_{,\rho} + \rho^2aY\Theta_{,\rho} + \rho a(2N+1)Y\Theta = 0 \tag{3.31b}$$

Solving Eq. (3.31) requires us to determine the temperature distribution along the radial direction. The axisymmetric, steady state heat conduction equation for an infinitely long hollow cylinder can be written as:

$$\frac{1}{r}\left(rk\vartheta_{,r}\right)_{,r} = 0 \qquad a \le r \le b \tag{3.32}$$

where k is the thermal conductivity varying according to Eq. (3.26) for the FGM cylinder. Rearranging Eq. (3.32), using the normalized radial coordinate and employing Eqs. (3.26) and (3.29), leads to:

$$\frac{1}{\rho}\left(\rho^{2N+1}\Theta_{,\rho}\right)_{,\rho} = 0 \qquad 1 \le \rho \le \imath \tag{3.33}$$

where $\imath = \frac{b}{a}$ is the aspect ratio of the hollow cylinder. The general solution is obtained as:

$$\Theta = C_1 \rho^{-2N} + C_2, \quad N \neq 0 \tag{3.34a}$$

$$\Theta = C_1 \ln(\rho) + C_2, \quad N = 0 \tag{3.34b}$$

where C_1 and C_2 are integration constants to be determined by thermal boundary conditions. The following general thermal boundary conditions are considered for Eq. (3.33) [32]:

$$\begin{aligned} A_{11}\Theta(1) + A_{12}\Theta'(1) &= f_1 \\ A_{21}\Theta(\imath) + A_{22}\Theta'(\imath) &= f_2 \end{aligned} \tag{3.35}$$

in which, A_{ij} is the Robin-type thermal boundary condition coefficients, and f_1 and f_2 are known functions on the inner and outer radii. Equations (3.34) are substituted into Eq. (3.35) and are solved for the integration constants C_1 and C_2. When $N \neq 0$:

$$\begin{aligned} C_1 &= \frac{A_{11}f_2 - A_{21}f_1}{A_{11}(A_{21}\imath - 2NA_{22})\imath^{-2N-1} - A_{21}(A_{11} - 2NA_{12})} \\ C_2 &= \frac{f_1}{A_{11}} - \frac{(A_{11} - 2NA_{12})(A_{11}f_2 - A_{21}f_1)}{A_{11}^2(A_{21}\imath - 2NA_{12})\imath^{-2N-1} - A_{21}\left(A_{11}^2 - 2NA_{12}A_{11}\right)} \end{aligned} \tag{3.36}$$

and when $N = 0$:

$$\begin{aligned} C_1 &= \frac{A_{11}f_2 - A_{21}f_1}{A_{11}(A_{21}\ln(\imath) + A_{22}\imath^{-1}) - A_{21}A_{12}} \\ C_2 &= \frac{f_1}{A_{11}} - \frac{A_{12}(A_{11}f_2 - A_{21}f_1)}{A_{11}^2(A_{21}\ln(\imath) + A_{22}\imath^{-1}) - A_{21}A_{12}A_{11}} \end{aligned} \tag{3.37}$$

Using the aforementioned temperature distribution, the coupled governing differential equations (3.31) can be solved. First, Eq. (3.31) is converted to the following new differential equations about ς with constant coefficients by introducing a variable substitution, $\rho = e^{\varsigma}$:

$$\ddot{u} + 2N\dot{u} + (2N\delta - \alpha)u + \ddot{\Psi} + (2N - v)\dot{\Psi} = ae^{\varsigma}\dot{\Theta} + (2N + 1 - \eta)ae^{\varsigma}\Theta - \omega_n ae^{3\varsigma} \tag{3.38a}$$

$$\ddot{u} + (2N + v)\dot{u} + 2Nvu - \lambda\ddot{\Psi} - 2N\lambda\dot{\Psi} = -aYe^{\varsigma}\dot{\Theta} - (2N + 1)aYe^{\varsigma}\Theta \tag{3.38b}$$

where the overdot stands for differentiation with respect to ς and

$$\omega_n = \frac{\rho_{d_0}\omega^2 a^2}{c_{330}} \tag{3.39}$$

Eliminating $\ddot{\Psi}$ between the two equations of (3.38) and solving for $\dot{\Psi}$ leads to:

$$\begin{aligned}\dot{\Psi} &= \frac{1+\lambda}{\nu\lambda}\ddot{u} + \frac{2N(1+\lambda)+\nu}{\nu\lambda}\dot{u} + \frac{2N(\delta\lambda+\nu)-\alpha\lambda}{\nu\lambda}u + \frac{a(Y-\lambda)}{\nu\lambda}e^{\varsigma}\dot{\Theta} \\ &+ \frac{(2N+1)Y-(2N+1-\eta)\lambda}{\nu\lambda}ae^{\varsigma}\Theta + \frac{a\omega_n}{\nu}e^{3\varsigma}\end{aligned} \tag{3.40}$$

Equation (3.40) and its derivative are substituted into the first equation of (3.38) to give us the following decoupled differential equation for u:

$$a_3\dddot{u} + a_2\ddot{u} + a_1\dot{u} + a_0 u = d_3 e^{3\varsigma} + d_2 e^{\varsigma} + d_1 e^{(-2N+1)\varsigma} \tag{3.41}$$

in which,

$$\begin{aligned}a_3 &= \frac{1+\lambda}{\nu\lambda},\ a_2 = \frac{4N(1+\lambda)}{\nu\lambda},\ a_1 = \frac{2N(\delta\lambda+\nu)+4N^2(\lambda+1)-\alpha\lambda-\nu^2}{\nu\lambda}, \\ a_0 &= \frac{4N^2(\nu+\delta\gamma)-2N(\alpha\lambda+\nu^2)}{\nu\lambda},\ d_3 = -\frac{a\omega_n(3+2N)}{\nu}, \\ d_2 &= -\frac{aY(2N+1)(2N+1-\nu)-a\lambda(2N+1-\eta)(1+2N)}{\nu\lambda}C_2, \\ d_1 &= -\frac{aY(1-\nu)+a\lambda(\eta-2N\nu-1)}{\nu\lambda}C_1\end{aligned} \tag{3.42}$$

The solution of Eq. (3.41) can be exactly obtained as follows

$$u = Ae^{m_1\varsigma} + Be^{m_2\varsigma} + Ce^{m_3\varsigma} + K_1 e^{(-2N+1)\varsigma} + K_2 e^{\varsigma} + K_3 e^{3\varsigma} \tag{3.43}$$

where $m_i\,(i=1,2,3)$ are the roots of the characteristic equation of Eq. (3.41); A, B, and C are the constants of integration determined by the boundary conditions; and $K_i\,(i=1,2,3)$ are obtained as:

$$\begin{aligned}K_1 &= \frac{d_1}{(-2N+1)^3 a_3 + (-2N+1)^2 a_2 + (-2N+1)a_1 + a_0} \\ K_2 &= \frac{d_2}{a_3+a_2+a_1+a_0},\ K_3 = \frac{d_3}{27a_3+9a_2+3a_1+a_0}\end{aligned} \tag{3.44}$$

The final solution for u in terms of ρ is:

$$u = A\rho^{m_1} + B\rho^{m_2} + C\rho^{m_3} + K_1\rho^{-2N+1} + K_2\rho + K_3\rho^3 \tag{3.45}$$

Substituting Eq. (3.43) into (3.40) and doing the integration leads to the following expression for Ψ:

$$\Psi = b_1 A\rho^{m_1} + b_2 B\rho^{m_2} + b_3 C\rho^{m_3} + D + K_5\rho^{-2N} + K_6\rho^{-2N+1} + K_7\rho + K_8\rho^3 \tag{3.46}$$

where D is a new integration constant and

$$\begin{aligned}
b_i &= \frac{1}{v\lambda}\left((1+\lambda)m_i + 2N(1+\lambda) + v + \frac{1}{m_i}(2N(v+\delta\lambda) - \alpha\lambda)\right) \\
K_4 &= \frac{1}{v\lambda}\left(K_1(1+\lambda+v) + \frac{K_1}{-2N+1}(2N(\delta\lambda+v) - \alpha\lambda) + \frac{C_1 a(Y - \lambda(1-\eta))}{-2N+1}\right) \\
K_5 &= \frac{1}{v\lambda}(K_2((2N+1)(1+\lambda) + v) + K_2(2N(\delta\lambda+v) - \alpha\lambda) \\
&\quad + C_2 a(Y(2N+1) - \lambda(2N+1-\eta)) \\
K_6 &= \frac{1}{v\lambda}\left(K_3((2N+3)(\lambda+1) + v) + \frac{K_3}{3}(2N(\delta\lambda+v) - \alpha\lambda) + \frac{\omega_n a\lambda}{3}\right)
\end{aligned} \tag{3.47}$$

The following non-dimensional stresses, displacement, electric potential, magnetic potential and magnetic induction are used for convenience in this paper:

$$\Sigma_{rr} = \frac{\sigma_{rr}}{c_{330}}, \Sigma_{\theta\theta} = \frac{\sigma_{\theta\theta}}{c_{330}}, u_1 = \frac{u}{a}, \Phi_1 = \frac{\Phi}{a}, \Psi_1 = \frac{\Psi}{a}, B_{r_1} = \frac{B_r}{d_{330}} \tag{3.48}$$

Using Eqs. (3.23), (3.45), (3.46), and (3.48), we achieve:

$$\Sigma_{rr} = \imath^{-2N}\frac{\rho^{2N-1}}{a}\begin{pmatrix} A\rho^{m_1}(m_1+\delta+b_1 m_1) + B\rho^{m_2}(m_2+\delta+b_2 m_2) \\ + C\rho^{m_3}(m_3+\delta+b_3 m_3) \\ + \rho^{-2N+1}(K_1(-2N+1) + K_1\delta + K_4(-2N+1) - aC_1) \\ + \rho(K_2(1+\delta) + K_5 - aC_2) + \rho^3(3K_3 + K_3\delta + 3K_6) \end{pmatrix} \tag{3.49a}$$

$$\Sigma_{\theta\theta} = \imath^{-2N}\frac{\rho^{2N-1}}{a}\begin{pmatrix} A\rho^{m_1}(m_1\delta+\alpha+vb_1 m_1) + B\rho^{m_2}(m_2\delta+\alpha+vb_2 m_2) \\ + C\rho^{m_3}(m_3\delta+\alpha+vb_3 m_3) \\ + \rho^{-2N+1}(K_1\delta(-2N+1) + K_1\alpha + K_4 v(-2N+1) - \eta aC_1) \\ + \rho(K_2(\delta+\alpha) + vK_5 - \eta aC_2) + \rho^3(K_3(3\delta+\alpha) + 3K_6 v) \end{pmatrix} \tag{3.49b}$$

$$\Psi_1 = \frac{b_1}{a} A\rho^{m_1} + \frac{b_2}{a} B\rho^{m_2} + \frac{b_3}{a} C\rho^{m_3} + \frac{D}{a} + \frac{K_4}{a}\rho^{-2N+1} + \frac{K_5}{a}\rho + \frac{K_6}{a}\rho^3 \quad (3.49c)$$

$$B_{r_1} = \imath^{-2N} \frac{\rho^{2N-1}}{a} \left(\begin{array}{c} A\rho^{r_1}(r_1 + v - \lambda b_1 r_1) + B\rho^{r_2}(r_2 + v - \lambda b_2 r_2) \\ + C\rho^{r_3}(r_3 + v - \lambda b_3 r_3) + \rho^{-2N+1}(K_1(-2N+1) + vK_1 + C_1 aY) \\ + \rho(K_2(1+v) - \lambda K_5 + C_2 Ya) + \rho^3(K_3(3+v) - 3\lambda K_6) \end{array} \right) \quad (3.49d)$$

$$u_1 = \frac{A}{a}\rho^{m_1} + \frac{B}{a}\rho^{m_2} + \frac{C}{a}\rho^{m_3} + \frac{K_1}{a}\rho^{-2N+1} + \frac{K_2}{a}\rho + \frac{K_3}{a}\rho^3 \quad (3.49e)$$

The following boundary conditions are assumed for thermomagnetoelastic analysis of the rotating FGM cylinder:

$$\Sigma_{rr}(1) = \Sigma_{rri}, \Psi_1(1) = \Psi_{1i} \quad (3.50a)$$

$$\Sigma_{rr}(\imath) = \Sigma_{rro}, \Psi_1(\imath) = \Psi_{1o} \quad (3.50b)$$

Using Eqs. (3.49) and (3.50), the four unknown A, B, C, and D are obtained by solving the following linear algebraic system of equations:

$$I \begin{Bmatrix} A \\ B \\ C \\ D \end{Bmatrix} = J \quad (3.51)$$

where I is a 4×4 nontrivial matrix with the following arrays:

$$\begin{aligned}
& I_{11} = \imath^{-2N} \frac{m_1(1+b_1)+\delta}{a}, \; I_{12} = \imath^{-2N} \frac{m_2(1+b_2)+\delta}{a} \\
& I_{13} = \imath^{-2N} \frac{m_3(1+b_3)+\delta}{a}, \; I_{14} = 0 \\
& I_{21} = \frac{b_1}{a}, \; I_{22} = \frac{b_2}{a}, \; I_{23} = \frac{b_3}{a}, I_{24} = \frac{1}{a} \\
& I_{31} = \imath^{m_1-1} \frac{m_1(1+b_1)+\delta}{a}, \; I_{32} = \imath^{m_2-1} \frac{m_2(1+b_2)+\delta}{a} \\
& I_{33} = \imath^{m_3-1} \frac{m_3(1+b_3)+\delta}{a}, \; I_{34} = 0 \\
& I_{41} = \imath^{m_1} \frac{b_1}{a}, \; I_{42} = \imath^{m_2} \frac{b_2}{a}, \; I_{43} = \imath^{m_3} \frac{b_3}{a}, \; I_{44} = \frac{1}{a}
\end{aligned} \quad (3.52)$$

and the vector J has the following components:

$$
\begin{aligned}
J_1 &= \Sigma_{rri} - \frac{\iota^{-2N}}{a}(K_1(-2N+1+\delta)+K_4(-2N+1)-aC_1+K_2(1+\delta)+K_5-aC_2 \\
&\quad +K_3(3+\delta)+3K_6) \\
J_2 &= \Psi_{1i} - \frac{K_4}{a} - \frac{K_5}{a} - \frac{K_6}{a} \\
J_3 &= \Sigma_{rro} - \frac{1}{a}((K_1(-2N+1+\delta)+K_4(-2N+1)-aC_1)\iota^{-2N}+K_2(1+\delta)+K_5-aC_2 \\
&\quad +K_3(3+\delta)+3K_6)\iota^2 \\
J_4 &= \Psi_{1o} - \frac{K_4}{a}\iota^{-2N+1} - \frac{K_5}{a}\iota - \frac{K_6}{a}\iota^3
\end{aligned}
\tag{3.53}
$$

For brevity the expressions for A, B, C, and D are not presented here.

3.2.2 Solution for Thermo-Magnetoelectroelastic Homogeneous Cylinder

An analytical solution for thermal analysis of a magnetoelectroelastic hollow cylinder is obtained in this part. The cylinder is assumed to be orthotropic and homogeneous. The governing differential equations are obtained using Eqs. (3.23), (3.25), (3.26), (3.28), (3.29) and (3.39) into (3.24) and setting $N = 0$ as follows:

$$
r^2u_{,rr} + ru_{,r} - \alpha u + r^2\Phi_{,rr} + r(1-\beta)\Phi_{,r} + r^2\Psi_{,rr} + r(1-v)\Psi_{,r} - r^2\Theta_{,r} - r(1-\eta)\Theta + \omega_n\frac{r^3}{a^2} = 0 \tag{3.54a}
$$

$$
r^2u_{,rr} + r(1+\beta)u_{,r} - r^2\gamma\Phi_{,rr} - r\gamma\Phi_{,r} - r^2\zeta\Psi_{,rr} - r\zeta\Psi_{,r} + r^2X\Theta_{,r} + rX\Theta = 0 \tag{3.54b}
$$

$$
r^2u_{,rr} + r(1+v)u_{,r} - r^2\zeta\Phi_{,rr} - r\zeta\Phi_{,r} - r^2\lambda\Psi_{,rr} - r\lambda\Psi_{,r} + r^2Y\Theta_{,r} + rY\Theta = 0 \tag{3.54c}
$$

Using the normalized radial coordinate $\rho = \frac{r}{a}$ and then changing the variable of $\rho = e^{\varsigma}$, the following differential equations with constant coefficients are obtained:

$$
\ddot{u} - \alpha u + \ddot{\Phi} - \beta\dot{\Phi} + \ddot{\Psi} - v\dot{\Psi} = ae^{\varsigma}\dot{\Theta} + (1-\eta)ae^{\varsigma}\Theta - \omega_n ae^{3\varsigma} \tag{3.55a}
$$

$$
\ddot{u} + \beta\dot{u} - \gamma\ddot{\Phi} - \zeta\ddot{\Phi} = -Xae^{\varsigma}\dot{\Theta} - Xae^{\varsigma}\Theta \tag{3.55b}
$$

$$\ddot{u} + v\dot{u} - \zeta\ddot{\Phi} - \lambda\ddot{\Phi} = -Yae^{\varsigma}\dot{\Theta} - Yae^{\varsigma}\Theta \tag{3.55c}$$

We eliminate Φ between Eqs. (3.55) and reach the two differential equations about u and Ψ as:

$$\begin{aligned}(1+\gamma)u - \alpha\gamma u + \beta u - \beta\gamma\Phi + (\gamma-\zeta)\Psi - v\gamma\Psi = ae^{\varsigma}\Theta(\gamma - X)\\ + ae^{\varsigma}\Theta(\gamma(1-\eta) - X) - \Omega a\gamma e^{3\varsigma}\end{aligned} \tag{3.56a}$$

$$(\gamma-\zeta)\ddot{u} + (v\gamma - \beta\zeta)\dot{u} + \left(\zeta^2 - \lambda\gamma\right)\ddot{\Psi} = ae^{\varsigma}\dot{\Theta}(X\zeta - Y\gamma) + ae^{\varsigma}\Theta(X\zeta - Y\gamma) \tag{3.56b}$$

Eliminating Ψ between Eqs. (3.56) and considering the temperature distribution according to Eq. (3.34b) leads to the following fourth-order ordinary differential equation:

$$a_4\ddot{\ddot{u}} + a_2\ddot{u} = (b_1C_2 + b_2C_1)e^{\varsigma} + b_1C_1te^{\varsigma} + b_3e^{3\varsigma} \tag{3.57}$$

in which,

$$\begin{aligned}
a_4 &= \frac{1}{v\gamma - \beta\zeta}\left(1 + \gamma + \frac{(\gamma-\zeta)^2}{\lambda\gamma - \zeta^2}\right), \quad a_2 = -\left(\frac{\alpha\gamma + \beta^2}{v\gamma - \beta\zeta} + \frac{v\gamma - \beta\zeta}{\lambda\gamma - \zeta^2}\right)\\
b_1 &= a\left(\frac{X\zeta - Y\gamma}{\zeta^2 - \lambda\gamma} + \frac{\gamma(1-\eta) + (\beta-1)X}{v\gamma - \beta\zeta} - \frac{\gamma-\zeta}{v\gamma - \beta\zeta}\frac{X\zeta - Y\gamma}{\zeta^2 - \lambda\gamma}\right)\\
b_2 &= a\left(2\frac{X\zeta - Y\gamma}{\zeta^2 - \lambda\gamma} + \frac{\gamma(3-2\eta) + (2\beta-3)X}{v\gamma - \beta\zeta} - 3\frac{\gamma-\zeta}{v\gamma - \beta\zeta}\frac{X\zeta - Y\gamma}{\zeta^2 - \lambda\gamma}\right)\\
b_3 &= -\frac{9a\gamma\omega_n}{v\gamma - \beta\zeta}
\end{aligned} \tag{3.58}$$

The solution of Eq. (3.57) can be written as:

$$u = A + B\ln(\rho) + C\rho^m + D\rho^{-m} + (K_1\ln(\rho) + K_2)\rho + K_3\rho^3 \tag{3.59}$$

where

$$\begin{aligned}
m &= \sqrt{-\frac{a_2}{a_4}}(m \in R), \ K_1 = \frac{b_1C_1}{a_4 + a_2}, \ K_2 = \frac{b_1C_2 + b_2C_1}{a_4 + a_2} - \frac{b_1C_1(4a_4 + 2a_2)}{(a_4 + a_2)^2},\\
K_3 &= \frac{b_5}{81a_4 + 9a_2}
\end{aligned} \tag{3.60}$$

In Eq. (3.59), A, B, C, and D are integration constants. Using the second equation of (3.56), Ψ can be found as:

$$\Psi = F + E\,\ln(\rho) + Bc_0(\ln(\rho))^2 + Cc_1\rho^m + Dc_2\rho^{-m} + (c_3\ln(\rho) + c_4)\rho + K_3c_5\rho^3 \tag{3.61}$$

where E and F are new integration constants and

$$\begin{aligned}
c_0 &= \frac{\nu\gamma - \beta\zeta}{2(\lambda\gamma - \zeta^2)}, c_1 = \frac{1}{m(\lambda\gamma - \zeta^2)}(m(\gamma - \zeta) + \nu\gamma - \beta\zeta),\\
c_2 &= \frac{1}{m(\lambda\gamma - \zeta^2)}(m(\gamma - \zeta) - \nu\gamma + \beta\zeta),\\
c_3 &= \frac{1}{\lambda\gamma - \zeta^2}(K_1(\gamma(1+\nu) - \zeta(1+\beta)) - aC_1(X\zeta - Y\gamma)),\\
c_4 &= \frac{1}{\lambda\gamma - \zeta^2}\begin{pmatrix} (2K_1 + K_2)(\gamma - \zeta) + (K_2 + K_1)(\nu\gamma - \beta\zeta)\\ -aC_2(X\zeta - Y\gamma) - K_1(\gamma(1+\nu) - \zeta(1+\beta)) \end{pmatrix},\\
c_5 &= \frac{1}{3(\lambda\gamma - \zeta^2)}(3(\gamma - \zeta) + \nu\gamma - \beta\zeta)
\end{aligned} \tag{3.62}$$

Considering the expressions for u and Ψ, one can obtain the following expression for Φ using Eq. (3.55c):

$$\Phi = H + G\ln(\rho) + Bl_0(\ln(\rho))^2 + Cl_1\rho^m + Dl_2\rho^{-m} + (l_3\ln(\rho) + l_4)\rho + K_3l_5\rho^3 \tag{3.63}$$

where G and H are integration constants and

$$\begin{aligned}
l_0 &= \frac{1}{2\gamma}(\beta - 2\zeta c_0), l_1 = \frac{1}{m\gamma}(m + \beta - m\zeta c_1), l_2 = \frac{1}{m\gamma}(m + \beta - m\zeta c_2),\\
l_3 &= \frac{1}{\gamma}(K_1(1+\beta) - \zeta c_3 + XaC_1),\\
l_4 &= \frac{1}{\gamma}(K_1 + K_2(1+\beta) - \zeta(c_3 + c_4) + XaC_2), l_5 = \frac{1}{3\gamma}(3 + \beta - 3\zeta c_5)
\end{aligned} \tag{3.64}$$

Since we have six boundary conditions while there are eight integration constants, we need two more complimentary equations. These equations are acquired by inserting Eqs. (3.59), (3.61), and (3.63) into Eq. (3.55a) as follows:

$$B = 0, \alpha A + \nu E + \beta G = 0 \tag{3.65}$$

Using the non-dimensional parameters (3.48), we have:

$$\Sigma_{rr} = \frac{\rho^{-1}}{a}(A\delta + G + E) + \frac{\rho^{m-1}}{a}(m(1 + l_1 + c_1) + \delta)C + \frac{\rho^{-m-1}}{a}(-m(1 + l_2 + c_2))D + \frac{\ln(\rho)}{a}(K_1(1+\delta) + l_3 + c_3 - C_1 a) + \frac{1}{a}(K_1 + (1+\delta)K_2 + l_3 + l_4 + c_3 + c_4 - C_2 a) + \frac{\rho^2}{a}(3(1 + l_5 + c_5) + \delta)K_3 \tag{3.66a}$$

$$\Sigma_{\theta\theta} = \frac{\rho^{-1}}{a}(A\alpha + G\beta + Ev) + \frac{\rho^{m-1}}{a}(m(\delta + \beta l_1 + vc_1) + \alpha)C + \frac{\rho^{-m-1}}{a}(-m(\delta + \beta l_2 + vc_2) + \alpha)D + \frac{\ln(\rho)}{a}(K_1(\delta+\alpha) + \beta l_3 + vc_3 - \eta C_1 a) + \frac{1}{a}(\delta K_1 + (\delta + \alpha)K_2 + \beta(l_3 + l_4) + v(c_3 + c_4) - \eta q C_2) + \frac{\rho^2}{a}(3(\delta + \beta l_5 + vc_5) + \alpha)K_3 \tag{3.66b}$$

$$\Phi_1 = \frac{H}{a} + \frac{G}{a}\ln(\rho) + \frac{C}{a}l_1\rho^{-m} + \frac{D}{a}l_2\rho^{-m} + (l_3\ln(\rho) + l_4)\frac{\rho}{a} + \frac{K_3 l_5}{a}\rho^3 \tag{3.66c}$$

$$\Psi_1 = \frac{F}{a} + \frac{E}{a}\ln(\rho) + \frac{C}{a}c_1\rho^{m} + \frac{D}{a}c_2\rho^{-m} + (c_3\ln(\rho) + c_4)\frac{\rho}{a} + \frac{K_3 c_5}{a}\rho^3 \tag{3.66d}$$

The following boundary conditions in this thermo-magnetoelectroelastic analysis are:

$$\Sigma_{rr}(1) = \Sigma_{rri},\ \Phi_1(1) = \Phi_{1i},\ \Psi_1(1) = \Psi_{1i} \tag{3.67a}$$

$$\Sigma_{rr}(\imath) = \Sigma_{rro},\ \Phi_1(\imath) = \Phi_{1o},\ \Psi_1(\imath) = \Psi_{1o} \tag{3.67b}$$

Employing Eqs. (3.65) through (3.67), we can obtain a linear algebraic equation for the integration constants as:

$$I\begin{Bmatrix} A \\ C \\ D \\ E \\ F \\ G \\ H \end{Bmatrix} = J \tag{3.68}$$

where I is a 7×7 nontrivial matrix with the following arrays:

$$
\begin{aligned}
& I_{11} = \frac{\delta}{a}, I_{12} = \frac{1}{a}(m(1+l_1+c_1)+\delta), I_{13} = \frac{1}{a}(-m(1+l_2+c_2)+\delta), I_{14} = \frac{1}{a}, \\
& I_{15} = 0, I_{16} = \frac{1}{a}, I_{17} = 0 \\
& I_{21} = 0, I_{22} = \frac{l_1}{a}, I_{23} = \frac{l_2}{a}, I_{24} = 0, I_{25} = 0, I_{26} = 0, I_{27} = \frac{1}{a} \\
& I_{31} = 0, I_{32} = \frac{c_1}{a}, I_{33} = \frac{c_2}{a}, I_{34} = 0, I_{35} = \frac{1}{a}, I_{36} = 0, I_{37} = 0 \\
& I_{41} = \alpha, I_{42} = 0, I_{43} = 0, I_{44} = v, I_{45} = 0, I_{46} = \beta, I_{47} = 0 \\
& I_{51} = \frac{\delta}{a}\iota^{-1}, I_{52} = \frac{\iota^{m-1}}{a}(m(1+l_1+c_1)+\delta), I_{53} = \frac{\iota^{-m-1}}{a}(-m(1+l_2+c_2)+\delta), \\
& I_{54} = \frac{\iota^{-1}}{a}, I_{55} = 0, I_{56} = \frac{\iota^{-1}}{a}, I_{57} = 0 \\
& I_{61} = 0, I_{62} = \frac{l_1}{a}\iota^{m}, I_{63} = \frac{l_2}{a}\iota^{-m}, I_{64} = 0, I_{65} = 0, I_{66} = \frac{\ln(\iota)}{a}, I_{67} = \frac{1}{a} \\
& I_{71} = 0, I_{72} = \frac{c_1}{a}\iota^{m}, I_{73} = \frac{c_2}{a}\iota^{-m}, I_{74} = \frac{\ln(\iota)}{a}, I_{75} = \frac{1}{a}, I_{76} = 0, I_{77} = 0
\end{aligned}
\tag{3.69}
$$

and the components of vector J are:

$$
\begin{aligned}
& J_1 = \Sigma_i - \frac{1}{a}(K_1 + (1+\delta)K_2 + l_3 + l_4 + c_3 + c_4 - C_2 a) - \frac{1}{a}(3(1+l_5+c_5)+\delta)K_3 \\
& J_2 = \Phi_{1i} - \frac{l_4}{a} - \frac{K_3 l_5}{a}, J_3 = \Psi_{1i} - \frac{c_4}{a} - \frac{K_3 c_5}{a}, J_4 = 0 \\
& J_5 = \Sigma_o - \frac{\ln(\iota)}{a}(K_1(1+\delta) + l_3 + c_3 - C_1 a) - \frac{1}{a}(K_1 + (1+\delta)K_2 + l_3 + l_4 + c_3 + c_4 - C_2 a) \\
& \quad - \frac{\iota^2}{a}(3(1+l_5+c_5)+\delta)K_3 \\
& J_6 = \Phi_{1o} - \frac{\iota}{a}(l_3 \ln(\iota) + l_4) - \frac{K_3 l_5}{a}\iota^3, J_7 = \Psi_{1o} - \frac{\iota}{a}(c_3 \ln(\iota) + c_4) - \frac{K_3 c_5}{a}\iota^3
\end{aligned}
\tag{3.70}
$$

Solving Eq. (3.68) renders the integration constants and completes the analytical solution for the thermo-magnetoelectroelastic hollow cylinder.

3.2.3 Benchmark Results

In this section, the numerical results are presented to show the multiphysical behavior of the hollow thermomagnetoeleastic functionally graded cylinder and the homogeneous thermo-magnetoelectroelastic orthotropic cylinder. The outer surface of the FGM cylinder is assumed to be $BaTiO_3/CoFe_2O_4$ with material properties given in Table 3.2, and these material properties are taken for the homogenous orthotropic cylinder as well. The following mechanical and magnetic boundary conditions are assumed for the thermomagnetoelastic FGM hollow cylinder:

$$\Sigma_{rr}(1) = -1, \Psi_1(1) = 1$$

$$\Sigma_{rr}(\imath) = 0, \Psi_1(\imath) = 0$$

The non-dimensional temperature on the inner and outer surfaces of the FGM hollow cylinder are assigned to be Θ_a and Θ_b, respectively. For the thermomagnetoelastic analysis, the radial and hoop stresses, magnetic potential, magnetic induction, and temperature distribution are given for a variety of non-dimensional values of inner temperature, inertial effects, aspect ratios and non-homogeneity parameters.

The influence of the thermal boundary condition on the distribution of stresses, magnetic induction, and magnetic potential along the radial direction for an FGM thermomagnetoelastic hollow cylinder are shown in Fig. 3.4a–e. The non-homogeneity parameter, angular velocity, and aspect ratio are $N = 1$, $\omega_n = 1$, and $\imath = 4$, respectively. The electromagnetic boundary conditions in above

Table 3.2 Material properties of $BaTiO_3$/$CoFe_2O_4$ [33, 34]

Properties	$BaTiO_3/CoFe_2O_4$
$c_{330}\left(\frac{N}{m^2}\right)$	2.96×10^{11}
$c_{110}\left(\frac{N}{m^2}\right)$	2.86×10^{11}
$c_{130}\left(\frac{N}{m^2}\right)$	1.70×10^{11}
$e_{310}\left(\frac{C}{m^2}\right)$	-4.4
$e_{330}\left(\frac{C}{m^2}\right)$	18.6
$d_{310}\left(\frac{N}{A\,m}\right)$	580.3
$d_{330}\left(\frac{N}{A\,m}\right)$	699.7
$\varepsilon_{330}\left(\frac{C^2}{N\,m^2}\right)$	9.3×10^{11}
$g_{330}\left(\frac{N\,s^2}{C^2}\right)$	3.0×10^{-12}
$\mu_{330}\left(\frac{N\,s^2}{VC}\right)$	1.57×10^{-4}
$\beta_{110}\left(\frac{N}{km^2}\right)$	4.395×10^{6}
$\beta_{330}\left(\frac{N}{km^2}\right)$	4.560×10^{6}
$\gamma_{10}\left(\frac{C}{km^2}\right)$	-13.0×10^{-5}
$\tau_{10}\left(\frac{N}{A\,m\,K}\right)$	6.0×10^{-3}

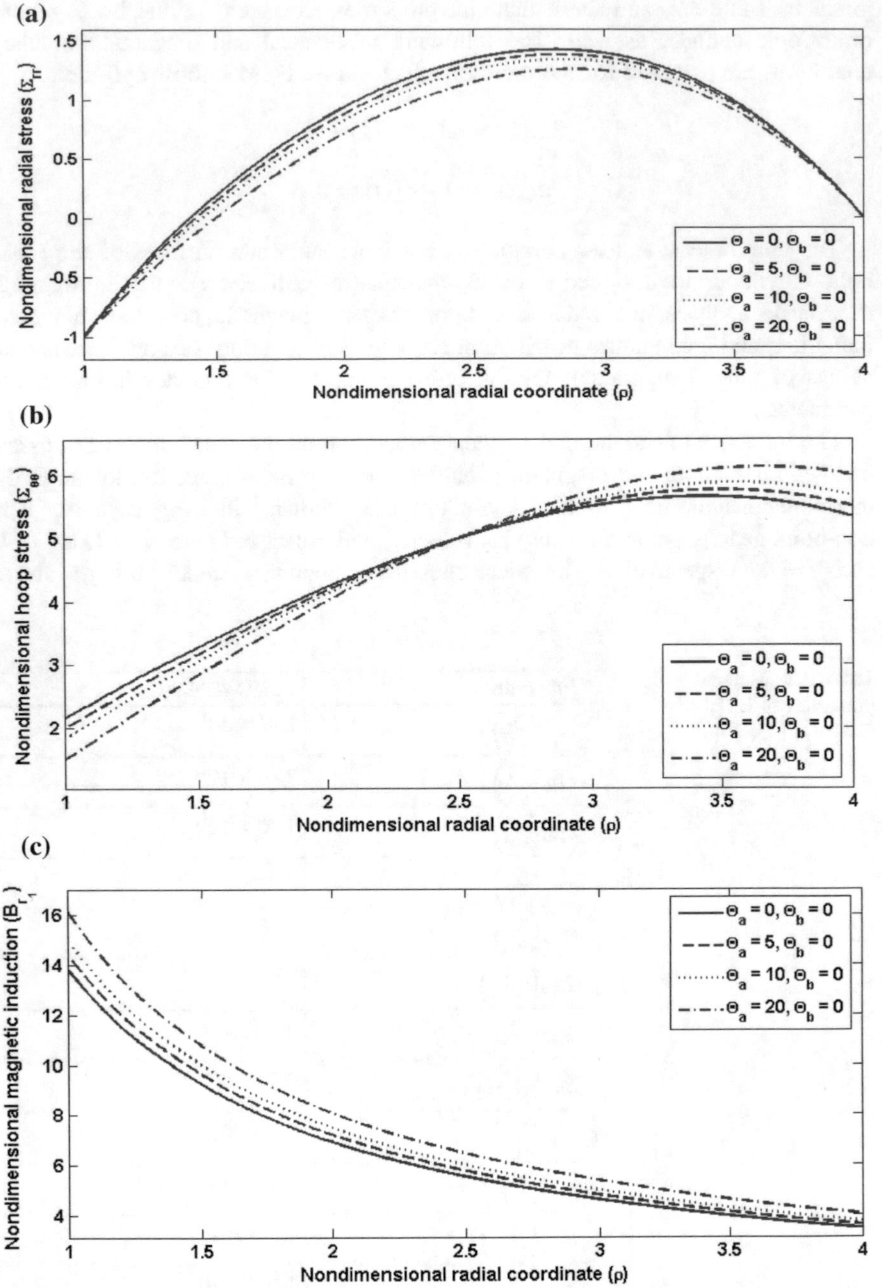

Fig. 3.4 **a** Radial stress, **b** Hoop stress, **c** Magnetic induction, **d** Total magnetic potential distribution, **e** Zoomed-in magnetic potential distribution for different thermal boundary conditions, $\omega_n = 1$, $N = 1$, $\imath = 4$. [Reproduced from [35] with permission from Taylor & Francis Ltd.]

(d)

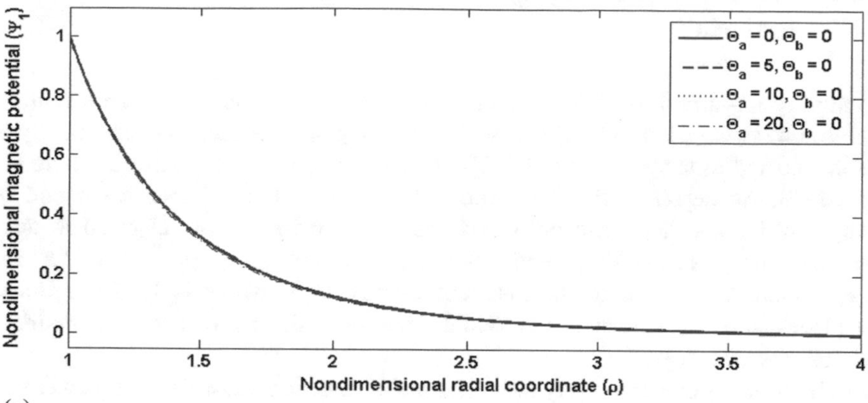

(e)

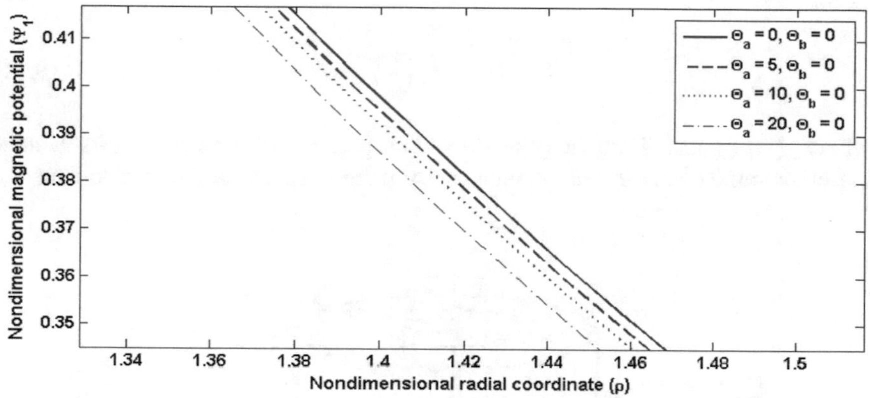

Fig. 3.4 (continued)

equations are used. The outer non-dimensional temperature of the FGM is assumed to be zero, while the non-dimensional temperature on the inner surface is increased gradually.

Increasing the temperature on the inner surface changes the radial stress distribution monotonically as shown in Fig. 3.4a and shifts the transition point of the radial stress, in which the radial stress changes from negative to positive, toward the outer surface of the cylinder. As depicted in Fig. 3.4b, the value of hoop stress decreases before $\rho = 2.4958$ and increases after this point when the inner temperature is increased. Furthermore, the non-dimensional magnetic induction in Fig. 3.4c increases when the inner temperature is augmented. Although the magnetic potential does not change significantly when altering the inner temperature in Fig. 3.4d, the zoomed-in examination clarifies that it decreases slightly when the inner temperature increases as shown in Fig. 3.4e. It should be mentioned that the results are similar to those reported in [36] when the inner and outer temperatures are set to be zero.

3.3 Thermal Stress Analysis of Heterogeneous Smart Materials

A similar approach is used in this section to analyze heterogeneous smart materials. An infinitely long, hollow, FGPM cylinder rotating at a constant angular velocity ω is considered as shown in Fig. 3.5. The cylinder is poled and graded in the radial direction. As depicted, the inner and outer radii of the cylinder are a and b, respectively. The inner and outer surfaces of the cylinder are subjected to temperature change θ, electric potential ϕ, and pressure P. Subscripts "a" and "b" are used to indicate the load on the inner and outer surfaces, respectively. The cylinder is placed in a constant magnetic field H_0 acting in the z direction of cylindrical coordinate system (r, θ, z).

The material properties vary along the radial direction according to a power law as follows:

$$\chi(r) = \chi_0 \left(\frac{r}{b}\right)^{2N} \tag{3.71}$$

where $\chi(r)$, χ_0, and N are the general material property of the cylinder, its value at the outer surface, and the non-homogeneity parameter, respectively. The constitu-

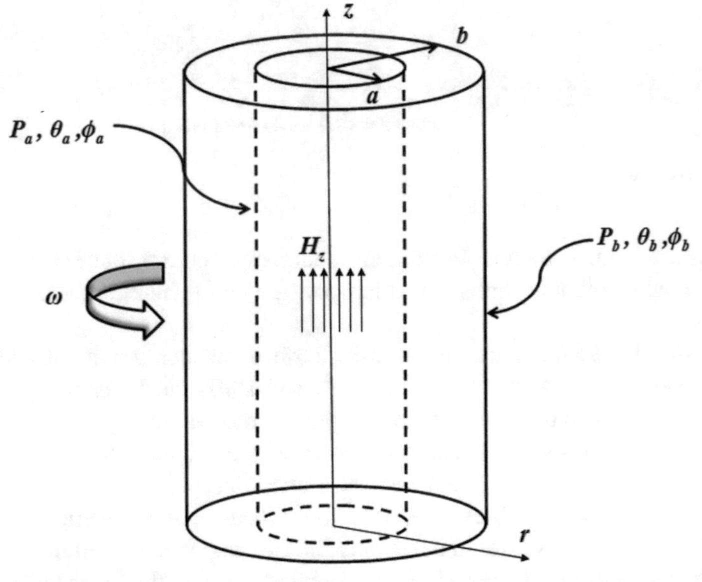

Fig. 3.5 FGPM rotating hollow cylinder and its boundary conditions [Reproduced from [37] with permission from IOP Publishing]

tive equations for a linear piezoelectric material are given in Eq. (3.1). Furthermore, the quasi-stationary electric field and the linear strain-displacement are in accordance with Eqs. (3.5) and (3.6).

The non-zero components of strains and electric fields for the current axisymmetric, plane strain problem are given as follows:

$$\begin{aligned} \bar{\varepsilon}_{rr} &= u_{,r}, \quad \bar{\varepsilon}_{\theta\theta} = \frac{u}{r} \\ E_r &= -\phi_{,r} \end{aligned} \tag{3.72}$$

where u is the radial displacement; r and θ are the radial and circumferential coordinates. For the cylindrically orthotropic piezoelectric material polarized in the radial direction, substituting Eq. (3.72) into Eq. (3.1) leads to:

$$\sigma_{rr} = c_{33}u_{,r} + c_{13}\frac{u}{r} + e_{33}\phi_{,r} - \beta_1\theta \tag{3.73a}$$

$$\sigma_{\theta\theta} = c_{13}u_{,r} + c_{11}\frac{u}{r} + e_{31}\phi_{,r} - \beta_3\theta \tag{3.73b}$$

$$D_r = e_{33}u_{,r} + e_{31}\frac{u}{r} - \varepsilon_{33}\phi_{,r} + \gamma_1\theta \tag{3.73c}$$

where $c_{mn} = C_{ijkl}$, $e_{mk} = e_{pij}(i,j,k,l,p = 1,2,3; m,n = 1,2,\ldots,6)$, $\beta_1 = \beta_{11}$, and $\beta_3 = \beta_{33}$.

The equation of motion in the rotating electromagnetic medium and Maxwell equation in the absence of electric charge under axisymmetric loading are expressed as:

$$\sigma_{rr,r} + \frac{1}{r}(\sigma_{rr} - \sigma_{\theta\theta}) + f_z = \rho_d u_{,tt} \tag{3.74a}$$

$$D_{r,r} + \frac{1}{r}D_r = 0 \tag{3.74b}$$

where ρ_d is the mass density and t stands for time; f_z and h_z are defined as Lorentz`s force and perturbation of magnetic field, respectively, which for a constant magnetic field H_0 can be written as follows [20]:

$$\begin{aligned} f_z &= H_0^2\left(\mu\left(u_{,r} + \frac{1}{r}u\right)\right)_{,r} \\ h_z &= -H_0\left(u_{,r} + \frac{1}{r}u\right) \end{aligned} \tag{3.75}$$

in which, μ is magnetic permeability.

3.3.1 Solution Procedures

Substituting Eqs. (3.71), (3.73), and (3.75) into Eq. (3.74) and considering $u_{tt} = -r\omega^2$, the two coupled governing differential equations for the problem are obtained:

$$\begin{aligned} & r^2\left(c_{330}+\mu_0 H_0^2\right)u_{,rr}+r(2N+1)\left(c_{330}+\mu_0 H_0^2\right)u_{,r} \\ & +\left(2Nc_{130}-c_{110}+\mu_0 H_0^2(2N-1)\right)u+r^2 e_{330}\varphi_{,rr} \\ & +r((2N+1)e_{330}-e_{310})\varphi_{,r}-r^2\beta_{10}\theta_{,r} \\ & -r((2N+1)\beta_{10}-\beta_{30})\theta+\rho_{d0}\omega^2 r^3=0 \end{aligned} \tag{3.76a}$$

$$\begin{aligned} & r^2 e_{330}u_{,rr}+r((2N+1)e_{330}+e_{310})u_{,r}+2Ne_{310}u-r^2\varepsilon_{330}\varphi_{,rr} \\ & -r(2N+1)\varepsilon_{330}\varphi_{r}+r^2\gamma_{10}\theta_{,r}+r(2N+1)\gamma_{10}\theta=0 \end{aligned} \tag{3.76b}$$

in which, ω is the constant angular velocity of the cylinder; $c_{110}, c_{130}, c_{330}, e_{130}, e_{330}, \varepsilon_{330}, \rho_{d0}, \beta_{10}, \beta_{30}, \gamma_{10}$, and μ_0 represent the corresponding values of $c_{11}, c_{13}, c_{33}, e_{13}, e_{33}, \varepsilon_{33}, \rho_d, \beta_1, \beta_3, \gamma_1$, and μ at the outer surface of the cylinder, respectively. To simplify the analysis, we introduce the following non-dimensional parameters:

$$\alpha=\frac{c_{110}}{c_{330}},\beta=\frac{e_{310}}{e_{330}},\gamma=\frac{\varepsilon_{330}c_{330}}{e_{330}^2},\delta=\frac{c_{130}}{c_{330}},\eta=\frac{\beta_{30}}{\beta_{10}},X=\frac{\gamma_{10}c_{330}}{e_{330}\beta_{10}},\Omega=\frac{\mu_0 H_0^2}{c_{330}} \tag{3.77}$$

as well as a new electric potential and temperature change as follows:

$$\Phi=\frac{e_{330}}{c_{330}}\phi,\quad \Theta=\frac{\beta_{110}}{c_{330}}\theta \tag{3.78}$$

Therefore, Eq. (8) can be rewritten in the following form:

$$\begin{aligned} & r^2(1+\Omega)u_{,rr}+r(2N+1)(1+\Omega)u_{,r} \\ & +(2N\delta-\alpha+\Omega(2N-1))u+r^2\Phi_{,rr}+r(2N+1-\beta)\Phi_{,r} \\ & -r^2\Theta_{,r}-r(2N+1-\eta)\Theta+\frac{\rho_{d0}\omega^2 r^3}{c_{330}}=0 \end{aligned} \tag{3.79a}$$

$$\begin{aligned} & r^2 u_{,rr}+r(2N+1+\beta)u_{,r}+2N\beta u-r^2\gamma\Phi_{,rr} \\ & -r(2N+1)\gamma\Phi_{,r}+r^2 X\Theta_{,r}+r(2N+1)X\Theta=0 \end{aligned} \tag{3.79b}$$

Using the normalized radial coordinate $\rho=\frac{r}{a}$ and employing the chain rule for differentiation, Eq. (3.79) is reduced to:

$$\begin{aligned}&\rho^2(1+\Omega)u_{,\rho\rho}+\rho(2N+1)(1+\Omega)u_{,\rho}\\&\quad+(2N\delta-\alpha+\Omega(2N-1))u+\rho^2\Phi_{,\rho\rho}+\rho(2N+1-\beta)\Phi_{,\rho}\\&\quad-\rho^2 a\Theta_{,\rho}-\rho(2N+1-\eta)\Theta+\frac{\rho_{d0}\omega^2\rho^3a^3}{c_{330}}=0\end{aligned} \tag{3.80a}$$

$$\begin{aligned}&\rho^2u_{,\rho\rho}+(2N+1+\beta)\rho u_{,\rho}+2N\beta u-\rho^2\gamma\Phi_{,\rho\rho}\\&\quad-\rho(2N+1)\gamma\Phi_{,\rho}+\rho^2aX\Theta_{,\rho}+\rho a(2N+1)X\Theta=0\end{aligned} \tag{3.80b}$$

To solve Eq. (3.80), the temperature distribution along the radial direction of the hollow cylinder must be obtained. The axisymmetric, steady state heat conduction equation for an infinitely long hollow cylinder can be represented as:

$$\frac{1}{r}\left(rk\theta_{,r}\right)_{,r}=0 \qquad a\leq r\leq b \tag{3.81}$$

in which, k is the thermal conductivity varying according to Eq. (3.70). Rearranging Eq. (3.81), using the normalized radial coordinate and employing Eqs. (3.70) and (3.78) leads to:

$$\frac{1}{\rho}\left(\rho^{2N+1}\Theta_{,\rho}\right)_{,\rho}=0 \qquad 1\leq\rho\leq\imath \tag{3.82}$$

where $\imath=\frac{b}{a}$ is the aspect ratio of the hollow cylinder. The general solution for $N\neq0$ is obtained as follows:

$$\Theta=C_1\rho^{-2N}+C_2 \tag{3.83}$$

in which, C_1 and C_2 are integration constants to be determined by thermal boundary conditions. The general boundary conditions for Eq. (3.82) are [32]:

$$A_{11}\Theta(1)+A_{12}\Theta'(1)=f_1 \tag{3.84a}$$

$$A_{21}\Theta(\imath)+A_{22}\Theta'(\imath)=f_2 \tag{3.84b}$$

where A_{ij} are the Robin-type boundary condition coefficients, and f_1 and f_2 are known functions on the inner and outer radii. Substituting Eq. (3.83) into Eq. (3.84) gives us the following integration constant C_1 and C_2:

$$\begin{aligned}C_1&=\frac{A_{11}f_2-A_{21}f_1}{A_{11}(A_{21}\vartheta-2NA_{22})\vartheta^{-2N-1}-A_{21}(A_{11}-2NA_{12})}\\C_2&=\frac{f_1}{A_{11}}-\frac{(A_{11}-2NA_{12})(A_{11}f_2-A_{21}f_1)}{A_{11}^2(A_{21}\vartheta-2NA_{12})\vartheta^{-2N-1}-A_{21}\left(A_{11}^2-2NA_{12}A_{11}\right)}\end{aligned} \tag{3.85}$$

Having the temperature distribution, we can solve the coupled governing equations (3.80). Making a variable change of $\rho = e^{\varsigma}$, Eq. (3.80) is converted to the new differential equations with constant coefficients in the following form:

$$\begin{aligned}&(1+\Omega)u + 2N(1+\Omega)u + (2N\delta - \alpha + \Omega(2N-1))u + \Phi + (2N-\beta)\Phi \\ &= ae^{\varsigma}\Theta + (2N+1-\eta)e^{\varsigma}a\Theta - \omega_n a e^{3\varsigma}\end{aligned} \tag{3.86a}$$

$$\ddot{u} + (2N+\beta)\dot{u} + 2N\beta u - \gamma\ddot{\Phi} - 2\gamma N\dot{\Phi} = -ae^{\varsigma}X\dot{\Theta} - (2N+1)Yae^{\varsigma}\Theta \tag{3.86b}$$

in which, the superposed dot represents differentiation with respect to ς and

$$\omega_n = \frac{\rho_{d0}\omega^2 a^2}{c_{330}} \tag{3.87}$$

By eliminating $\ddot{\Phi}$ between the two equations in Eq. (3.87) and solving for $\dot{\Phi}$, we obtain:

$$\begin{aligned}\dot{\Phi} = &\frac{(1+\Omega)\gamma+1}{\beta\gamma}\ddot{u} + \frac{2N(1+(1+\Omega)\gamma)+\beta}{\beta\gamma}\dot{u} \\ &+ \frac{2N(\delta\gamma+\beta) - \alpha\gamma + \Omega\gamma(2N-1)}{\beta\gamma}u + \frac{a(X-\gamma)}{\beta\gamma}e^{\varsigma}\dot{\Theta} \\ &+ \frac{(2N+1)X - (2N+1-\eta)\gamma}{\beta\gamma}ae^{\varsigma}\Theta + \frac{a\omega_n}{\beta}e^{3\varsigma}\end{aligned} \tag{3.88}$$

Substituting Eq. (3.88) and its derivative into (3.86a), we obtain the following decoupled differential equation about u:

$$a_3\dddot{u} + a_2\ddot{u} + a_1\dot{u} + a_0 u = d_3 e^{3\varsigma} + d_2 e^{\varsigma} + d_1 e^{(-2N+1)\varsigma} \tag{3.89}$$

where

$$\begin{aligned}a_3 &= \frac{(1+\Omega)\gamma+1}{\beta\gamma}, \quad a_2 = \frac{4N(1+\gamma(1+\Omega))}{\beta\gamma}, \\ a_1 &= \frac{2N(\gamma\delta+\beta) + 4N^2((1+\Omega)\gamma+1) + \gamma(\Omega(2N-1)-\alpha) - \beta^2}{\beta\gamma}, \\ a_0 &= \frac{4N^2(\gamma\delta+\beta) + 2N\big(\Omega\gamma(2N-1) - \alpha\gamma - \beta^2\big)}{\beta\gamma}, \\ d_3 &= -\frac{a\omega_n(3+2N)}{\beta}, \quad d_2 = \frac{aX(2N+1)(2N+1-\beta) - a\gamma(2N+1-\eta)(1+2N)}{\beta\gamma}C_2, \\ d_1 &= \frac{aX(1-\beta) + a\gamma(\eta - 2N\beta - 1)}{\beta\gamma}C_1\end{aligned} \tag{3.90}$$

The solution of the differential equation (3.89) includes two parts, the general solution and the particular solution. The general solution is:

$$u_g = Ae^{m_1\varsigma} + Be^{m_2\varsigma} + Ce^{m_3\varsigma} \tag{3.91}$$

where, m_i $(i = 1, 2, 3)$ are the roots of the characteristic equation (3.89). Also, A, B, and C are constants of integration determined by the boundary conditions. The particular solutions can be written in the following form:

$$u_p = K_1 e^{(-2N+1)\varsigma} + K_2 e^{\varsigma} + K_3 e^{3\varsigma} \tag{3.92}$$

Substituting Eq. (3.92) into Eq. (3.89), we obtain:

$$\begin{aligned} K_1 &= \frac{d_1}{(-2N+1)^3 a_3 + (-2N+1)^2 a_2 + (-2N+1)a_1 + a_0}, \\ K_2 &= \frac{d_2}{a_3 + a_2 + a_1 + a_0}, \quad K_3 = \frac{d_3}{27a_3 + 9a_2 + 3a_1 + a_0} \end{aligned} \tag{3.93}$$

Consequently, the solution for u in terms of ρ is:

$$u = u_g + u_p = A\rho^{m_1} + B\rho^{m_2} + C\rho^{m_3} + K_1\rho^{-2N+1} + K_2\rho + K_3\rho^3 \tag{3.94}$$

Substituting Eq. (3.94) into Eq. (3.88) and integrating, Φ is obtained as:

$$\Phi = b_1 A\rho^{m_1} + b_2 B\rho^{m_2} + b_3 C\rho^{m_3} + D + K_4\rho^{-2N+1} + K_5\rho + K_6\rho^3 \tag{3.95}$$

in which D is a new integration constant and

$$\begin{aligned} b_i &= \frac{1}{\beta\gamma}\left(((1+\Omega)\gamma+1)m_i + 2N((1+\Omega)\gamma+1) + \beta + \frac{2N(\beta+\delta\gamma) - \alpha\gamma + \Omega\gamma(2N-1)}{m_i}\right) \\ K_4 &= \frac{1}{\beta\gamma}\left(K_1((1+\Omega)\gamma+1+\beta) + \frac{K_1}{-2N+1}(2N(\delta\gamma+\beta) + \gamma(\Omega(2N-1)-\alpha)) \right. \\ &\quad \left. + \frac{C_1 a(X-\gamma(1-\eta))}{-2N+1}\right) \\ K_5 &= \frac{1}{\beta\gamma}(K_2((2N+1)((1+\Omega)\gamma+1)+\beta) + K_2(2N(\delta\gamma+\beta) + \gamma(\Omega(2N-1)-\alpha)) \\ &\quad + C_2 a(X(2N+1) - \gamma(2N+1-\eta))) \\ K_6 &= \frac{1}{\beta\gamma}\left(\begin{array}{c} K_3((2N+3)((1+\Omega)\gamma+1)+\beta) \\ +\frac{K_3}{3}(2N(\delta\gamma+\beta) + \gamma(\Omega(2N-1)-\alpha)) + \frac{\omega_n a\gamma}{3} \end{array}\right) \end{aligned} \tag{3.96}$$

For convenience, we define the following non-dimensional stresses, displacement, electric potential, electric displacement, and perturbation of magnetic field as follows:

$$\Sigma_{rr} = \frac{\sigma_{rr}}{c_{330}}, \Sigma_{\theta\theta} = \frac{\sigma_{\theta\theta}}{c_{330}}, u_1 = \frac{u}{a}, \Phi_1 = \frac{\Phi}{a}, D_{r_1} = \frac{D_r}{e_{330}}, h_z^* = \frac{h_z}{H_0} \tag{3.97}$$

Using Eqs. (3.73), (3.94), (3.95), and (3.97), we obtain:

$$\Sigma_{rr} = \imath^{-2N}\frac{\rho^{2N-1}}{a}\begin{pmatrix} A\rho^{m_1}(m_1+\delta+b_1m_1)+B\rho^{m_2}(m_2+\delta+b_2m_2) \\ +C\rho^{m_3}(m_3+\delta+b_3m_3) \\ +\rho^{-2N+1}(K_1(-2N+1)+K_1\delta+K_4(-2N+1)-aC_1) \\ +\rho(K_2+K_2\delta+K_5-aC_2)+\rho^3(3K_3+K_3\delta+3K_6) \end{pmatrix} \tag{3.98a}$$

$$\Sigma_{\theta\theta} = \imath^{-2N}\frac{\rho^{2N-1}}{a}\begin{pmatrix} A\rho^{m_1}(m_1\delta+\alpha+\beta b_1m_1)+B\rho^{m_2}(m_2\delta+\alpha+\beta b_2m_2) \\ +C\rho^{m_3}(m_3\delta+\alpha+\beta b_3m_3) \\ +\rho^{-2N+1}(K_1\delta(-2N+1)+K_1\alpha+K_4\beta(-2N+1)-\eta aC_1) \\ +\rho(K_2\delta+K_2\alpha+\beta K_5-\eta aC_2)+\rho^3(3K_3\delta+K_3\alpha+3K_6\beta) \end{pmatrix} \tag{3.98b}$$

$$D_{r_1} = \imath^{-2N}\frac{\rho^{2N-1}}{a}\begin{pmatrix} A\rho^{m_1}(m_1+\beta-\gamma b_1m_1)+B\rho^{m_2}(m_2+\beta-\gamma b_2m_2) \\ +C\rho^{m_3}(m_3+\beta-\gamma b_3m_3) \\ +\rho^{-2N+1}(K_1(-2N+1)+K_1\beta-K_4\gamma(-2N+1)-\eta aC_1) \\ +\rho(K_2+\beta K_2-\gamma K_5+XaC_2)+\rho^3(3K_3+K_3\beta-3K_6\gamma) \end{pmatrix} \tag{3.98c}$$

$$h_z^* = -\frac{1}{a}\begin{pmatrix} A\rho^{m_1-1}(m_1+1)+B\rho^{m_2-1}(m_2+1)+C\rho^{m_3-1}(m_3+1) \\ +2K_1\rho^{-2N}(-N+1)+2K_2+4K_3\rho^2 \end{pmatrix} \tag{3.98d}$$

$$u_1 = \frac{1}{a}A\rho^{m_1} + \frac{1}{a}B\rho^{m_2} + \frac{1}{a}C\rho^{m_3} + \frac{K_1}{a}\rho^{-2N+1} + \frac{K_2}{a}\rho + \frac{K_3}{a}\rho^3 \tag{3.98e}$$

$$\Phi_1 = \frac{b_1}{a}A\rho^{m_1} + \frac{b_2}{a}B\rho^{m_2} + \frac{b_3}{a}C\rho^{m_3} + \frac{1}{a}D + \frac{K_4}{a}\rho^{-2N+1} + \frac{K_5}{a}\rho + \frac{K_6}{a}\rho^3 \tag{3.98f}$$

The boundary conditions according to Fig. 3.5 can be expressed in the following form:

$$\Sigma_{rr}(1) = \Sigma_{rri}, \quad \Phi_1(1) = \Phi_{1i} \tag{3.99a}$$

$$\Sigma_{rr}(\imath) = \Sigma_{rro}, \quad \Phi_1(\imath) = \Phi_{1o} \tag{3.99b}$$

Applying Eqs. (3.98) and (3.99), we reach a set of four linear algebraic equations for the four unknowns, A, B, C, and D, as follows:

$$I\begin{Bmatrix} A \\ B \\ C \\ D \end{Bmatrix} = J \tag{3.100}$$

in which, I is a 4×4 nontrivial matrix with the following arrays:

$$\begin{aligned}
&I_{11} = \iota^{-2N}\frac{m_1(1+b_1)+\delta}{a}, I_{12} = \iota^{-2N}\frac{m_2(1+b_2)+\delta}{a}, \\
&I_{13} = \iota^{-2N}\frac{m_3(1+b_3)+\delta}{a}, I_{14} = 0, \\
&I_{21} = \frac{b_1}{a}, I_{22} = \frac{b_2}{a}, I_{23} = \frac{b_3}{a}, I_{24} = \frac{1}{a}, \\
&I_{31} = \iota^{m_1-1}\frac{m_1(1+b_1)+\delta}{a}, I_{32} = \iota^{m_2-1}\frac{m_2(1+b_2)+\delta}{a}, \\
&I_{33} = \iota^{m_3-1}\frac{m_3(1+b_3)+\delta}{a}, I_{34} = 0, \\
&I_{41} = \iota^{m_1}\frac{b_1}{a}, I_{42} = \iota^{m_2}\frac{b_2}{a}, I_{43} = \iota^{m_3}\frac{b_3}{a}, I_{44} = \frac{1}{a}
\end{aligned} \tag{3.101}$$

and the vector J has the following components:

$$J_{11} = \Sigma_{rri} - \frac{1}{a}\iota^{-2N}\begin{pmatrix} K_1(-2N+1)+K_1\delta+K_4(-2N+1) \\ -aC_1+K_2+K_2\delta+K_5-aC_2+3K_3+K_3\delta+3K_6 \end{pmatrix} \tag{3.102}$$

Finally, solving Eq. (3.100) completes our analysis. For brevity the expressions for A, B, C, and D have been omitted here.

3.3.2 *Benchmark Results*

Numerical examples of the analytical results for the multiphysical response of the FGPM hollow cylinder are presented graphically in this section. The outer surface of the cylinder is taken to be PZT-4 with its material properties listed in Table 3.3. Two mechanical and electrical boundary conditions are assumed here to illustrate the behavior of the FGPM cylinder:

Case 1:	Case 2:
$\Sigma_{rr}(1) = -1, \Phi_1(1) = 1$	$\Sigma_{rr}(1) = -1, \Phi_1(1) = 0$
$\Sigma_{rr}(\iota) = 0, \Phi_1(\iota) = 0$	$\Sigma_{rr}(\iota) - 0, \Phi_1(\iota) = 0$

Table 3.3 Material properties of the outer surface of the FGPM cylinder

Properties	PZT-4 (outer surface)
$c_{330}\left(\frac{\mathrm{N}}{\mathrm{m}^2}\right)$	115×10^9
$c_{110}\left(\frac{\mathrm{N}}{\mathrm{m}^2}\right)$	139×10^9
$c_{130}\left(\frac{\mathrm{N}}{\mathrm{m}^2}\right)$	74.3×10^9
$e_{310}\left(\frac{\mathrm{C}}{\mathrm{m}^2}\right)$	-5.20
$e_{330}\left(\frac{\mathrm{C}}{\mathrm{m}^2}\right)$	15.1
$\varepsilon_{330}\left(\frac{\mathrm{C}^2\mathrm{N}}{\mathrm{m}^2}\right)$	5.62×10^{-9}
$\rho_{d_0}\left(\frac{\mathrm{kg}}{\mathrm{m}^3}\right)$	7.5×10^3
$\beta_{10}\left(\frac{\mathrm{NK}}{\mathrm{m}^2}\right)$	1.0089×10^6
$\beta_{30}\left(\frac{\mathrm{NK}}{\mathrm{m}^2}\right)$	0.8439×10^6
$\gamma_{10}\left(\frac{\mathrm{CK}}{\mathrm{m}^2}\right)$	-2.5×10^{-5}

Although the formulation for thermal analysis includes a general form of thermal boundary conditions, the numerical results are obtained for the FGPM hollow cylinder with assigned non-dimensional temperature Θ_a and Θ_b on the inner and outer surfaces, respectively. The stresses, electric potential, electric displacement, perturbation of magnetic field and temperature distribution are given for various non-dimensional magnetic fields, inertial effects, aspect ratios and non-homogeneity parameters as well as different non-dimensional values of inner and outer temperature.

Figure 3.6a–d show the effects of non-dimensional magnetic field Ω on the distribution of stresses, electric displacement, and electric potential along the radial direction of the hollow FGPM cylinder with $N = -1$, angular velocity $\omega_n = 1$, and aspect ratio $\iota = 4$ under Case 1 boundary conditions in the equations above without any thermal disturbance. The results for $\Omega = 0$ are completely similar to those reported in (Babaei and Chen 2008), which validates our solution procedure.

3.4 Effect of Hygrothermal Excitation on One-Dimensional Smart Structures

The constitutive, potential field, and conservation equations for solving the uncoupled hygrothermomagnetoelectroelastic problems are presented in this section. The geometry of an infinitely long MEE cylinder is depicted in Fig. 3.7. The cylinder is radially polarized and magnetized and rotating at a constant angular velocity ω about the z-axis of cylindrical coordinate system (r, θ, z). The inner and outer radii of the cylinder are a and b, respectively. The MEE cylinder rests on the

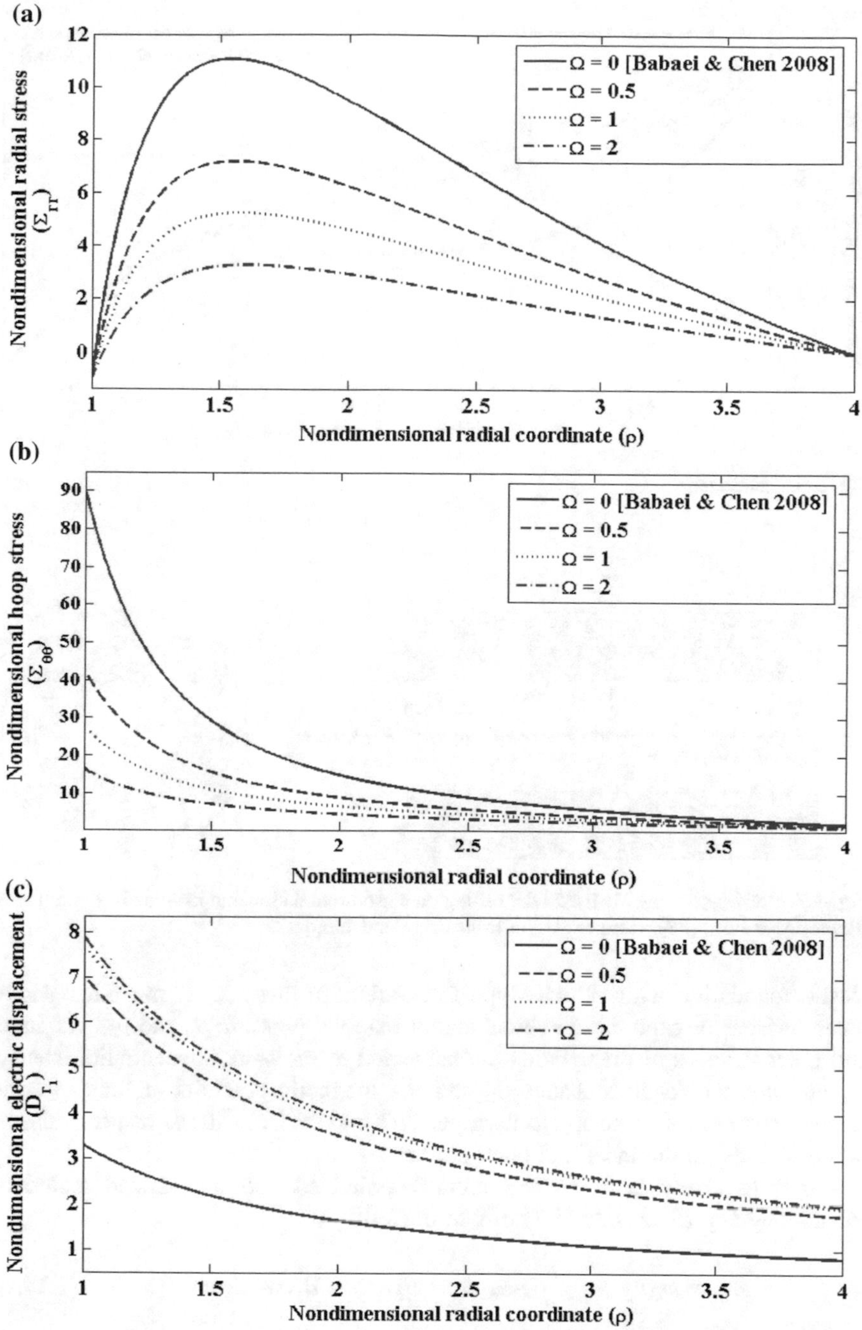

Fig. 3.6 **a** Radial stress for different Ω, $\omega_n = 1$, $N = -1$, $\iota = 4$, case 1, **b** Hoop stress for different Ω, $\omega_n = 1$, $N = -1$, $\iota = 4$, case 1, **c** Electric displacement for different Ω, $\omega_n = 1$, $N = -1$, $\iota = 4$, case 1, **d** Electric potential for different Ω, $\omega_n = 1$, $N = -1$, $\iota = 4$, case 1. [Reproduced from [37] with permission from IOP Publishing]

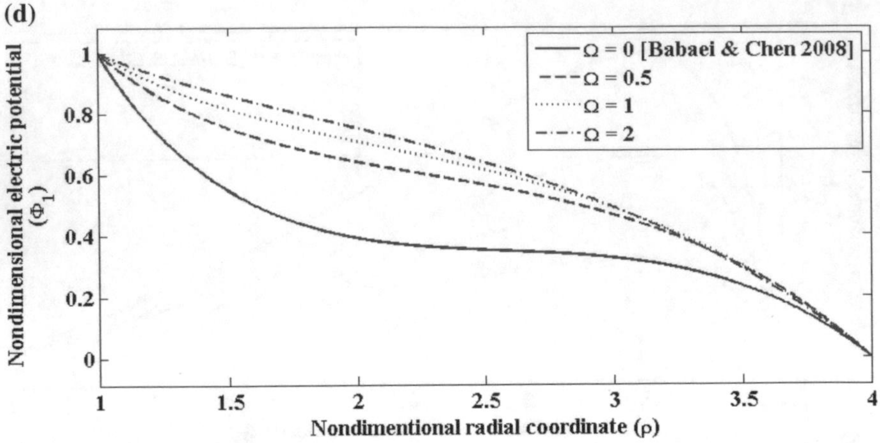

Fig. 3.6 (continued)

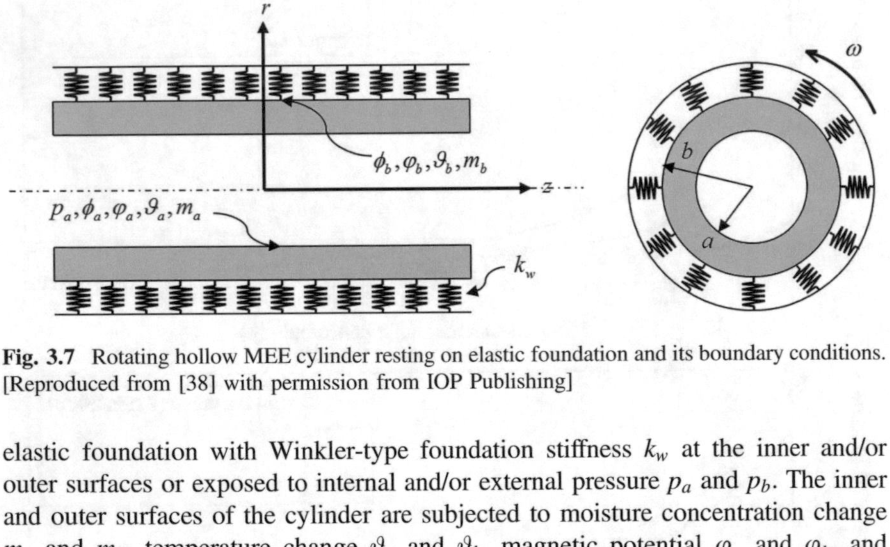

Fig. 3.7 Rotating hollow MEE cylinder resting on elastic foundation and its boundary conditions. [Reproduced from [38] with permission from IOP Publishing]

elastic foundation with Winkler-type foundation stiffness k_w at the inner and/or outer surfaces or exposed to internal and/or external pressure p_a and p_b. The inner and outer surfaces of the cylinder are subjected to moisture concentration change m_a and m_b, temperature change ϑ_a and ϑ_b, magnetic potential φ_a and φ_b, and electric potential ϕ_a and ϕ_b. Subscripts "*a*" and "*b*" are used, respectively, to indicate loads on the inner and outer surfaces.

For transversely isotropic and radially polarized and magnetized materials, substituting Eq. (3.22) into (3.7) results in [39]:

$$\sigma_{rr} = c_{33}u_{,r} + c_{13}\frac{u}{r} + e_{33}\phi_{,r} + d_{33}\varphi_{,r} - \beta_1\vartheta - \zeta_1 m \tag{3.103a}$$

$$\sigma_{\theta\theta} = c_{13}u_{,r} + c_{11}\frac{u}{r} + e_{31}\phi_{,r} + d_{31}\varphi_{,r} - \beta_3\vartheta - \zeta_3 m \tag{3.103b}$$

$$\sigma_{zz} = c_{13}u_{,r} + c_{12}\frac{u}{r} + e_{31}\phi_{,r} + d_{31}\varphi_{,r} - \beta_3\vartheta - \zeta_3 m \tag{3.103c}$$

$$D_r = e_{33}u_{,r} + e_{31}\frac{u}{r} - \in_{33}\phi_{,r} - g_{33}\varphi_{,r} + \gamma_1\vartheta + \chi_1 m \tag{3.103d}$$

$$B_r = d_{33}u_{,r} + d_{31}\frac{u}{r} - g_{33}\phi_{,r} - \mu_{33}\varphi_{,r} + \tau_1\vartheta + \upsilon_1 m \tag{3.103e}$$

where, $c_{mn} = C_{ijkl}$, $e_{ml} = e_{kij}$, $d_{ml} = d_{kij}$, $\beta_m = \beta_{ij}$, and $\xi_m = \xi_{ij}\,(i,j,k,l = 1,2,3;$ $m,n = 1,2,\ldots,6)$. In the absence of body force, free charge density, and current density, the equation of motion and Maxwell's electromagnetic equations for the axisymmetric, infinitely long cylinder are written as Eq. (3.24).

3.4.1 Solution Procedure

The solution procedure for hollow and solid MEE cylinders is given in this section similar to those reported in [40] for magnetoelectroelastic cylinders. Substituting Eqs. (3.103) and (3.25) into Eq. (3.24) results in the following coupled ordinary differential equations in terms of moisture concentration, temperature change, magnetic potential, electrical potential, and displacement:

$$\begin{aligned} &c_{33}r^2u_{,rr} + c_{33}ru_{,r} - c_{11}u + e_{33}r^2\varphi_{,rr} + (e_{33} - e_{31})r\varphi_{,r} \\ &+ d_{33}r^2\phi_{,rr} + (d_{33} - d_{31})r\phi_{,r} - \beta_1 r^2\vartheta_{,r} \\ &-(\beta_1 - \beta_3)r\vartheta - \xi_1 r^2 m_{,r} - (\xi_1 - \xi_3)rm + \rho_d r^3\omega^2 = 0 \end{aligned} \tag{3.104a}$$

$$\begin{aligned} &e_{33}r^2u_{,rr} + (e_{31} + e_{33})ru_{,r} - \in_{33} r^2\varphi_{,rr} - \in_{33} r\phi_{,r} \\ &-g_{33}r^2\varphi_{,rr} - g_{33}r\varphi_{,r} + \gamma_1 r^2\vartheta_{,r} + \gamma_1 r\vartheta + \chi_1 r^2 m_{,r} + \chi_1 rm = 0 \end{aligned} \tag{3.104b}$$

$$\begin{aligned} &d_{33}r^2u_{,rr} + (d_{31} + d_{33})ru_{,r} - g_{33}r^2\phi_{,rr} - g_{33}r\phi_{,r} - \mu_{33}r^2\varphi_{,rr} \\ &-\mu_{33}r\varphi_{,r} + \tau_1 r^2\vartheta_{,r} + \tau_1 r\vartheta + \upsilon_1 r^2 m_{,r} + \upsilon_1 rm = 0 \end{aligned} \tag{3.104c}$$

The non-dimensional parameters are introduced in Eq. (3.28) and new electric potential, magnetic potential, temperature change, and moisture concentration change as:

$$\Phi = \frac{e_{33}}{c_{33}}\phi,\ \Psi = \frac{d_{33}}{c_{33}}\varphi,\ \Theta = \frac{\beta_1}{c_{33}}\vartheta,\ M = \frac{\xi_1}{c_{33}}m \tag{3.105}$$

Using Eqs. (3.28) and (3.105), Eq. (3.104) can be rewritten in the following form:

$$\begin{aligned}&r^2u_{,rr}+ru_{,r}-\alpha u+r^2\Phi_{,rr}+(1-\beta)r\Phi_{,r}+r^2\Psi_{,rr}+(1-v)r\Psi_{,r}\\&-r^2\Theta_{,r}-(1-\eta)r\Theta-r^2M_{,r}-(1-\varsigma)rM+\Omega\frac{r^3}{a^2}=0\end{aligned}\tag{3.106a}$$

$$\begin{aligned}&r^2u_{,rr}+(1+v)ru_{,r}-\zeta r^2\Phi_{,rr}-\zeta r\Phi_{,r}-\lambda r^2\Psi_{,rr}-\lambda r\Psi_{,r}\\&+Yr^2\Theta_{,r}+Yr\Theta+Wr^2M_{,r}+WrM=0\end{aligned}\tag{3.106b}$$

$$\begin{aligned}&r^2u_{,rr}+(1+\beta)ru_{,r}-\gamma r^2\Phi_{,rr}-\gamma r\Phi_{,r}-\zeta r^2\Psi_{,rr}-\zeta r\Psi_{,r}\\&+Xr^2\Theta_{,r}+Xr\Theta+Vr^2M_{,r}+VrM=0\end{aligned}\tag{3.106c}$$

in which,

$$\Omega=\frac{\rho_d\omega^2a^2}{c_{33}}\tag{3.107}$$

The following set of second-order coupled ordinary differential equations with constant coefficients is obtained by using the non-dimensional radial coordinate $\rho=\frac{r}{a}$ and then changing variable ρ with s by $\rho=e^s$ as follows:

$$\begin{aligned}&\ddot{u}-\alpha u+\ddot{\Phi}-\beta\dot{\Phi}+\ddot{\Psi}-v\dot{\Psi}=ae^s\dot{\Theta}+(1-\eta)ae^s\Theta\\&+ae^s\dot{M}+(1-\varsigma)ae^sM-\Omega ae^{3s}\end{aligned}\tag{3.108a}$$

$$\ddot{u}+\beta\dot{u}-\gamma\ddot{\Phi}-\zeta\ddot{\Psi}=-Xae^s\dot{\Theta}-Xae^s\Theta-Vae^s\dot{M}-Vae^sM\tag{3.108b}$$

$$\ddot{u}+v\dot{u}-\zeta\ddot{\Phi}-\lambda\ddot{\Psi}=-Yae^s\dot{\Theta}-Yae^s\Theta-Wae^s\dot{M}-Wae^sM\tag{3.108c}$$

in which, the overdot stands for differentiation with respect to s. For uncoupled hygrothermomagnetoelectroelastic problems, temperature and moisture concentration distributions are obtained separately by solving heat conduction and moisture diffusion equations. The axisymmetric and steady state Fourier heat conduction and Fickian moisture diffusion equations for an infinitely long hollow cylinder are written as [17]:

$$\frac{1}{r}\left(rk^T\theta_{,r}\right)_{,r}=0 \qquad (\textit{Heat conduction equation})\tag{3.109a}$$

$$\frac{1}{r}\left(rk^Cm_{,r}\right)_{,r}=0 \qquad (\textit{Moisture diffusion equation})\tag{3.109b}$$

where, k^T and k^C are thermal conductivity and moisture diffusivity coefficients, respectively. Solving Eq. (3.109) and using the non-dimensional radial coordinate as well as the new parameters defined in Eq. (3.105) result in the following temperature and moisture concentration distribution along the radial direction for a homogenous cylinder:

$$\Theta = C_1 \ln(\rho) + C_2 \tag{3.110a}$$

$$M = C_3 \ln(\rho) + C_4 \tag{3.110b}$$

where C_1, C_2, C_3, and C_4 are integration constants which are determined by satisfying the hygrothermal boundary conditions. The non-dimensional temperature change on the inner and outer surfaces of the hollow cylinder are, respectively, assumed to be Θ_a and Θ_b. Similarly, non-dimensional moisture concentration change are assigned to be M_a and M_b on the inner and outer surfaces. The integration constants according to the hygrothermal boundary conditions can be obtained as:

$$C_1 = \frac{\Theta_b - \Theta_a}{\ln(\imath)}, \quad C_2 = \Theta_a, \quad C_3 = \frac{M_b - M_a}{\ln(\imath)}, \quad C_4 = M_a \tag{3.111}$$

in which, $\imath = \frac{b}{a}$ is the aspect ratio of the hollow cylinder. It should be mentioned that for solid cylinders as well as hollow cylinders with uniform temperature and moisture concentration rise, we have $C_1 = C_3 = 0$. For more general hygrothermal boundary conditions, one may refer to the Robin-type boundary conditions considered in [17, 32]. It is worth noting that temperature and moisture concentration have similar effects on magnetoelectroelastic responses in uncoupled hygrothermomagnetoelectroelasticity according to Eqs. (3.108) through (3.111). The solution of Eq. (3.108) can be found analytically by successive decoupling method [40]. Eliminating Φ between equations in (3.108) leads to the following two ordinary differential equations about u and Ψ:

$$\begin{aligned}
&(1+\gamma)\dddot{u} - (\beta^2 + \alpha\gamma)\dot{u} + (\gamma - \zeta)\dddot{\Psi} + (\beta\zeta - \nu\gamma)\ddot{\Psi} \\
&= (\gamma - X)ae^s\ddot{\Theta} + (\gamma(2-\eta) + X(\beta - 2))ae^s\dot{\Theta} + (\gamma(1-\eta) + X(\beta - 1))ae^s\Theta \\
&\quad + (\gamma - V)ae^s\ddot{M} + (\gamma(2-\varsigma) + V(\beta - 2))ae^s\dot{M} \\
&\quad + (\gamma(1-\varsigma) + V(\beta - 1))ae^sM - 3a\Omega\gamma e^{3s}
\end{aligned} \tag{3.112a}$$

$$\begin{aligned}
&(\gamma - \zeta)\ddot{u} + (\nu\gamma - \beta\zeta)\dot{u} + (\zeta^2 - \lambda\gamma)\ddot{\Psi} = (X\zeta - Y\gamma)ae^s\dot{\Theta} \\
&\quad + (X\zeta - Y\gamma)ae^s\Theta \times (V\zeta - W\gamma)ae^s\dot{M} + (V\zeta - W\gamma)ae^sM
\end{aligned} \tag{3.112b}$$

By eliminating Ψ between Eqs. (3.112) and considering the temperature change and moisture concentration change distributions according to Eq. (3.110), the following ordinary differential equation with constant coefficients for radial displacement u is achieved:

$$a_2\dddot{u} + a_1\dot{u} = (b_1C_2 + b_2C_1 + b_4C_4 + b_5C_3)e^s + (b_1C_1 + b_4c_3)se^s + b_5e^{3s} \tag{3.113}$$

where,

$$
\begin{aligned}
a_1 &= \frac{(v\gamma-\beta\zeta)^2}{\zeta^2-\lambda\gamma}-(\beta^2+\alpha\gamma), \quad a_2 = 1+\gamma-\frac{(\gamma-\zeta)^2}{\zeta^2-\lambda\gamma} \\
b_1 &= a\left(\gamma(1-\eta)+X(\beta-1)-\frac{(X\zeta-Y\gamma)(\beta\zeta-v\gamma+\gamma\zeta)}{\zeta^2-\lambda\gamma}\right) \\
b_2 &= a\left(\gamma(2-\eta)+X(\beta-2)-\frac{(X\zeta-Y\gamma)(\beta\zeta-v\gamma+2\gamma-2\zeta)}{\zeta^2-\lambda\gamma}\right) \\
b_3 &= a\left(\gamma(1-\varsigma)+V(\beta-1)-\frac{(V\zeta-W\gamma)(\beta\zeta-v\gamma+\gamma-\zeta)}{\zeta^2-\lambda\gamma}\right) \\
b_4 &= a\left(\gamma(2-\varsigma)+V(\beta-2)-\frac{(V\zeta-W\gamma)(\beta\zeta-v\gamma+2\gamma-2\zeta)}{\zeta^2-\lambda\gamma}\right) \\
b_5 &= -3a\gamma\Omega
\end{aligned}
\tag{3.114}
$$

The solution of Eq. (3.113) in terms of variable ρ can be expressed as:

$$u = A + C\rho^m + D\rho^{-m} + (K_1\ln(\rho)+K_2)\rho + K_3\rho^3 \tag{3.115}$$

in which,

$$m = i\sqrt{\frac{a_1}{a_2}} \qquad \left(m \in R \quad and \quad i=\sqrt{-1}\right) \tag{3.116}$$

and A, C, and D are integration constants. Employing Eqs. (3.113) and (3.115), the following expression for Ψ can be obtained:

$$\Psi = F + E\ln(\rho) + Cc_1\rho^m + Dc_2\rho^{-m} + (c_3\ln(\rho)+c_4)\rho + K_3c_5\rho^3 \tag{3.117}$$

where, E and F are new integration constants and:

$$
\begin{aligned}
c_1 &= \frac{1}{m(\lambda\gamma-\zeta^2)}(m(\gamma-\zeta)+v\gamma-\beta\zeta), c_2 = \frac{1}{m(\lambda\gamma-\zeta^2)}(m(\gamma-\zeta)-v\gamma+\beta\zeta) \\
c_3 &= \frac{1}{\lambda\gamma-\zeta^2}(K_1(\gamma(1+v)-\zeta(1+\beta))-aC_1(X\zeta-Y\gamma)-aC_3(V\zeta-W\gamma)) \\
c_4 &= \frac{1}{\lambda\gamma-\zeta^2}(K_2(\gamma-\zeta)+(K_2-K_1)(v\gamma-\beta\zeta)+a(C_1-C_2)(X\zeta-Y\gamma) \\
&\quad + a(C_3-C_4)(V\zeta-W\gamma)) \\
c_5 &= \frac{1}{3(\lambda\gamma-\zeta^2)}(3(\gamma-\zeta)+v\gamma-\beta\zeta)
\end{aligned}
\tag{3.118}
$$

Using Eq. (3.108b) and considering Eqs. (3.115) and (3.117), Φ can be expressed by:

$$\Phi = H + G\ln(\rho) + Cl_1\rho^m + Dl_2\rho^{-m} + (l_3\ln(\rho) + l_4)\rho + K_3 l_5 \rho^3 \tag{3.119}$$

in which, G and H are integration constants and:

$$\begin{aligned}
l_1 &= \frac{1}{m\gamma}(m + \beta - m\zeta c_1), l_2 = \frac{1}{m\gamma}(m - \beta - m\zeta c_2) \\
l_3 &= \frac{1}{\gamma}(K_1(1+\beta) - \zeta c_3 + aXC_1 + aVC_3) \\
l_4 &= \frac{1}{\gamma}(K_2 + (K_2 - K_1)\beta - c_4\zeta + aX(C_2 - C_1) + aV(C_4 - C_3)) \\
l_5 &= \frac{1}{3\gamma}(3 + \beta - 3\zeta c_5)
\end{aligned} \tag{3.120}$$

For convenience, the following non-dimensional stresses, displacement, electric potential, and magnetic potential are used:

$$\Sigma_{rr} = \frac{\sigma_{rr}}{c_{33}}, \Sigma_{\theta\theta} = \frac{\sigma_{\theta\theta}}{c_{33}}, \Sigma_{zz} = \frac{\sigma_{zz}}{c_{33}}, U = \frac{u}{a}, \Phi_1 = \frac{\Phi}{a}, \Psi_1 = \frac{\Psi}{a} \tag{3.121}$$

Then, we reach:

$$\begin{aligned}
\Sigma_{rr} = {} & \frac{\rho^{-1}}{a}(\delta A + G + E) + \frac{\rho^{m-1}}{a}(m(1 + l_1 + c_1) + \delta)C \\
& + \frac{\rho^{m-1}}{a}(-m(1 + l_2 + c_2) + \delta)D + \frac{\ln(\rho)}{a}(K_1(1+\delta) + l_3 + c_3 - a(C_1 + C_3)) \\
& + \frac{1}{a}(K_1 + (1+\delta)K_2 + l_3 + l_4 + c_3 + c_4 - a(C_2 + C_4)) + \frac{\rho^2}{a}(3(1 + l_5 + c_5) + \delta)K_3
\end{aligned} \tag{3.122a}$$

$$\begin{aligned}
\Sigma_{\theta\theta} = {} & \frac{\rho^{-1}}{a}(\delta A + \beta G + vE) + \frac{\rho^{m-1}}{a}(m(\delta + \beta l_1 + vc_1) + \delta)C \\
& + \frac{\rho^{m-1}}{a}(-m(\delta + \beta l_2 + vc_2) + \delta)D + \frac{\ln(\rho)}{a}(K_1(\delta + \alpha) + \beta l_3 + vc_3 - a(\eta C_1 + \varsigma C_3)) \\
& + \frac{1}{a}(\delta K_1 + (\delta + \alpha)K_2 + \beta(l_3 + l_4) + v(c_3 + c_4) - a(\eta C_2 - \varsigma C_4)) \\
& + \frac{\rho^2}{a}(3(\delta + \beta l_5 + vc_5) + \alpha)K_3
\end{aligned} \tag{3.122b}$$

$$
\begin{aligned}
\Sigma_{zz} &= \frac{\rho^{-1}}{a}(\delta^* A + \beta G + vE) + \frac{\rho^{m-1}}{a}(m(\delta + \beta l_1 + vc_1) + \delta^*)C \\
&+ \frac{\rho^{m-1}}{a}(-m(\delta + \beta l_2 + vc_2) + \delta^*)D + \frac{\ln(\rho)}{a}(K_1(\delta + \delta^*) + \beta l_3 + vc_3 - a(\eta C_1 + \varsigma C_3)) \\
&+ \frac{1}{a}(\delta K_1 + (\delta + \delta^*)K_2 + \beta(l_3 + l_4) + v(c_3 + c_4) - a(\eta C_2 + \varsigma C_4)) \\
&+ \frac{\rho^2}{a}(3(\delta + \beta l_5 + vc_5) + \delta^*)K_3
\end{aligned}
\tag{3.122c}
$$

$$
\Psi_1 = \frac{F}{a} + \frac{E}{a}\ln(\rho) + \frac{C}{a}c_1\rho^m + \frac{D}{a}C_2\rho^{-m} + (c_3\ln(\rho) + c_4)\frac{\rho}{a} + \frac{K_3 c_5}{a}\rho^3 \tag{3.122d}
$$

$$
\Phi_1 = \frac{H}{a} + \frac{G}{a}\ln(\rho) + \frac{C}{a}l_1\rho^m + \frac{D}{a}l_2\rho^{-m} + (l_3\ln(\rho) + l_4)\frac{\rho}{a} + \frac{K_3 l_5}{a}\rho^3 \tag{3.122e}
$$

$$
U = \frac{A}{a} + \frac{C}{a}\rho^m + \frac{D}{a}\rho^{-m} + \left(\frac{K_1}{a}\ln(\rho) + \frac{K_2}{a}\right)\rho + \frac{K_3}{a}\rho^3 \tag{3.122f}
$$

We have seven integration constants in the aforementioned equations; however, there exist only six boundary conditions in the magnetoelectroelastic medium. Therefore, one complimentary equation is needed which is obtained by substituting Eqs. (3.115), (3.117), and (3.119) into Eq. (3.108):

$$
\alpha A + vE + \beta G = 0 \tag{3.123}
$$

The integration constants are obtained in the following subsections for hollow and solid cylinders. Although the solution procedure for this hygrothermomagnetoelectroelastic analysis under steady-state condition is the same as [40], the work is a pioneer in such emerging multiphysical analysis. The analytical solutions given in Eq. (3.122) could be employed for the design of MEE structures as well as a benchmark solution for verification of the other analytical and numerical results which will be used later for the multiphysical problem.

To consider the effect of temperature and moisture dependency of elastic coefficients on the magnetoelectroelastic response, the elastic coefficients are expressed in the following form [41, 42]:

$$
C_{ij} = C_{ij0}(1 + \alpha^* \vartheta + \beta^* M) \tag{3.124}
$$

in which, C_{ij0} is an elastic coefficient at stress-free temperature and moisture concentration; α^* and β^* are empirical material constants for temperature and moisture dependency. In the current work, the temperature and moisture dependency is only considered for uniform temperature and moisture concentration rise to avoid dealing with non-linear–problems [43].

3.4.2 MEE Hollow Cylinder

The MEE hollow cylinder may be exposed to different hygrothermomagnetoelectroelastic boundary conditions. The hygrothermal boundary conditions were considered earlier; here we specify other magnetoelectroelastic boundary conditions. Since the hollow cylinder may be simulated with or without Winkler-type elastic foundation on the inner and outer surfaces, different non-dimensional elastic boundary conditions could be considered as follows [44, 45]:

- Internal and external pressure:

$$\Sigma_{rr}(1) = \Sigma_{rra} \; and \; \Sigma_{rr}(\imath) = \Sigma_{rrb} \tag{3.125a}$$

- Internal pressure and outer surface elastic foundation:

$$\Sigma_{rr}(1) = \Sigma_{rra} \; and \; \Sigma_{rr}(\imath) = -K_W U_b \tag{3.125b}$$

- Inner surface elastic foundation and external pressure:

$$\Sigma_{rr}(1) = K_W U_a \; and \; \Sigma_{rr}(\imath) = \Sigma_{rrb} \tag{3.125c}$$

- Elastic foundation on both the inner and outer surfaces:

$$\Sigma_{rr}(1) = K_W U_a \, and \; \Sigma_{rr}(\imath) = -K_W U_b \tag{3.125d}$$

Moreover, the magnetoelectric boundary conditions are specified as:

$$\Phi_1(1) = \Phi_{1a}, \quad \Psi_1(1) = \Psi_{1a} \tag{3.126a}$$

$$\Phi_1(\imath) = \Phi_{1b}, \quad \Psi_1(\imath) = \Psi_{1b} \tag{3.126b}$$

in which, $K_W = \frac{ak_w}{c_{33}}$ and subscripts "a" and "b" are associated with quantities on the inner and outer surfaces. It is worth to note that the Winkler-type elastic foundation changes the type of mechanical boundary conditions and does not affect the governing differential equations. Employing Eqs. (3.122), (3.124), (3.125), and (3.126) lead to the following linear algebraic equation for integration constants as follows:

$$I \begin{Bmatrix} A \\ C \\ D \\ E \\ F \\ G \\ H \end{Bmatrix} = J \tag{3.127}$$

where, I is a 7×7 matrix and J is a 7×1 vector and their components are given in [38]. Once the integration constants are obtained by solving Eq. (3.127), the analytical solution for hollow MEE cylinder under hygrothermal loading is eventually obtained.

3.4.3 MEE Solid Cylinder

The solution procedure for an MEE solid cylinder with outer radius b is analogous to an MEE hollow cylinder; however, the new non-dimensional radial coordinate $\rho = \frac{r}{b} (0 \leq \rho \leq 1)$ is needed to enable us to use previously obtained expressions for stresses, displacement, electric potential, and magnetic potential. The following new non-dimensional displacement, electric potential, and magnetic potential are defined for the MEE solid cylinder:

$$U = \frac{u}{b}, \quad \Phi_1 = \frac{\Phi}{b}, \quad \Psi_1 = \frac{\Psi}{b} \tag{3.128}$$

As the displacement, electric potential, and magnetic potential should be finite at the axis of symmetry of solid cylinders, we can deduce from Eqs. (3.115), (3.117), (3.119), and (3.123):

$$A = C = E = G = 0 \tag{3.129}$$

Therefore, one can obtain the following results for the MEE solid cylinder from Eqs. (3.115), (3.117), (3.119), and (3.122):

$$\Sigma_{rr} = \frac{\rho^{-m-1}}{b}(-m(1 + l_2 + c_2) + \alpha)D + \frac{1}{b}((1 + \delta)K_2 + l_4 + c_4 - b(C_2 + C_4)) + \frac{\rho^2}{b}(3(1 + l_5 + c_5) + \delta)K_3 \tag{3.130a}$$

$$\Sigma_{\theta\theta} = \frac{\rho^{-m-1}}{b}(-m(\delta + \beta l_2 + \nu c_2) + \alpha)D + \frac{1}{b}((\delta + \alpha)K_2 + \beta l_4 + \nu c_4 - b(\eta C_2 - \varsigma C_4)) + \frac{\rho^2}{b}(3(\delta + \beta l_5 + \nu c_5) + \alpha)K_3 \tag{3.130b}$$

$$\Sigma_{zz} = \frac{\rho^{-m-1}}{b}(-m(\delta + \beta l_2 + \nu c_2) + \delta^*)D + \frac{1}{b}((\delta + \delta^*)K_2 + \beta l_4 + \nu c_4 - b(\eta C_2 + \varsigma C_4)) + \frac{\rho^2}{b}(3(\delta + \beta l_5 + \nu c_5) + \delta^*)K_3 \tag{3.130c}$$

$$\Psi_1 = \frac{F}{b} + \frac{D}{b} c_2 \rho^{-m} + \frac{c_4}{b} \rho + \frac{K_3 c_5}{b} \rho^3 \tag{3.130d}$$

$$\Phi_1 = \frac{H}{b} + \frac{D}{b} l_2 \rho^{-m} + \frac{l_4}{b} \rho + \frac{K_3 l_5}{b} \rho^3 \tag{3.130e}$$

$$U = \frac{D}{b} \rho^{-m} + \frac{K_2}{b} \rho + \frac{K_3}{b} \rho^3 \tag{3.130f}$$

where all constants are similar to those defined for the hollow cylinder except the following parameters:

$$\begin{aligned}
& C_1 = C_3 = 0 \\
& K_1 = 0, K_2 = \frac{b_1 C_2 + b_3 C_4}{a_2 + a_1} \\
& c_3 = 0, c_4 = \frac{1}{\lambda\gamma - \zeta^2} (K_2(\gamma - \zeta) + K_2(v\gamma - \beta\zeta) - bC_2(X\zeta - Y\gamma) - bC_4(V\zeta - W\gamma)) \\
& l_3 = 0, l_4 = \frac{1}{\gamma} (K_2(1 + \beta) - c_4\zeta + bXC_2 + bVC_4)
\end{aligned} \tag{3.131}$$

Furthermore, it is worth recalling that the temperature and moisture concentration remain constant through the radial direction of axisymmetric solid cylinders in the steady-state condition according to the heat conduction and moisture diffusion equations. The constant values can be determined by the non-dimensional temperature and moisture concentration at the outer surface of cylinders. To obtain the integration constants in Eq. (3.130), the magnetoelectroelastic boundary conditions are required. The outer surface of the solid cylinder could be exposed to external pressure or rest on a Winkler-type elastic foundation; the following elastic boundary conditions can be expressed accordingly:

- External pressure:

$$\Sigma_{rr}(1) = \Sigma_{rrb} \tag{3.132a}$$

- Elastic foundation on the outer surface:

$$\Sigma_{rr}(1) = -K_W U_b \tag{3.132b}$$

The electromagnetic boundary conditions are:

$$\Phi_1(1) = \Phi_{1b} \; and \; \Psi_1(1) = \Psi_{1b} \tag{3.133}$$

Employing Eqs. (3.130) and (3.132a, 3.132b), the following linear algebraic equation can be established to obtain the integration constants:

$$P\left\{\begin{matrix} D \\ F \\ H \end{matrix}\right\} = Q \tag{3.134}$$

in which, P is a 3×3 matrix and Q is a 3×1 vector with components defined in [38]. Solving Eq. (3.134) gives the integration constants and complete the analytical solution. According to Eq. (3.130), the value of displacement, electric potential, and magnetic potential are always finite along the central axis of solid cylinder; however, the stress components are singular for $m = \dfrac{a_1}{a_2} > -1$ depending on the material properties of the MEE cylinder [40].

3.4.4 Benchmark Results

The numerical results for uncoupled hygrothermomagnetoelectroelastic behavior of transversely isotropic hollow and solid cylinders are presented graphically in this section. The numerical results include radial, hoop, and axial stresses as well as electrical and magnetic potentials. There does not exist any experimental results for such multifield analysis; however, due to the application of smart wood structures with piezoelectric/piezomagnetic sensors and actuators in different environmental conditions, such multiphysical experiment is feasible. These theoretical results reveal the possible interaction of different physical fields when studying the structural behavior of smart materials.

The material properties of MEE cylinders are given in Table 3.4 according to the material properties of an adaptive wood made of $BaTiO_3/CoFe_2O_4$ [34, 46, 47].

The effect of hygrothermal boundary conditions on the multiphysical responses of an MEE hollow cylinder is depicted in Fig. 3.8. The cylinder is assumed to be under internal pressure and the traction-free boundary condition exists on the outer surface. The following magnetoelectroelastic boundary conditions are considered:

$$\Sigma_{rr}(1) = -1, \;\; \Phi_1(1) = 1, \;\; \Psi_1(1) = 1$$

$$\Sigma_{rr}(\imath) = 0, \;\; \Phi_1(\imath) = 0, \;\; \Psi_1(\imath) = 0$$

The aspect ratio, inner radius, and non-dimensional angular velocity of the rotating MEE hollow cylinder are $\imath = 4$, $a = 1$, and $\Omega = 1$, respectively. The non-dimensional moisture concentration and temperature on the inner surface are kept at be zero, while the moisture concentration and temperature rise on the outer surface are Θ_b and M_b. Since the effects of moisture concentration and temperature on the multiphysical responses are similar, the same values are taken for Θ_b and M_b. As depicted in Fig. 3.11, the results are quite close to those reported in [36] for the magnetoelectroelastic response of rotating MEE hollow cylinders in the absence of elastic foundation and hygrothermal loading ($\Theta_a = \Theta_b = 0, M_a = M_b = 0$).

Table 3.4 Material properties of an adapative wood made of $BaTiO_3$/$CoFe_2O_4$

$c_{33}\left(\frac{N}{m^2}\right)$	2.695×10^{11}
$c_{11}\left(\frac{N}{m^2}\right)$	2.86×10^{11}
$c_{13}\left(\frac{N}{m^2}\right)$	1.705×10^{11}
$c_{12}\left(\frac{N}{m^2}\right)$	1.73×10^{11}
$e_{31}\left(\frac{C}{m^2}\right)$	-4.4
$e_{33}\left(\frac{C}{m^2}\right)$	18.6
$d_{31}\left(\frac{N}{Am}\right)$	580.3
$d_{31}\left(\frac{N}{Am}\right)$	699.7
$\in_{33}\left(\frac{C^2}{Nm^2}\right)$	9.3×10^{-11}
$g_{33}\left(\frac{Ns^2}{C^2}\right)$	3.0×10^{-12}
$\mu_{33}\left(\frac{Ns^2}{C^2}\right)$	1.57×10^{-4}
$\alpha_1^T = \alpha_2^T = \alpha_3^T\left(\frac{1}{K}\right)$	1×10^{-5}
$\beta_1^C\left(\frac{m^3}{kg}\right)$	0
$\beta_2^C = \beta_3^C\left(\frac{m^3}{kg}\right)$	1.1×10^{-4}
$\gamma_1\left(\frac{C}{km^2}\right)$	-13.0×10^{-5}
$\chi_1\left(\frac{Cm}{kg}\right)$	0
$\tau_1\left(\frac{N}{AmK}\right)$	6.0×10^{-3}
$\upsilon_1\left(\frac{Nm^2}{Akg}\right)$	0

Figure 3.8a–c illustrate the effect of hygrothermal loading Θ_b and M_b on the distribution of stresses through the thickness. The greater hygrothemal loadings on the outer surface result in greater absolute values of maximum radial stress. Increasing the outer hygrothermal loading amplifies the hoop stress on the inner surface and lessens the hoop stress on the outer surface. Since electroelastic cylinders have been observed to fail at a critical hoop stress [48], the effect of hygrothermal loading on the hoop stress is noteworthy for design and manufacturing of magnetoelectroelastic cylinders. Furthermore, the hygrothermal loading generally decreases the axial stresses through the thickness of the MEE cylinder. The effect of hygrothermal loading on the distribution of electric and magnetic potentials is shown in Fig. 3.8d and e. As depicted in Fig. 3.8d, the electric potential distribution shows double concavities in the absence of temperature and moisture concentration; however, applying the hygrothermal loading on the outer surface leads to the disappearance of one concavity and increases the maximum

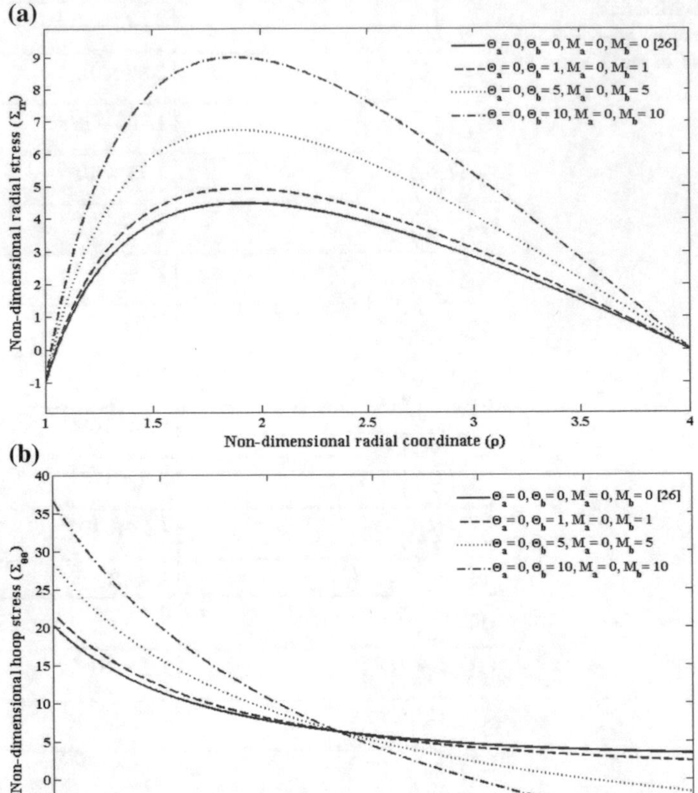

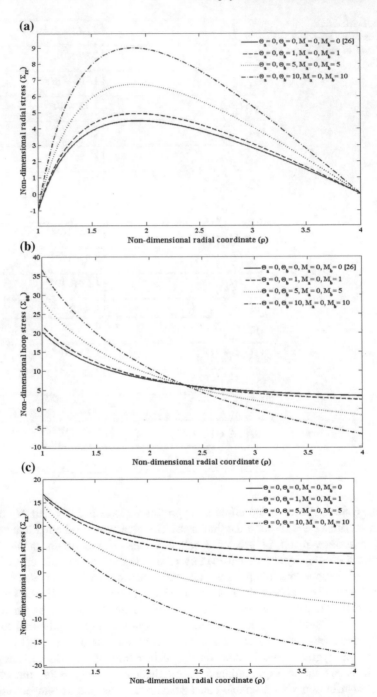

Fig. 3.8 Effect of hygrothermal loading on the distribution of: **a** radial stress, **b** hoop stress, **c** axial stress, **d** electric potential, **e** magnetic potential, **f** temperature, and **g** moisture concentration [38]

(d)

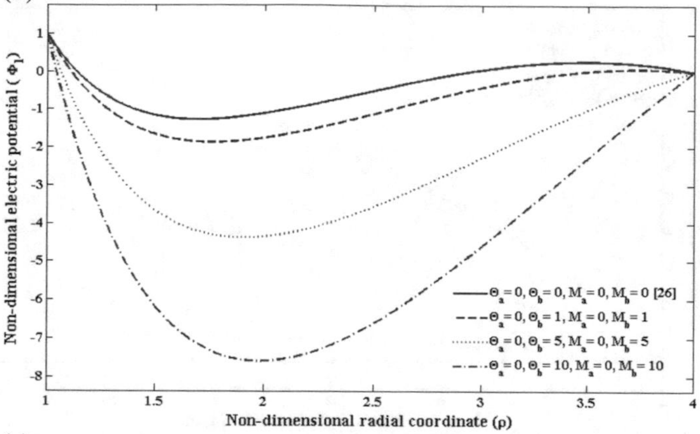

(e)

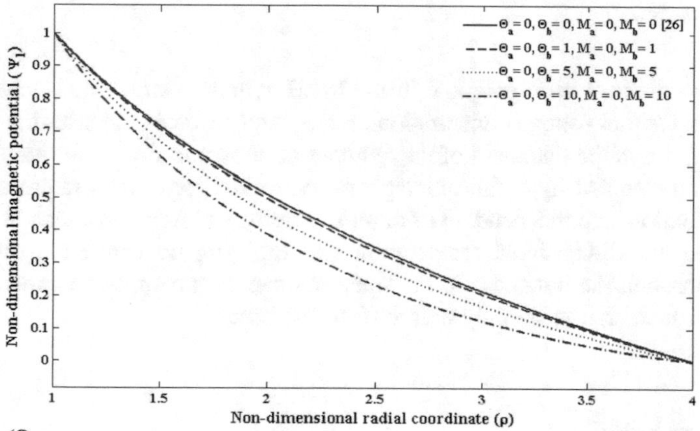

(f)

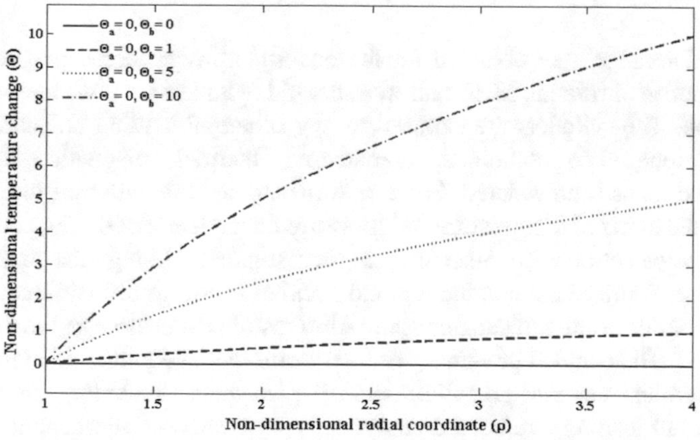

Fig. 3.8 (continued)

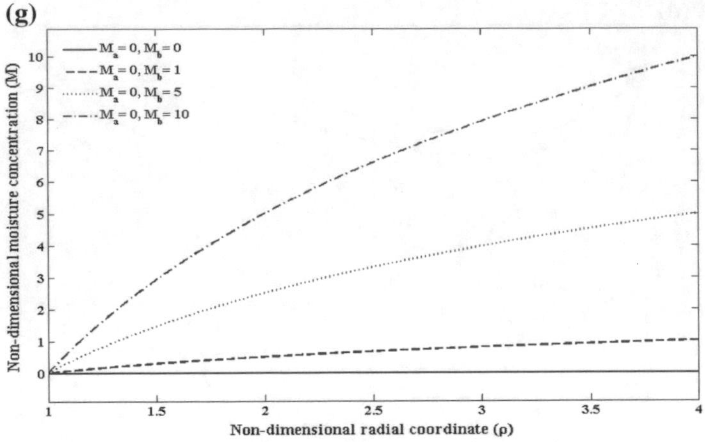

Fig. 3.8 (continued)

absolute value of electric potential in the MEE cylinder. Moreover, increasing the temperature and moisture concentration on the outer surface decreases the magnetic potential through the thickness of the cylinder as shown in Fig. 3.8e. As mentioned earlier, in uncoupled hygrothermomagnetoelectroelasticity, temperature and moisture distributions are independent of other multiphysical fields. As seen in Fig. 3.8f and g, the non-dimensional temperature and moisture concentration distribution increase through the thickness in the same manner as the applied temperature and moisture concentration on the outer surface increases.

3.5 Remarks

Analytical solutions are obtained for the uncoupled hygrothermomagnetoelectroelastic response of rotating MEE hollow and solid cylinders on a Winkler-type elastic foundation. The cylinders are exposed to hygrothermal loading and assumed to be infinitely long. The combined hygroscopic, thermal, magnetic, electric, and mechanical loads is considered. For a uniform temperature and moisture concentration rise, the effect of temperature and moisture dependency of elastic coefficients on the magnetoelectroelastic response is investigated. Using the axisymmetric, steady-state Fourier heat conduction and Fickian moisture diffusion equations, the radial temperature and moisture concentration distributions through the thickness are determined. The coupled governing ordinary differential equations in terms of magnetic potential, electric potential, and displacement, including the effects of hygrothermal loading, are solved analytically by a successive decoupling method. Numerical results are calculated to reveal the effect of hygrothermal boundary conditions, elastic foundation, and temperature and moisture dependency of elastic coefficients on the multiphysical responses of the hollow and solid cylinders. The investigation reveals that the coupling effects of magneto-electro-elastic fields cannot be ignored when the material properties exhibit piezomagnetic/piezoelectric effects simultaneously. Although the governing and constitutive equations of the magnetic and electric potentials are similar to each other, their distributions are not the same due to the different coupling coefficients. Furthermore, it is seen that imposing a proper magnetic field can reduce the hoop stress in a rotating FGPM cylinder, and as a result can make the smart structures more reliable. Finally, the investigation shows that moisture concentration and temperature have similar effects on the multiphysical responses of an MEE cylinder in uncoupled hygrothermomagnetoelectroelasticity. It is observed that hygrothermal loading can change the radial stress, hoop stress, axial stress, and electric potential significantly for both hollow and solid MEE cylinders. It is worth mentioning that a theoretical micromechanical model or a computational homogenization technique can be used to obtain the effective properties of smart materials to be used in the closed-form solutions obtained in this chapter for multi-physical analysis of smart materials and structures [49, 50].

References

1. Krzhizhanovskaya VV, Sun S (2007) Simulation of multiphysics multiscale systems: introduction to the ICCS'2007 workshop. In: Proceedings of the 7th international conference on computational science, Part I: ICCS 2007. Springer, Beijing, China, pp 755–761
2. Babaei MH (2009) Multiphysics analysis of functionally graded piezoelectrics. Mechanical Engineering Department, University of New Brunswick, Ph.D. thesis
3. Altay G, Dökmeci MC (2008) Certain hygrothermopiezoelectric multi-field variational principles for smart elastic laminae. Mech Adv Mater Struct 15(1):21–32
4. Qin QH (2001) Fracture mechanics of piezoelectric materials
5. Ballato A (2001) Modeling piezoelectric and piezomagnetic devices and structures via equivalent networks. IEEE Trans Ultrason Ferroelectr Freq Control 48(5):1189–1240
6. Nan C-W (1994) Magnetoelectric effect in composites of piezoelectric and piezomagnetic phases. Phys Rev B 50(9):6082–6088
7. Wang Y et al (2010) Multiferroic magnetoelectric composite nanostructures. NPG Asia Mater 2:61–68
8. Tzou HS, Lee HJ, Arnold SM (2004) Smart materials, precision sensors/actuators, smart structures, and structronic systems. Mech Adv Mater Struct 11(4–5):367–393
9. Robert GL (1997) Recent developments in smart structures with aeronautical applications. Smart Mater Struct 6(5):R11
10. Javanbakht M, Shakeri M, Sadeghi SN (2009) Dynamic analysis of functionally graded shell with piezoelectric layers based on elasticity. Proc Inst Mech Eng Part C: J Mech Eng Sci 223 (9):2039–2047
11. Yas MH, Shakeri M, Khanjani M (2011) Layer-wise finite-element analysis of a functionally graded hollow thick cylinder with a piezoelectric ring. Proc Inst Mech Eng Part C: J Mech Eng Sci 225(5):1045–1060
12. Kapuria S, Yasin MY (2010) Active vibration control of piezoelectric laminated beams with electroded actuators and sensors using an efficient finite element involving an electric node. Smart Mater Struct 19(4):045019
13. Gabbert U, Ringwelski S (2014) Active vibration and noise control of a car engine: modeling and experimental validation. In: Belyaev AK, Irschik H, Krommer M (eds) Mechanics and model-based control of advanced engineering systems. Springer Vienna, Vienna, pp 123–135

14. Roederer AG, Jensen NE, Crone GAE (1996) Some European satellite-antenna developments and trends. IEEE Antennas Propag Mag 38(2):9–21
15. Barman S.L.a.G.P., (2003) Damage detection of structures based on spectral methods using piezoelectric materials. In: Proceeding of 4th international workshop on structural health monitoring, Stanford University
16. Bein T, Nuffer J (2006) Application of piezoelectric materials in transportation industry. In: Global symposium on innovative solution for the advancement of the transport industry
17. Sih GC, Michopoulos JG, Chou SC (1986) Coupled diffusion of temperature and moisture. In: Sih GC, Michopoulos JG, Chou SC (eds) Hygrothermoelasticity. Springer Netherlands: Dordrecht, pp 17–45
18. Smittakorn W (2001) A theoretical and experimental study of adaptive wood composites. Ph. D. thesis
19. Parton VZ, Kudryavtsev BA (1988) Electromagnetoelasticity: piezoelectrics and electrically conductive solids. Gordon and Breach Science Publishers
20. Kraus JD (1984) Electromagnetics. McGraw Hill
21. Chen P, Shen Y (2007) Propagation of axial shear magneto–electro-elastic waves in piezoelectric–piezomagnetic composites with randomly distributed cylindrical inhomogeneities. Int J Solids Struct 44(5):1511–1532
22. Eslami MR, Hetnarski RB (2009) Heat Conduction problems. In: Thermal stresses—advanced theory and applications. Springer Netherlands, Dordrecht, pp 132–218
23. Biot MA (1956) Thermoelasticity and irreversible thermodynamics. J Appl Phys 27(3):240–253
24. Heyliger WSPR (2000) A discrete-layer model of laminated hygrothermopiezoelectric plates. Mech Compos Mater Struct 7(1):79–104
25. Aboudi J, Williams TO (2000) A coupled micro–macromechanical analysis of hygrothermoelastic composites. Int J Solids Struct 37(30):4149–4179
26. Chandrasekharaiah DS (1998) Hyperbolic thermoelasticity: a review of recent literature. Appl Mech Rev 51(12):705–729
27. Reddy JN, Chin CD (1998) thermomechanical analysis of functionally graded cylinders and plates. J Therm Stresses 21(6):593–626
28. Reddy J (2011) A general nonlinear third-order theory of functionally graded plates. Int J Aerosp Lightweight Struct (IJALS) 1(1)
29. Kurimoto M et al (2010) Application of functionally graded material for reducing electric field on electrode and spacer interface. IEEE Trans Dielectr Electr Insul 17(1):256–263
30. Pompe W et al (2003) Functionally graded materials for biomedical applications. Mater Sci Eng, A 362(1):40–60
31. Carrera E, Soave M (2011) Use of functionally graded material layers in a two-layered pressure vessel. J Press Vessel Technol 133(5):051202
32. Poultangari R, Jabbari M, Eslami M (2008) Functionally graded hollow spheres under non-axisymmetric thermo-mechanical loads. Int J Press Vessels Pip 85(5):295–305
33. Tang T, Yu W (2009) Micromechanical modeling of the multiphysical behavior of smart materials using the variational asymptotic method. Smart Mater Struct 18(12):125026
34. Challagulla K, Georgiades A (2011) Micromechanical analysis of magneto-electro-thermo-elastic composite materials with applications to multilayered structures. Int J Eng Sci 49(1):85–104
35. Akbarzadeh AH, Chen ZT (2014) Thermo-magneto-electro-elastic responses of rotating hollow cylinders. Mech Adv Mater Struct 21(1):67–80
36. Babaei M, Chen Z (2008) Analytical solution for the electromechanical behavior of a rotating functionally graded piezoelectric hollow shaft. Arch Appl Mech 78(7):489–500
37. Akbarzadeh AH, Babaei MH, Chen ZT (2011) The thermo-electromagnetoelastic behavior of a rotating functionally graded piezoelectric cylinder. Smart Mater Struct 20(6):065008
38. Akbarzadeh AH, Chen ZT (2012) Magnetoelectroelastic behavior of rotating cylinders resting on an elastic foundation under hygrothermal loading. Smart Mater Struct 21(12):125013

39. Yu J, Ma Q, Su S (2008) Wave propagation in non-homogeneous magneto-electro-elastic hollow cylinders. Ultrasonics 48(8):664–677
40. Babaei M, Chen Z (2008) Exact solutions for radially polarized and magnetized magnetoelectroelastic rotating cylinders. Smart Mater Struct 17(2):025035
41. Youssef HM (2005) Generalized thermoelasticity of an infinite body with a cylindrical cavity and variable material properties. J Therm Stresses 28(5):521–532
42. Adams DF, Miller AK (1977) Hygrothermal microstresses in a unidirectional composite exhibiting inelastic material behavior. J Compos Mater 11(3):285–299
43. Babaei M, Akhras G (2011) Temperature-dependent response of radially polarized piezoceramic cylinders to harmonic loadings. J Intell Mater Syst Struct 22(7):645–654
44. Ying J, Lü CF, Chen WQ (2008) Two-dimensional elasticity solutions for functionally graded beams resting on elastic foundations. Compos Struct 84(3):209–219
45. Kiani Y, Bagherizadeh E, Eslami M (2011) Thermal buckling of clamped thin rectangular FGM plates resting on Pasternak elastic foundation (Three approximate analytical solutions). ZAMM-J Appl Math Mech (Z Angew Math Mech) 91(7):581–593
46. Smittakorn W, Heyliger PR (2007) An adaptive wood composite: theory. Wood Fiber Sci 33 (4):595–608
47. Hou P-F, Leung AY (2004) The transient responses of magneto-electro-elastic hollow cylinders. Smart Mater Struct 13(4):762
48. Galic D, Horgan C (2003) The stress response of radially polarized rotating piezoelectric cylinders. Trans Am Soc Mech Eng J Appl Mech 70(3):426–435
49. Shi J, Akbarzadeh AH (2019) Architected cellular piezoelectric metamaterials: Thermo-electro-mechanical properties. Acta Materialia 163:91–121
50. Kalamkarov AL, Andrianov IV, Danishevs'Kyy VV (2009) Asymptotic homogenization of composite materials and structures. Appl. Mech. Rev. 62(3):669–676

Chapter 4
Coupled Thermal Stresses in Advanced Smart Materials

4.1 Functionally Graded Materials

Besides being the snack of choice of the Chinese Giant panda, the bamboo plant also represents a near-perfect natural example of a functionally graded material [1]. This type of materials can be defined as a composite characterized by a spatially varying microstructure [2]. The properties of the functionally graded material (FGM) vary gradually along a given spatial axis throughout the material. This is usually accomplished by continuously and gradually alternating the presence of the reinforcement and matrix materials in creating composites, essentially using biomimicry inspired by naturally occurring examples such as bamboo [2]. Most species of bamboo plant have hollow culms with varying structural characteristics between the inner and outer peripheries. The fibres at the outermost layer are more numerous and have a compact, circular cross-section compared to the larger and elliptically shaped fibres at the inner layer. This variation in the microstructure leads to a tensile strength of 160 and 45 kg/mm^2 at the outer and inner peripheries, respectively [1].

The principle advantage of FGMs is the possibility of "combining" advantages and desired properties based on the constituent materials used. For example, using a metal and a ceramic correctly in an FGM would incorporate the heat and corrosion resistance of the ceramic as well as the mechanical strength of the metal [2]. Moreover, FGMs have smoothly varying material layers instead of the abrupt layer changes of typical composites. This gradual variation leads to a reduction of stress concentration at layer interfaces, reduces creep and failure, and increases the life of the material [2, 3].

In certain materials such as quartz and tourmaline, the piezoelectric effect occurs naturally. However, even when the piezoelectric effect is not naturally present, it can be induced through an electric polarization process. Barium titanate ($BaTiO_3$), polyvinylidene fluoride (PVDF) and other polycrystalline materials can be provided with piezoelectric properties by excessive heating and exposure to a strong DC field (higher than 2000 V/mm). This process aligns the molecular dipoles of the material

Z. T. Chen and A. H. Akbarzadeh, *Advanced Thermal Stress Analysis of Smart Materials and Structures*, Structural Integrity 10,
https://doi.org/10.1007/978-3-030-25201-4_4

according to the direction of the applied field [2]. FGMs can be used in this context as more efficient piezoelectric materials with less mechanical stress developed throughout the material [2]. Typical applications of piezoelectric materials that can be enhanced through the use of FGMs include: microelectromechanical systems, accelerometers, acoustic, pressure, and monitoring sensors, as well as precision position control among others [2, 3]. Furthermore, FGMs were found to have inherent thermal characteristics due to their unique nature. By correctly choosing the degree of variation of the composite materials, it is possible to essentially control (increase or decrease) the temperature of the material when exposed to excessive thermal stress [4]. For example, a spacecraft re-entering earth's atmosphere is subjected to a temperature gradient of approximately 1000 °C from the outer surface to the inside of the vessel. The design of an FGM with outer ceramic properties and inner thermally conductive properties, as well as a properly selected degree of gradation could greatly reduce the temperature within the walls of the spacecraft and reduce the risk of failure [2, 4].

Knowing this, the intelligent design of smart graded materials can greatly improve performance in extreme thermal conditions and in electrical applications, benefitting the fields and industries that could potentially rely on these materials. The very nature of functionally graded materials make them inherently useful since they offer the property-blending benefits of typical composites without the dangers of stress concentration at layer interfaces. Their potential as smart materials transcends any one scientific discipline, instead blurring the lines between mechanical, electrical, and thermal applications for use in structural design, electrical machinery and aerospace technologies. Manufacturing advanced materials like FGMs using smart, responsive, piezoelectric materials generates a powerful tool that is not only more efficient, but also more useful than traditional materials.

This chapter seeks to further this potential by examining the response of FGPMs to thermal stresses. Within the context of various governing thermoelasticity theories, analyses are carried out in order to graphically compare the results with those from previous authors. In Sect. 4.2, a generalized theory is used to quantify the behaviour of a homogeneous piezoelectric rod under thermal stress. Section 4.3 employs another generalized theory to analyze a functionally graded piezoelectric cylinder undergoing thermal shock. Finally, a functionally graded piezoelectric rod's response to a heat source is examined through coupled, uncoupled and generalized theories of thermoelasticity in Sect. 4.4.

4.2 Hyperbolic Coupled Thermopiezoelectricity in One-Dimensional Rod

In this section, the dynamic response of a thermospiezoelectric rod is analyzed on the basis of the Lord and Shulman theory of generalized thermoelasticity [5]. The rod, which is assumed to be made of a homogeneous material, is subject to a moving heat source travelling along its length. Solutions for the displacement, temperature and electric potential are analytically obtained from three coupled, dynamic, governing different equations for the given problem [5]. The equations are firstly solved in the Laplace domain through successive decoupling, then the time-dependent dynamic solutions are attained through a numerical inversion in the Laplace domain. Finally, through numerical examples, the present results are juxtaposed with those previously reported in the literature to substantiate the conclusions.

In many applications, piezoelectric materials are used under conditions of high temperature [2, 6]. For this reason, pyroelectricity, which plays a crucial role in thermopiezoelectric media and showcases the potential of piezoelectric materials to generate electricity out of temperature changes, is incorporated in the present analysis.

4.2.1 Introduction

In analysis, the classical theory of thermoelasticity yields unrealistic results; thermal wave speeds are found to be infinite, and the thermal and elastic fields are independent of each other [5]. Vernotte [7] and Cattaneo [8] introduced a hyperbolic non-Fourier heat conduction theory which included a "relaxation time" to address the first issue of infinite speed of heat propagation. The concept of this addition will be explored in Sect. 4.3. Biot [9] investigated the idea of coupling the thermal and elastic fields using the first law of thermodynamics and successfully proposed the aptly named coupled thermoelasticity theory. Following these two breakthroughs, more generalized thermoelasticity theories have been developed which amend both the thermal wave velocity and the coupling effect problems. One such theory arose from Lord and Shulman (L-S) [10] who used the hyperbolic heat conduction theory to define a more generalized classical thermoelasticity theory. Using the same hyperbolic theory, Chandrasekharaiah [11] proposed the generalized thermopiezoelasticity, essentially extending the work of LS to a thermopiezoelectric theory. This theory will be seen in Sect. 4.3 as well.

Using the generalized Lord-Shulman theory of thermoelasticity, the investigation of a finite thermopiezoelectric rod was performed by He et al. [12]. However, in this case the electric displacement in the rod was considered to be independent of time in order to simplify the analysis. As such, the time histories of both electric displacement and electric potential were omitted from the research. Furthermore, the equation for entropy was stated in terms of strain, temperature, and electric field instead of electric displacement. This leads to an inconsistency whereby the variables used in the entropy equation differ from those used in the equations for stress and electric field [5]. When the pyroelectric constant of the material becomes large, the effect of this discrepancy is magnified, and there is also no evidence of thermal wavefronts present in the temperature distributions found. Finally, the solutions proposed by He et al. for the displacement and temperature of the rod only contain four integration constants, arising from two boundary conditions for temperature and two for displacement. This system implies that there is no mathematical option to have the rod electroded at its ends in order to create specified voltage conditions [5].

We now investigate the problem of a thermopiezoelectric rod of finite length subjected to a moving heat source. In the following analysis, the simplifications in [12] will not be included. As such, six integration constants are utilized, allowing the ends of the rod to experience preset conditions of voltage. Results, now including stress and electric displacement distributions, and thermal wavefronts, are obtained and compared graphically to those based on the simplified problem.

4.2.2 Homogeneous Rod Problem

Consider a thermopiezoelectric rod of length L, lying on the horizontal z-axis situated in a one-dimensional coordinate system, as shown in Fig. 4.1. The left end of the rod is located at the origin of the coordinate system. The boundary conditions of the ends of the rod are as follows:

$$w(z=0,t)=w_0, \quad w(z=L,t)=w_L \tag{4.1a, b}$$

$$\phi(z=0,t)=\phi_0, \quad \phi(z=L,t)=\phi_L \tag{4.1c, d}$$

$$\begin{aligned} \delta_{10}T(z=0,t)+\delta_{20}\frac{\partial T}{\partial z}(z=0,t)=\Theta_0, \\ \delta_{1L}T(z=L,t)+\delta_{2L}\frac{\partial T}{\partial z}(z=L,t)=\Theta_L \end{aligned} \tag{4.1e, f}$$

where w, ϕ, T, and t are, respectively, the displacement, electric potential, absolute temperature and time. δ_{10}, δ_{1L}, δ_{20}, δ_{2L}, w_0, w_L, ϕ_0, ϕ_L, Θ_0 and Θ_L are arbitrary constants and $\frac{\partial}{\partial z}$ stands for partial differentiation with respect to z.

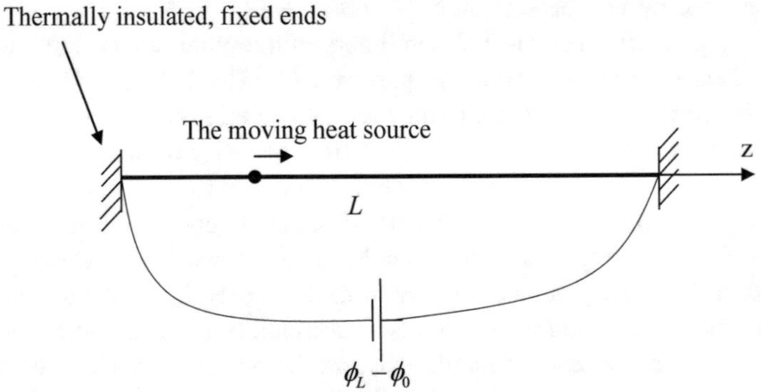

Fig. 4.1 A thermopiezoelectric rod subjected to a moving heat source

The initial conditions are given below:

$$w(z, t = 0) = 0, \quad \frac{\partial w}{\partial t}(z, t = 0) = 0 \tag{4.2a, b}$$

$$\phi(z, t = 0) = 0, \quad \frac{\partial \phi}{\partial t}(z, t = 0) = 0 \tag{4.2c, d}$$

$$T(z, t = 0) = T_0, \quad \frac{\partial T}{\partial t}(z, t = 0) = 0 \tag{4.2e, f}$$

where T_0 is the initial temperature of the rod.

4.2.2.1 Fundamental and Governing Equations

The constitutive equations for linear thermopiezoelectric media are [11]:

$$\sigma_{ij} = C_{ijkl}\varepsilon_{kl} - e_{ijk}E_k - \beta_{ij}\theta \tag{4.3a}$$

$$\rho S = \frac{\rho C_E}{T_0}\theta + \beta_{ij}\varepsilon_{ij} + p_i E_i \tag{4.3b}$$

$$D_i = e_{ijk}\varepsilon_{jk} + \in_{ij} E_j + p_i\theta \tag{4.3c}$$

$$\left(1 + \tau\frac{\partial}{\partial t}\right) q_i = -K_{ij}\theta_{,j} \tag{4.3d}$$

where σ_{ij}, S, D_i and q_i are stress, entropy, electric displacement and heat flux, respectively. C_{ijkl}, e_{ijk}, β_{ij}, p_i, $\in_{ij}$, τ, C_E, K_{ij} and ρ are, respectively, elastic and piezoelectric constants, thermal moduli, pyroelectric and dielectric constants, thermal relaxation, specific heat, coefficient of thermal conductivity, and density. Here, a subscript comma denotes the partial differentiation with respect to the variable that follows it and θ is the temperature change $(\theta = T - T_0)$.

The equations of energy and motion, as well as the Coulomb equation and linear strain-displacement, in the absence of electric current and free charge, are:

$$\rho\left(S_{,t}T_0 - R\right) + q_{i,i} = 0 \tag{4.4a}$$

$$\sigma_{ij,j} + \rho F_i = \rho u_{i,tt} \tag{4.4b}$$

$$D_{i,i} = 0, \quad E_i = -\phi_{,i}(i, j = 1, 2, 3) \tag{4.4c, d}$$

$$\varepsilon_{ij} = \frac{1}{2}\left(u_{i,j} + u_{j,i}\right) \tag{4.4e}$$

where F_i, R and u_i are the external body force, heat source intensity and displacement components. We can see that for a problem in which the only dimension is z, the remaining displacement and electric field components are $u_3 = w$ and $E_3 = E_z$. Likewise, the only components of heat flux, stress and electric displacement that are non-zero are in the z direction, and any spatial differentiations are zero except those with respect to z. Thus, Eqs. (4.3) and (4.4) are reduced to:

$$\sigma_{zz} = c_{33}\varepsilon_{zz} - e_{33}E_3 - \beta_3\theta \tag{4.5a}$$

$$\rho S = \frac{\rho C_E}{T_0}\theta + \beta_3\varepsilon_{zz} + p_3E_z \tag{4.5b}$$

$$D_z = e_{33}\varepsilon_{zz} + \in_3 E_z + p_3\theta \tag{4.5c}$$

$$\left(1 + \tau\frac{\partial}{\partial t}\right)q_z = -K_3\theta_{,z} \tag{4.5d}$$

and

$$\rho\left(S_{,t}T_0 - R\right) + q_{z,z} = 0 \tag{4.6a}$$

$$\sigma_{zz,z} = \rho w_{tt} \tag{4.6b}$$

$$D_{z,z} = 0, \quad E_z = -\phi_{,z} \tag{4.6c, d}$$

$$\varepsilon_{zz} = w_{,z} \tag{4.6e}$$

in which, C_{3333}, e_{333}, β_{33}, $\in_{33}$ and k_{33} are replaced by c_{33}, e_{33}, β_3, $\in_3$ and k_3 for convenience. By using Eqs. (4.5) and (4.6d, e), Eq. (4.6a–c) can be reduced to the following three governing differential equations of the stated problem:

$$c_{33}w_{,zz} + e_{33}\phi_{,zz} - \beta_3\theta_{,z} = \rho w_{,tt} \tag{4.7a}$$

$$e_{33}w_{,zz} - \in_{33}\phi_{,zz} + p_3\theta_{,z} = 0 \tag{4.7b}$$

$$K_3\theta_{,ij} + \left(1 + \tau\frac{\partial}{\partial t}\right)\left(\rho R - \rho C_E\theta_{,t} - \beta_3T_0w_{,zt} + p_3T_0\phi_{,zt}\right) = 0 \tag{4.7c}$$

4.2.3 Solution Procedure

4.2.3.1 Solution in Laplace Domain

Using the following relations, we can normalize the governing equations in order to make the solution procedure more convenient:

$$\bar{z} = c_0\eta_0 z, \quad \bar{w} = c_0\eta_0 w, \quad \bar{\phi} = \frac{\in_3}{e_{33}L}\phi, \quad \bar{R} = \frac{R}{K_3 T_0 c_0^2 \eta_0^2} \tag{4.8a–d}$$

$$\bar{t} = c_0^2\eta_0 t, \quad \bar{\tau} = c_0^2\eta_0\tau \tag{4.8e, f}$$

$$\bar{\theta} = \frac{\theta}{T_0}, \quad \bar{\sigma}_{zz} = \frac{\sigma_{zz}}{c_{33}}, \quad \bar{D}_z = \frac{D_z}{e_{33}} \tag{4.8g–i}$$

where $c_0 = \sqrt{\frac{c_{33}}{\rho}}$ and $\eta_0 = \frac{\rho C_E}{K_3}$ are the propagation speed of an elastic wave in a homogeneous, purely elastic, linearly isotropic solid, and the reciprocal of thermal diffusivity, respectively. Using the above normalized values and after dropping the overbars for convenience, the non-dimensional form of Eq. (4.7) is:

$$w_{,zz} + \frac{c_0\eta_0 L e_{33}^2}{\in_3 c_{33}}\phi_{,zz} - \frac{\beta_3 T_0}{c_{33}}\theta_{,z} - w_{,tt} = 0 \tag{4.9a}$$

$$w_{,zz} - c_0\eta_0 L\phi_{,zz} + \frac{p_3 T_0}{e_{33}}\theta_{,z} = 0 \tag{4.9b}$$

$$\theta_{,zz} + \left(1 + \tau\tfrac{\partial}{\partial t}\right)\left(\rho R - \theta_{,t} - \tfrac{\beta_3}{\rho C_E} w_{,zt} + \tfrac{p_3 c_0 L e_{33}}{K_3 \in_3}\phi_{,zt}\right) = 0 \tag{4.9c}$$

In this problem, it is assumed that the intensity of the heat source has the following form on a non-dimensional basis:

$$R = R_0\delta(z - vt) \tag{4.10}$$

where R_0, δ, and v are the non-dimensional magnitude of the moving heat source, the Dirac delta function, and the non-dimensional velocity of the heat source, respectively. In order to simplify the solution procedure, we first find the solution of Eq. (4.9) in the Laplace domain. The Laplace transform of a function $f(t)$, in terms of its argument t, is defined by $\bar{f}(s)$ as follows:

$$L[f(t)] = f(s) = \int_0^\infty e^{-st} f(t)dt \quad (Re(s) > 0) \tag{4.11}$$

where s is the Laplace parameter and Re is the real part of its argument. Using the integral transform above and incorporating the initial conditions stated in Eq. (4.2), after dropping overbars for simplicity, the governing, non-dimensional equations of the rod in the Laplace domain are:

$$w_{,zz} - a_1 s^2 w + a_2 \phi_{,zz} - a_3 \theta_{,z} = 0 \tag{4.12a}$$

$$w_{,zz} - b_1 \phi_{,zz} + b_2 \theta_{,z} = 0 \tag{4.12b}$$

$$\theta_{,zz} - s(1+\tau s)\theta - d_1 s(1+\tau s) w_{,z} + d_2 s(1+\tau s)\phi_{,z} = -(1+\tau s) d e^{-\frac{s}{v}z} \tag{4.12c}$$

where

$$\begin{aligned} a_1 &= \frac{c_0^2 \rho}{c_{33}}, \quad a_2 = \frac{c_0 \eta_0 L e_{33}^2}{\in_{33}\, c_{33}}, \quad a_3 = \frac{\beta_3 T_0}{c_{33}} \\ b_1 &= c_0 \eta_0 L, \quad b_2 = \frac{p_3 T_0}{e_{33}} \\ d_1 &= \frac{\beta_3}{\rho C_E}, \quad d_2 = \frac{p_3 c_0 e_{33} L}{K_3 \in_3}, \quad d = \frac{\rho R_0}{v} \end{aligned} \tag{4.13}$$

Typically, a system of ordinary differential equations with constant coefficients is solved using its eigenvalues and eigenvectors [13]. For current problem, however, one unknown can be consecutively eliminated from the equations. This method is more straightforward, as seen below:

By eliminating $\theta_{,z}$ between Eqs. (4.12a) and (4.12b), we obtain the following equations for w and ϕ:

$$\left(1 + \frac{a_3}{b_2}\right) w_{,zz} - a_1 s^2 w + \left(a_2 - \frac{b_1 a_3}{b_2}\right)\phi_{,zz} = 0 \tag{4.14}$$

After solving for $\theta_{,z}$ using Eq. (4.12a) and using its derivative $(\theta_{,zz})$ and Eq. (4.12c), we can find θ in terms of other unknowns:

$$\theta = \frac{1}{a_3 s(1+\tau s)} w_{,zzz} - \left(d_1 + \frac{a_1 s}{a_3(1+\tau s)}\right) w_{,z} + \frac{a_2}{a_s s(1+\tau s)}\phi_{,zzz} + d_2 \phi_{,z} + \frac{d}{s} e^{-\frac{s}{v}z} \tag{4.15}$$

Using the above equations, we can eliminate θ in Eq. (4.12b) and obtain the second equation of w and ϕ:

$$\begin{aligned} &\frac{b_2}{a_3 s(1+\tau s)} w_{,zzzz} + \left(1 - b_2\left(d_1 + \frac{a_1 s}{a_3(1+\tau s)}\right)\right) w_{,zz} \\ &+ \frac{a_2 b_2}{a_3 s(1+\tau s)}\phi_{,zzzz} + (b_2 d_2 - b_1)\phi_{,zz} = \frac{d b_2}{v} e^{-\frac{s}{v}z} \end{aligned} \tag{4.16}$$

Similarly, ϕ can be eliminated between Eqs. (4.14) and (4.16). Then we can obtain the final differential equation of w:

$$f_4 w_{,zzzz} + f_2 w_{,zz} + f_0 w = f e^{-\frac{s}{v}z} \tag{4.17}$$

where

$$\begin{aligned} f_4 &= \frac{b_2}{a_3 s(1+\tau s)}\left(1 - a_2 \frac{b_2 + a_3}{a_2 b_2 - a_3 b_1}\right) \\ f_2 &= 1 - b_2\left(d_1 + \frac{a_1 s}{a_3(1+\tau s)}\right) + \frac{a_2 a_1 b_2^2 s}{a_3(a_2 b_2 - a_3 b_1)(1+\tau s)} \\ &\quad - (b_2 d_2 - b_1)\frac{b_2 + a_3}{a_2 b_2 - a_3 b_1} \\ f_0 &= (b_2 d_2 - d_1)\frac{a_1 b_2 s^2}{a_2 b_2 - a_3 b_1}, f = \frac{d b_2}{v} \end{aligned} \tag{4.18}$$

The solution of Eq. (4.17) can be expressed by the sum of its general and particular solutions, as:

$$w = C_i e^{\lambda_i z} + K e^{-\frac{s}{v}z} (i = 1, \ldots, 4) \tag{4.19}$$

where C_i are integration constants to be found using the boundary conditions and λ_i are corresponding characteristic roots of Eq. (4.17):

$$\begin{aligned} \lambda_1 = -\lambda_2 &= \sqrt{\frac{-f_2 + \sqrt{f_2^2 - 4 f_4 f_0}}{2 f_4}} \\ \lambda_3 = -\lambda_4 &= \sqrt{\frac{-f_2 - \sqrt{f_2^2 - 4 f_4 f_0}}{2 f_4}} \end{aligned} \tag{4.20}$$

Also, we can define K as:

$$K = \frac{f}{f_4\left(\frac{s}{v}\right)^4 + f_2\left(\frac{s}{v}\right)^2 + f_0} \tag{4.21}$$

Substituting Eq. (4.19) into Eqs. (4.14) and (4.12), we can find ϕ and θ, respectively, as follows:

$$\phi = j_i C_i e^{\lambda_i z} + j K e^{-\frac{s}{v}z} + C_5 z + C_6 \quad (i = 1, \ldots, 4) \tag{4.22a}$$

$$\theta = l_i C_i e^{\lambda_i z} + l K e^{-\frac{s}{v}z} + \tfrac{b_1}{b_2} C_5 + C_7 \quad (i = 1, \ldots, 4) \tag{4.22b}$$

where C_5, C_6, and C_7 are three new constants of integration and:

$$j_1 = j_2 = \frac{1}{a_2b_2 - a_3b_1}\left(\frac{a_1b_2s^2}{\lambda_1^2} - b_2 - a_3\right)$$

$$j_3 = j_4 = \frac{1}{a_2b_2 - a_3b_1}\left(\frac{a_1b_2s^2}{\lambda_3^2} - b_2 - a_3\right), j = \frac{1}{a_2b_2 - a_3b_1}\left(a_1b_2v^2 - b_2 - a_3\right)$$

$$\begin{aligned} l_1 = -l_2 &= \frac{\lambda_1}{b_2}(b_1j_1 - 1), l_3 = -l_4 = \frac{\lambda_3}{b_2}(b_1j_3 - 1), \\ &= \frac{s}{vb_2}(1 - b_1j) \end{aligned} \tag{4.23}$$

Substituting Eqs. (4.19) and (4.22) into Eq. (4.12c), we can obtain another auxiliary equation:

$$\left(d_2 - \frac{b_1}{b_2}\right)C_5 - C_7 = 0 \tag{4.24}$$

Having solved the governing different equations for displacement, electric potential and temperature of the rod, we can obtain equations for stress and electric displacement using Eqs. (4.5a, c):

$$\begin{aligned} \sigma_{zz} = {} & C_i\left(\lambda_i\left(1 + j_i\frac{e_{33}^2 c_0\eta_0 L}{\epsilon_3\, c_{33}}\right) - \frac{\beta_3 T_0}{c_{33}}l_i\right)e^{\lambda_i z} \\ & + K\left(-\frac{s}{v}\left(1 + j\frac{e_{33}^2 c_0\eta_0 L}{\epsilon_3\, c_{33}}\right) - \frac{\beta_3 T_0}{c_{33}}l\right)e^{-\frac{s}{v}z} \\ & + C_5\left(\frac{e_{33}^2 c_0\eta_0 L}{\epsilon_3\, c_{33}} - \frac{b_1}{b_2}\frac{\beta_3 T_0}{c_{33}}\right) - \frac{\beta_3 T_0}{c_{33}}C_7 \end{aligned} \tag{4.25a}$$

$$\begin{aligned} D_z = {} & C_i\left(\lambda_i(1 - j_i c_0\eta_0 L) + \frac{p_3 T_0}{e_{33}}l_i\right)e^{\lambda_i z} \\ & + K\left(-\frac{s}{v}(1 - j c_0\eta_0 L) + \frac{p_3 T_0}{e_{33}}l\right)e^{-\frac{s}{v}z} \\ & + C_5\left(\frac{p_3 T_0}{e_{33}}\frac{b_1}{b_2} - c_0\eta_0 L\right) + \frac{p_3 T_0}{e_{33}}C_7 \quad (i = 1, \ldots, 4) \end{aligned} \tag{4.25b}$$

After normalizing and transforming the boundary conditions in Eq. (1) into the Laplace domain, we encounter a linear system of equations with seven unknowns, $C_i (i = 1, \ldots, 7)$, as follows:

$$A_{7\times 7}C_{7\times 1} = B_{7\times 1} \quad (4.26)$$

The array of coefficients and the matrices $A_{7\times 7} = a_{ij}$ and $B_{7\times 1} = b_j(i,j = 1,\ldots,7)$ are given below:

$$a_{11} = a_{12} = a_{13} = a_{14} = 1, a_{15} = a_{16} = a_{17} = 0$$

$$b_1 = c_0\eta_0\frac{w_0}{s} - K$$

$$a_{21} = j_1, a_{22} = j_2, a_{23} = j_3, a_{24} = j_4, a_{25} = 0, a_{26} = 1, a_{27} = 0$$

$$b_2 = \frac{\in_3}{e_{33}L}\frac{\phi_0}{s} - jK$$

$$a_{31} = (\delta_{10} + \delta_{20}\lambda_1 c_0\eta_0)l_1, a_{32} = (\delta_{10} + \delta_{20}\lambda_2 c_0\eta_0)l_2, a_{33} = (\delta_{10} + \delta_{20}\lambda_3 c_0\eta_0)l_3,$$

$$a_{34} = (\delta_{10} + \delta_{20}\lambda_4 c_0\eta_0)l_4, a_{35} = \delta_{10}\frac{b_1}{b_2}, a_{36} = 0, a_{37} = \delta_{10}$$

$$b_3 = \left(\delta_{20}\frac{s}{v}c_0\eta_0 - \delta_{10}\right)lK + \frac{1}{s}\left(\frac{\Theta_0}{T_0} - \delta_{10}\right)$$

$$a_{41} = e^{\lambda_1 c_0\eta_0 L}, a_{42} = e^{\lambda_2 c_0\eta_0 L}, a_{43} = e^{\lambda_3 c_0\eta_0 L}, a_{44} = e^{\lambda_4 c_0\eta_0 L}, a_{45} = a_{46} = a_{47} = 0$$

$$b_4 = c_0\eta_0\frac{w_L}{s} - Ke^{-\frac{s}{v}c_0\eta_0 L}$$

$$a_{51} = a_{52} = a_{53} = a_{54} = 0, a_{55} = d_2 - \frac{b_1}{b_2}, a_{56} = 0, a_{57} = -1$$

$$b_5 = 0$$

$$a_{61} = j_1 e^{\lambda_1 c_0\eta_0 L}, a_{62} = j_2 e^{\lambda_2 c_0\eta_0 L}, a_{63} = j_3 e^{\lambda_3 c_0\eta_0 L}, a_{64} = j_4 e^{\lambda_4 c_0\eta_0 L},$$

$$a_{65} = c_0\eta_0 L, a_{66} = 1, a_{67} = 0$$

$$b_6 = \frac{\in_3}{e_{33}L}\frac{\phi_L}{s} - jKe^{-\frac{s}{v}c_0\eta_0 L}$$

$$a_{71} = (\delta_{10} + \delta_{20}\lambda_1 c_0\eta_0)l_1 e^{\lambda_1 c_0\eta_0 L}, a_{72} = (\delta_{10} + \delta_{20}\lambda_2 c_0\eta_0)l_2 e^{\lambda_2 c_0\eta_0 L},$$

$$a_{73} = (\delta_{10} + \delta_{20}\lambda_3 c_0\eta_0)l_3 e^{\lambda_3 c_0\eta_0 L}, a_{74} = (\delta_{10} + \delta_{20}\lambda_4 c_0\eta_0)l_4 e^{\lambda_4 c_0\eta_0 L},$$

$$a_{75} = \delta_{1L}\frac{b_1}{b_2}, a_{76} = 0, a_{77} = \delta_{1L}$$

$$b_7 = \left(\delta_{2L}\frac{s}{v}c_0\eta_0 - \delta_{1L}\right)lKe^{-\frac{s}{v}c_0\eta_0 L} + \frac{1}{s}\left(\frac{\Theta_L}{T_0} - \delta_{1L}\right)$$

Finally, we can solve Eq. (4.26), and find the unknowns. The next step in our procedure is the inversion of the results from the Laplace domain to the time domain, which is performed numerically.

4.2.3.2 Numerical Inversion of Laplace Transform

The analytical inversion of the solutions to the time domain is not straightforward. Therefore, a numerical technique is employed [14]. In this method, a function $(f(s))$ in the Laplace domain can be inverted at discrete time points t_j using the following formula:

$$f(t_j) = C(j)\left[-\frac{1}{2}Re\{f(a)\} + Re\left\{\sum_{k=0}^{N-1}(A(k) + iB(k))W^{jk}\right\}\right], j = 0, 1, 2, \ldots, N-1 \tag{4.27}$$

where

$$A(k) = \sum_{l=0}^{L} Re\left\{f\left(a + i(k + lN)\frac{2\pi}{T}\right)\right\}$$

$$B(k) = \sum_{l=0}^{L} Im\left\{f\left(a + i(k + lN)\frac{2\pi}{T}\right)\right\}$$

$$C(j) = \frac{2}{T}e^{aj\Delta t}, \Delta t = \frac{T}{N}$$

$$W = e^{i\frac{2\pi}{N}} \tag{4.28}$$

In the above equations, T, N and Δt are the total time over which the numerical inversion is performed, the number of time points to which the total time is divided, and the time increment, respectively; a is an arbitrary real number larger than the real parts of all singularities of $f(s)$; and Im denotes the imaginary part of a complex number. To minimize both discretization and truncation errors, the following constraints are defined [14]:

$$\begin{gathered} 5 \le aT \le 10 \\ 50 \le NL \le 5000 \end{gathered} \tag{4.29}$$

For our current solution, the above-mentioned parameters are chosen as follows:

$$\begin{aligned} aT &= 7.5 \\ L &= 10 \\ N &= 450 \end{aligned} \tag{4.29}$$

4.2.4 *Results and Discussion*

Let us now use an example to quantify the results found thus far. We consider a quartz thermopiezoelectric rod with the properties listed as follows [15, 16]:

$c_{33} = 8.674 \times 10^{10}\,\frac{\mathrm{N}}{\mathrm{m}^2}$	$\beta_3 = 1.16 \times 10^6\,\frac{\mathrm{N}}{\mathrm{m}^2\mathrm{K}}$
$\rho = 2.65 \times 10^3\,\frac{\mathrm{kg}}{\mathrm{m}^3}$	$C_E = 782\,\frac{\mathrm{J}}{\mathrm{kg\ K}}$
$e_{33} = 0.2\,\frac{\mathrm{C}}{\mathrm{m}^2}$	$\in_3 = 0.392 \times 10^{-10}\,\frac{\mathrm{F}}{\mathrm{m}}$
$p_3 = 4 \times 10^{-4}\,\frac{\mathrm{C}}{\mathrm{m}^2\mathrm{K}}$	$K_3 = 1.4\,\frac{\mathrm{W}}{\mathrm{mK}}$

Other numerical values used include: $R_0 = \frac{10}{\rho}$, $T_0 = 293$, $v = 0.5$, and $\tau = 0.05$. The non-dimensional length of the rod (L_N) is taken to be 1 [5]. Now, let us assume that the rod's ends are fixed and thermally insulated, such that no electric potential exists at either end. Under these conditions, the parameters in Eq. (4.1) reduce to the following:

$$w_0 = w_L = 0 \tag{4.31a}$$

$$\delta_{10} = \delta_{1L} = 0, \delta_{20} = \delta_{2L} = 1 \tag{4.31b, c}$$

$$\Theta_0 = \Theta_L = 0 \tag{4.31d}$$

$$\phi_0 = \phi_L = 0 \tag{4.31e}$$

Using these restrictions, the current solutions for displacement, temperature change, and stress are compared with those based on simplified assumptions in the figures that follow. When comparing the solution in this section with the simplified solution, one can note a few new and interesting phenomena.

In Fig. 4.2, the displacements along the length of the rod at non-dimensional time $t = 0.1848$ are plotted using both the solution found in this section and the simplified one. As shown, the current plot is negative and the peak displacement is slightly lower than that of the simplified solution. As can be expected, the extrema of the two solutions occur at the same location on the rod ($z = 0.19$). Finally, it can be seen that in the simplified solution, a significant portion of the rod (roughly $0.8 \leq z \leq 1$) has yet to respond to the thermal disturbance while in the non-simplified solution, the entire rod has responded.

Figure 4.3 depicts the temperature change at non-dimensional instant $t = 0.1848$ along the length of the rod. As shown, the thermal wavefront of the simplified solution is still travelling towards the right end of the rod. For the solution solved in this section, the wavefront is travelling in the opposite direction back towards the left end, implying a higher thermal wave speed for the current solution.

In Fig. 4.4, the time history of stress at a specific location on the rod ($z = 0.25$) is displayed. The main difference between the two solutions arises from the

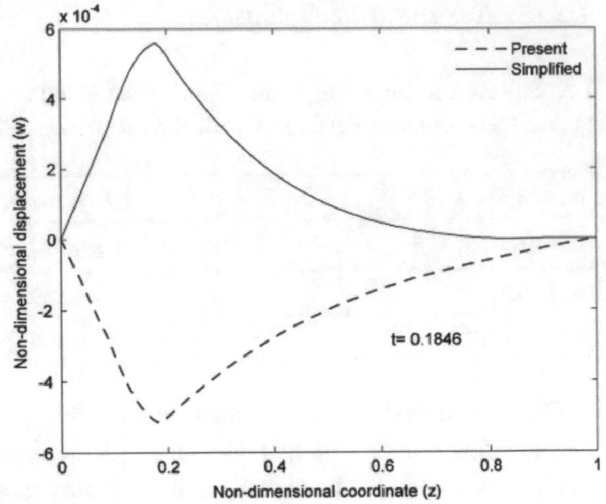

Fig. 4.2 Comparison of the displacement distributions at $t = 0.1848$ of the current solution [5] and simplified solution using the assumptions of [12]. [Reproduced from [5] with permission from IOP Publishing]

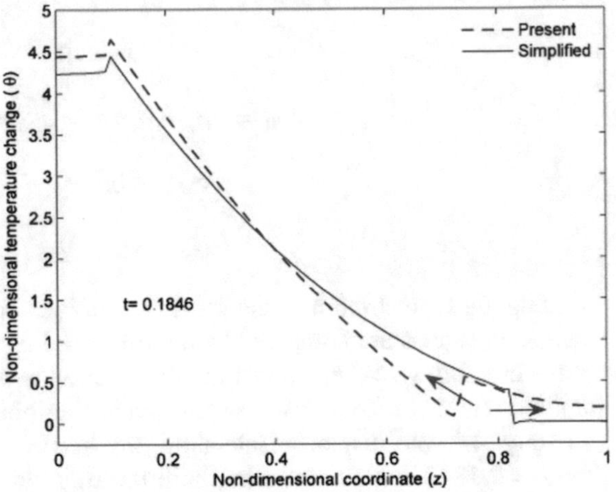

Fig. 4.3 Comparison of the temperature distributions at $t = 0.1848$ of the current solution [5] and that using the assumptions of [12]. [Reproduced from [5] with permission from IOP Publishing]

difference in displacement, as seen in Fig. 4.2. Also shown in Fig. 4.4 is the point in time at which the heat source exits the rod, after which both stress fields vibrate freely.

As seen in Fig. 4.5, the history of electric potential can be analyzed based on two periods of time; before and after the heat source finished travelling along the rod. The former is dominated by the thermal effects of the disturbance, while the latter is dominated by the coupling between electric potential and displacement. This coupling leads to the fluctuation seen after the heat source exits the rod. Additionally, the positivity of the electric potential is changed along its length before the heat source exits the rod.

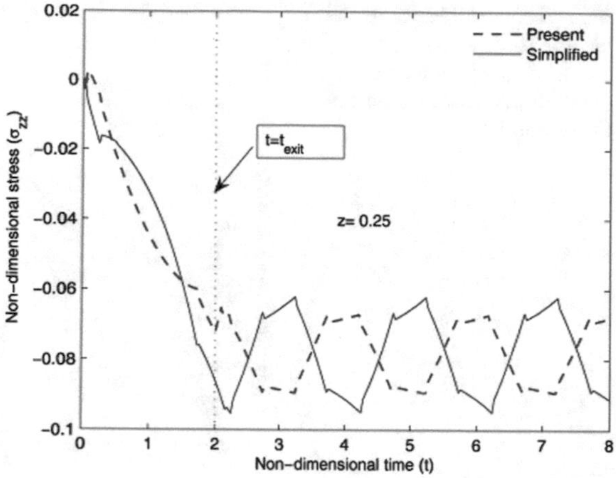

Fig. 4.4 Comparison of the stress history at $z = 0.25$ of the current solution [5] and that using the assumptions of [12]. [Reproduced from [5] with permission from IOP Publishing]

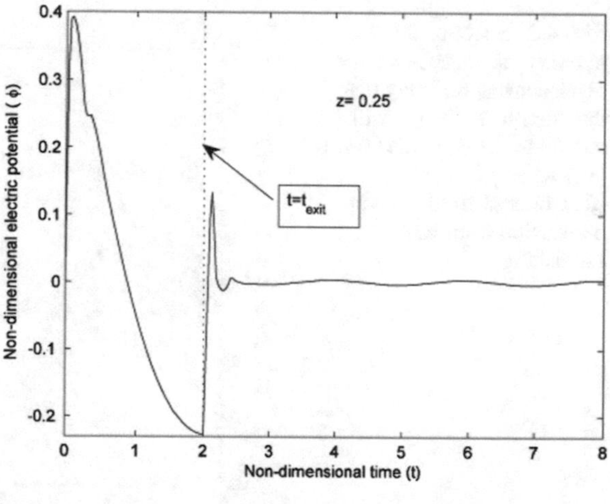

Fig. 4.5 Time history of electric potential at $z = 0.25$. [Reproduced from [5] with permission from IOP Publishing]

The plot of Fig. 4.6 shows that electric displacement is time dependent before the heat source exits the material, as was implied in the non-simplified solution [5]. Afterwards, the plot remains at a constant value, since for a one-dimensional problem the electric displacement is expected to remain spatially uniform along the length of the rod from Eq. (4.4c).

Given its non-simplified nature, the solution procedure in this section shows a more consistent and discrepancy-free approach to the hyperbolic theory of generalized thermoelasticity.

Finally, it should be noted that when the pyroelectric constant (p_3) decreases, the plots for both displacement and temperature change become equivalent to the simplified approach's plots. This relationship is demonstrated in Fig. 4.7 for displacement. Moreover, even when ($p_3 \rightarrow 0$), electric potential and electric displacement remain non-zero and time dependent along the rod, respectively.

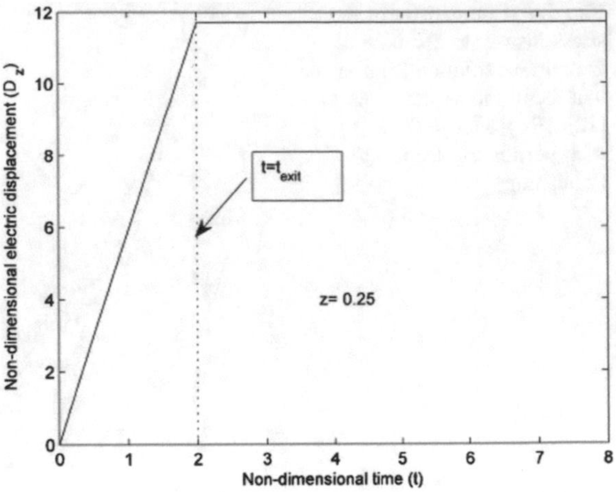

Fig. 4.6 Time history of the electric displacement at $z = 0.25$. [Reproduced from [5] with permission from IOP Publishing]

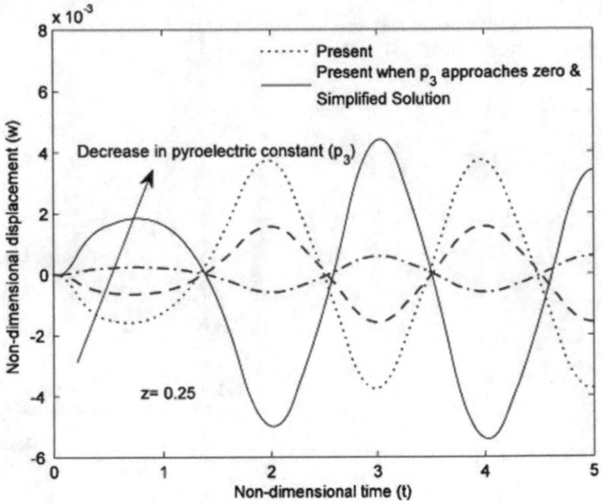

Fig. 4.7 Effect of the pyroelectric constant on the displacement verifying that the current results [5] will reduce to the simplified ones [12] when $p_3 \rightarrow 0$. [Reproduced from [5] with permission from IOP Publishing]

4.3 Hyperbolic Coupled Thermopiezoelectricity in Cylindrical Smart Materials

In the following section, we will analyze the transient response of a functionally graded, radially polarized hollow cylinder under dynamic axisymmetric loadings [17]. In this case, the Chandrasekharaiah hyperbolic theory of generalized thermo-piezoelectricity is used to simultaneously couple the displacement, temperature and electric fields under non-Fourier heat conduction. Moreover, all material properties (except thermal relaxation time) vary gradually throughout the material according to a volume fraction-based rule with varying degrees of non-homogeneity. The Galerkin finite element method is employed in the Laplace domain in order to solve the three coupled partial differential equations, after which a numerical method of the Laplace inversion restores the time variable. This section emphasizes the effects of the non-homogeneity index and the thermal relaxation time on the results.

4.3.1 Introduction

As previously stated at the beginning of this chapter, the benefits of functionally graded materials (FGMs) include lower stress concentrations at layer interfaces and improved residual stress distribution, among others [17]. When combined with the design of piezoelectric components, FGMs can improve the performance of advanced structures such as surface acoustic wave (SAW) sensors [18] and bimorph actuators [19]. This being said, piezoelectric devices operated at high temperatures require the coupling of temperature with elastic and electric fields. In order to model the response of piezoelectric media when exposed to different loading situations, it is imperative to choose a proper thermo-piezoelectricity theory. In the past, classic uncoupled theories were unable to quantify certain physical phenomena such as the wave-like behaviour of temperature and the effect of strain and electric field on the temperature distribution [20]. To account for these observations, generalized theories of thermo-piezoelectricity were developed such as Chandrasekharaiah's [11] work in extending the generalized Lord and Shulman theory of thermoelasticity into a thermo-piezoelectricity theory. Previous works in cylindrical piezoelectric structures include Babaei and Akhras [21] modeling the response of a radially polarized piezoceramic cylinder to harmonic loadings. Temperature dependence was incorporated in the analysis through material properties that fluctuated with temperature, pyroelectricity, and thermally-dependent dimensions.

Let us now introduce the problem solved in this section. The Chandrasekharaiah theory is used to analyze the transient response of a functionally graded piezoelectric hollow cylinder to an axisymmetric thermal shock [17]. The cylinder is assumed to be radially polarized and its materials vary in the radial direction, hence its functionally graded nature. The degree to which the properties vary is given according to a volume fraction rule and different non-homogeneity indices. The solutions for the displacement, electric potential and temperature, and the effects of the non-homogeneity indices and thermal relaxation time on the solutions, are displayed graphically.

4.3.2 Hollow Cylinder Problem

For this problem, we consider a long, hollow cylinder whose central axis is aligned with the z-direction of the coordinate system, as shown in Fig. 4.8 [17]. The cylinder, with inner and outer radii a and b respectively, is polarized and graded in

Fig. 4.8 An FGPM hollow cylinder and boundary condition locations. [Reproduced from [17] with permission from The Royal Society]

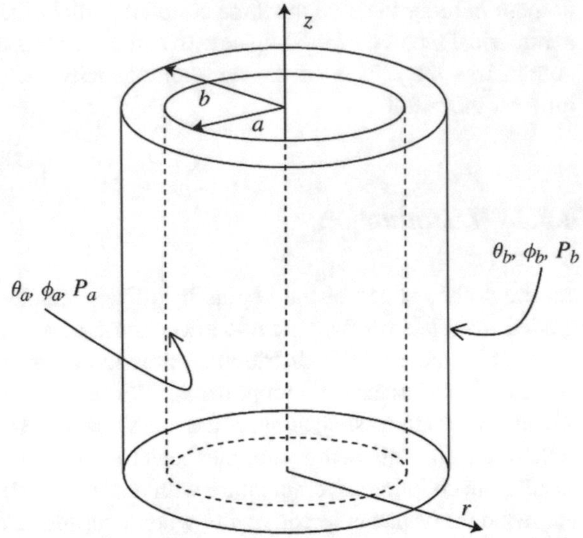

the radial direction. On each surface, the temperature change θ, electric potential ϕ, and pressure P are specified.

The variation of the material properties follows the volume fraction-based gradient rule as follows:

$$\chi(r) = \chi_a + \chi_{ba}\left(\frac{r-a}{b-a}\right)^{n_\chi} \tag{4.32a}$$

$$\chi_{ba} = \chi_b - \chi_a \tag{4.32b}$$

where χ is any property of the cylinder (except the thermal relaxation time, which is constant), the subscripts a and b denote the surface at which the property is considered, and n_χ is the non-homogeneity index of the corresponding material property, χ. Figure 4.9 shows how an arbitrary material property χ varies radially according to Eq. (4.32) with different non-homogeneity indices. When $n_\chi = 0$, the property is constant at the χ_b value, as shown in Fig. 4.9.

4.3.2.1 Fundamental and Governing Equations

The constitutive relations for a linear piezoelectric material and their variables, previously stated in Sect. 4.1, are repeated here for consistency [11]:

$$\sigma_{ij} = C_{ijkl}\varepsilon_{kl} - e_{ijk}E_k - \beta_{ij}\theta \tag{4.33a}$$

$$\rho S = \frac{\rho c}{T_0}\theta + \beta_{ij}\varepsilon_{ij} + \gamma_i E_i \tag{4.33b}$$

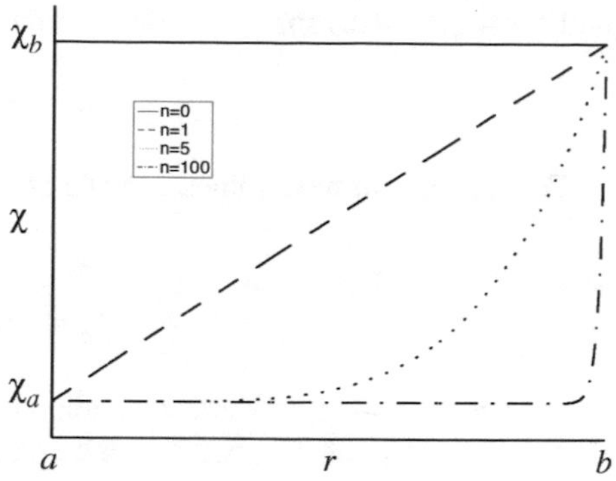

Fig. 4.9 Profiles of a material property, χ, in the radial direction; variations with different non-homogeneity indices. Solid line: $n_\chi = 0$, dashed line: $n_\chi = 1$, dotted line: $n_\chi = 5$, dash-dotted line: $n_\chi = 100$. [Reproduced from [17] with permission from The Royal Society]

$$D_i = e_{ijk}\varepsilon_{jk} + \in_{ij} E_j + \gamma_i\theta \tag{4.33c}$$

$$q_i + \tau q_{i,t} = -K_{ij}\theta_{,j} \quad i,j,k = 1,2,3 \tag{4.33d}$$

where σ_{ij}, S, D_i, and q_i are the stress, entropy, electric displacement and heat flux respectively. C_{ijkl}, e_{ijk}, β_{ij}, ρ, c, γ_i, are, respectively, the elastic and piezoelectric coefficients, thermal moduli, density, specific heat, and pyroelectric coefficient. $\in_{ij}$, τ, and K_{ij} are the dielectric coefficient, thermal relaxation time, and coefficient of thermal conductivity. Here, a comma subscript denotes partial differentiation with respect to what follows it. θ is the temperature change with respect to the initial temperature T_0, i.e. $\theta = T - T_0$, while ε_{jk} and E_k represent the strain and electric fields. In the absence of electric current, free charge and body force, the equations of energy and motion, as well as the Coulomb equation for the conservation of electric charge are [17]:

$$\rho S_{,t} T_0 + q_{i,i} = 0 \tag{4.34a}$$

$$\sigma_{ij,j} = \rho u_{i,tt} \tag{4.34b}$$

$$D_{i,i} = 0 \tag{4.34c}$$

where u_i are the displacement components. The linear strain-displacement relations are:

$$\varepsilon_{ij} = \frac{1}{2}\left(u_{i,j} + u_{j,i}\right) \tag{4.35}$$

and for a quasi-stationary electric field the following relation holds:

$$E_i = -\phi_{,i} \tag{4.36}$$

For the current axisymmetric plane strain problem:

$$\begin{aligned} u_z &= u_\theta = 0 \\ (\cdot)_{,z} &= (\cdot)_{,\theta} = 0 \end{aligned} \tag{4.37}$$

where θ is the azimuthal direction and $(\cdot)$ is an arbitrary variable. It should be stated that a symbol θ as a subscript corresponds to the azimuthal direction, while a regular lowercase θ represents temperature change. Considering Eq. (4.37), the non-zero components of strain and electric field, Eqs. (4.35) and (4.36), are:

$$\varepsilon_{rr} = u_{,r}, \varepsilon_{\theta\theta} = \frac{u}{r} \tag{4.38a, b}$$

$$E_r = -\phi_{,r} \tag{4.38c}$$

Here, u_r has been replaced by u for convenience. Therefore, the entropy and non-zero components of the stress, electric displacement and heat flux [Eq. (4.33)] are:

$$\sigma_{rr} = c_{33}u_{,r} + \frac{c_{13}}{r}u + e_{33}\phi_{,r} - \beta_1\theta, \sigma_{\theta\theta} = c_{13}u_{,r} + \frac{c_{11}}{r}u + e_{31}\phi_{,r} - \beta_3\theta \tag{4.39a, b}$$

$$\sigma_{zz} = c_{13}u_{,r} + \frac{c_{12}}{r}u + e_{31}\phi_{,r} - \beta_3\theta, \rho S = \beta_1 u_{,r} + \frac{\beta_3}{r}u - \gamma_1\phi_{,r} + \frac{\rho c}{T_0}\theta \tag{4.39c, d}$$

$$D_r = e_{33}u_{,r} + \frac{e_{31}}{r}u - \epsilon_{33}\,\phi_{,r} + \gamma_1\theta, q_r + \tau q_{r,t} = -K_{11}\theta_{,r} \tag{4.39e, f}$$

where $c_{\alpha\delta} = c_{ijkl}(i,j,k,l = 1,2,3; \alpha,\delta = 1,2,\ldots,6)$, $e_{m\alpha} = e_{mij}(m = 1,2,\ldots,6)$ and $\beta_1 = \beta_{11}$, $\beta_3 = \beta_{33}$. The notations are changed here for brevity and convenience. Similarly, Eq. (4.34) is reduced to the following set of equations:

$$\sigma_{rr,r} + \frac{\sigma_{rr} - \sigma_{\theta\theta}}{r} = \rho u_{,tt} \tag{4.40a}$$

$$\frac{1}{r}(rD_r)_{,r} = 0 \tag{4.40b}$$

$$\rho S_{,t}T_0 + \frac{1}{r}(rq_r)_{,r} = 0 \tag{4.40c}$$

We now introduce the normalized parameters as follows:

$$\begin{aligned}&\bar{r}=c^{*}\eta^{*}r,\bar{u}=c^{*}\eta^{*}u,\bar{t}=c^{*2}\eta^{*}t,\bar{\tau}=c^{*2}\eta^{*}\tau\\&\bar{\theta}=\frac{\theta}{T_0},\bar{\phi}=\frac{\in_{33a}}{e_{33a}b}\phi,\bar{\sigma}_{ij}=\frac{\sigma_{ij}}{c_{33a}},\bar{D}_i=\frac{D_i}{e_{33a}}\end{aligned}\tag{4.41}$$

where an overbar denotes a normalized value, $c^{*}=\sqrt{\frac{c_{33a}}{\rho_a}}$ represents the propagation speed of the elastic wave in a homogeneous, linearly elastic, isotropic medium, and $\eta^{*}=\frac{\rho_a c_a}{K_{11a}}$ is the reciprocal of the thermal diffusivity of the inner surface of the cylinder. By substituting Eq. (4.39) into Eq. (4.40), the normalized governing equations of the current problem are defined as follows:

$$\begin{aligned}&c_{33}u_{,rr}+\left(c_{33,r}+\frac{c_{33}}{r}\right)u_{,r}+\frac{1}{r}\left(c_{13,r}-\frac{c_{11}}{r}\right)u+\upsilon e_{33}\phi_{,rr}\\&+\upsilon\left(e_{33,r}+\frac{1}{r}(e_{33}-e_{31})\right)\phi_{,r}-\beta_1 T_0\theta_{,r}\\&+\left(-\beta_{1,r}+\frac{1}{r}(\beta_3-\beta_1)\right)T_0\theta=\rho c^{*2}u_{,tt}\end{aligned}\tag{4.42a}$$

$$\begin{aligned}&e_{33}u_{,rr}+\left(e_{33,r}+\frac{1}{r}(e_{31}+e_{33})\right)u_{,r}+\frac{e_{31,r}}{r}u-\upsilon\in_{33}\phi_{,rr}\\&-\upsilon\left(\in_{33,r}+\frac{\in_{33}}{r}\right)\phi_{,r}+\gamma_1 T_0\theta_{,r}+\left(\gamma_{1,r}+\frac{\gamma_1}{r}\right)T_0\theta=0\end{aligned}\tag{4.42b}$$

$$\begin{aligned}&-\left(1+\tau\frac{\partial}{\partial t}\right)\left(\rho c\theta_{,t}+\beta_1 u_{,rt}+\frac{\beta_3}{r}u_{,t}-\upsilon\gamma_1\phi_{,rt}\right)\\&+\eta^{*}\left(K_{11,r}+\frac{K_{11}}{r}\right)\theta_{,r}+\eta^{*}K_{11}\theta_{,rr}=0\end{aligned}\tag{4.42c}$$

where $\upsilon=\dfrac{e_{33a}bc^{*}\eta^{*}}{\in_{33a}}$ and overbars are once again omitted for convenience.

4.3.3 Solution Procedure

Equation (4.42) is a system of linear partial differential equations with variable coefficients. To solve this system, we will use the steps previously mentioned in the introduction to this section. Details of the solution procedure are presented in the following sections.

4.3.3.1 Solution in Laplace Domain

The Laplace transform and its inversion are defined as follows:

$$\begin{aligned} f(s) &= L[f(t)] = \int_0^{\infty} e^{-st} f(t)dt \\ f(t) &= L^{-1}[f(s)] = \frac{1}{2\pi i}\int_{\alpha-i\infty}^{\alpha+i\infty} e^{st} f(s)ds \end{aligned} \tag{4.43}$$

where α is an arbitrary real number greater than all real parts of the singularities of $f(s)$, s is the Laplace parameter, a tilde denotes a transformed function and i represents the imaginary unit. After applying the Laplace the transform to Eq. (4.42), and considering zero initial conditions, the governing equations in the Laplace domain become:

$$\begin{aligned} &c_{33}u_{,rr} + \left(c_{33,r} + \frac{c_{33}}{r}\right)u_{,r} + \frac{1}{r}\left(c_{13,r} - \frac{c_{11}}{r} - r\rho c^{*2}s^2\right)u + \upsilon e_{33}\phi_{,rr} \\ &+ \upsilon\left(e_{33,r} + \frac{1}{r}(e_{33} - e_{31})\right)\phi_{,r} - \beta_1 T_0\theta_{,r} + \left(-\beta_{1,r} + \frac{1}{r}(\beta_3 - \beta_1)\right)T_0\theta = 0 \end{aligned} \tag{4.44a}$$

$$\begin{aligned} &e_{33}u_{,rr} + \left(e_{33,r} + \frac{1}{r}(e_{31} + e_{33})\right)u_{,r} + \frac{e_{31,r}}{r}u - \upsilon \in_{33} \phi_{,rr} \\ &- \upsilon\left(\in_{33,r} + \frac{\in_{33}}{r}\right)\phi_{,r} + \gamma_1 T_0\theta_{,r} + \left(\gamma_{1,r} + \frac{\gamma_1}{r}\right)T_0\theta = 0 \end{aligned} \tag{4.44b}$$

$$-\imath\beta_1 u_{,r} - \frac{\imath\beta_3}{r}u + \imath\upsilon\gamma_1\phi_{,r} + \eta^* K_{11}\theta_{,rr} + \eta^*\left(K_{11,r} + \frac{K_{11}}{r}\right)\theta_{,r} - \imath\rho c\theta = 0 \tag{4.44c}$$

where $\imath = (1 + \tau s)s$. In Eq. (4.44), tildes are omitted for convenience.

4.3.3.2 Galerkin Finite Element Method

We now make use of a finite element method to solve Eq. (4.44). In order to discretize the governing equations, linear elements and shape functions are incorporated in the analysis method. The shape of the elements, the local coordinates and the shape functions are shown in Fig. 4.10 [17].

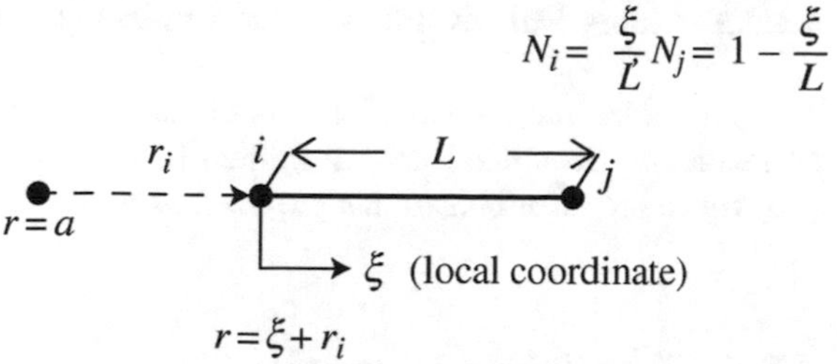

Fig. 4.10 The element, local coordinate and shape functions used in the current section [17]

The unknowns of the problem (displacement, electric potential, and temperature change) can then be approximated over the defined linear elements as follows:

$$(u, \phi, \theta)^{(e)} = N_k\left(u_k, \phi_{,k}, \theta_{,k}\right); k = i, j \tag{4.45}$$

Now, we apply the Galerkin method [22] to Eq. (4.44):

$$\int_0^L \left(\begin{array}{l} c_{33} u_{,\xi\xi} + \left(c_{33,\xi} + \frac{c_{33}}{r_i+\xi}\right) u_{,\xi} \\ + \frac{1}{r_i+\xi}\left(c_{13,\xi} - \frac{c_{11}}{r_i+\xi} - (r_i+\xi)\rho c^{*2} s^2\right) u + \upsilon e_{33} \phi_{,\xi\xi} \\ + \upsilon\left(e_{33,\xi} + \frac{1}{r_i+\xi}(e_{33} - e_{31})\right) \phi_{,\xi} - \beta_1 T_0 \theta_{,\xi} \\ + \left(-\beta_{1,\xi} + \frac{1}{r_i+\xi}(\beta_3 - \beta_1)\right) T_0 \theta \end{array} \right) N_k d\xi = 0 \tag{4.46a}$$

$$\int_0^L \left(\begin{array}{l} e_{33} u_{,\xi\xi} + \left(e_{33,\xi} + \frac{1}{r_i+\xi}(e_{31} + e_{33})\right) u_{,\xi} + \frac{e_{31,\xi}}{r_i+\xi} u - \upsilon \in_{33} \phi_{,\xi\xi} \\ -\upsilon\left(\in_{33,\xi} + \frac{\in_{33}}{r_i+\xi}\right) \phi_{,\xi} + \gamma_1 T_0 \theta_{,\xi} + \left(\gamma_{1,\xi} + \frac{\gamma_1}{r_i+\xi}\right) T_0 \theta \end{array} \right) N_k d\xi = 0 \tag{4.46b}$$

$$\int_0^L \left(-\iota \beta_1 u_{,\xi} - \iota \frac{\beta_3}{r_i+\xi} u + \iota \upsilon \gamma_1 \phi_{,\xi} + \eta^* \left(K_{11,\xi} + \frac{K_{11}}{r_i+\xi} \right) \theta_{,\xi} - \iota \rho c \theta \right) N_k d\xi = 0, k = i, j \tag{4.46c}$$

To decrease the order of the second order derivatives and obtain boundary conditions (the stresses on the boundaries), we now perform integration by parts on all of the second order derivatives in Eq. (4.46). After the integrations and sufficient manipulation, element matrices are found. These matriccs are then assembled into a system of linear algebraic equations which contain the variables at each node. Finally, all the unknowns can be obtained.

4.3.3.3 Numerical Inversion of the Laplace Transform

The procedure for numerical inversion used in this section is identical to that performed in Sect. 4.2.3 and is omitted here for brevity. The same discretization and truncation error constraints are used as well.

4.3.4 Results and Discussion

In this section, we analyze the effects of both the non-homogeneity index and thermal relaxation time on the response of the cylinder to thermal shock. All boundary conditions shown in Fig. 4.8 are assumed to be zero except the non-dimensional temperature at the inner surface of the cylinder, which is defined as $\theta_a = 10^4 t e^{-1000t}$ (t is non-dimensional). The non-dimensional inner and outer radii of the cylinder are taken to be 0.01 and 0.02, respectively [17]. The inner surface of the cylinder is made of Lead Zirconate Titanate while the outer surface is Cadmium Selenide. Their properties are listed in Table 4.1.

Figure 4.11a–c depicts varying non-homogeneity indices and the effects of this variation on the distribution of the displacement, electric potential, and temperature change respectively, with $t = 0.0011$ and $\tau = 0.05$. In this section, the non-homogeneity indices are assumed to be identical for every varying material

Table 4.1 Material properties of the inner and outer surfaces of the cylinder [17, 23–25]

Properties	Lead Zirconate Titanate (inner surface)	Cadmium Selenide (outer surface)
c_{33} (GPa)	117	83.6
c_{13} (GPa)	53	39.3
c_{11} (GPa)	126	74.1
$e_{33}\left(\frac{C}{m^2}\right)$	23.3	0.347
$e_{31}\left(\frac{C}{m^2}\right)$	−6.5	−0.16
$\beta_1 \times 10^6\left(\frac{N}{Km^2}\right)$	1.41	0.551
$\beta_3 \times 10^6\left(\frac{N}{Km^2}\right)$	1.97	0.621
$\epsilon_{33} \times 10^{-11}\left(\frac{C^2}{Nm^2}\right)$	1300	9.03
$\gamma_1 \times 10^{-6}\left(\frac{C}{Km^2}\right)$	−5.48	−2.94
$\rho \times 10^3\left(\frac{kg}{m^3}\right)$	7.87	5.68
$c \times 10^3\left(\frac{J}{kg\ K}\right)$	0.33	0.46
$K_{11}\left(\frac{W}{mK}\right)$	50	12.9

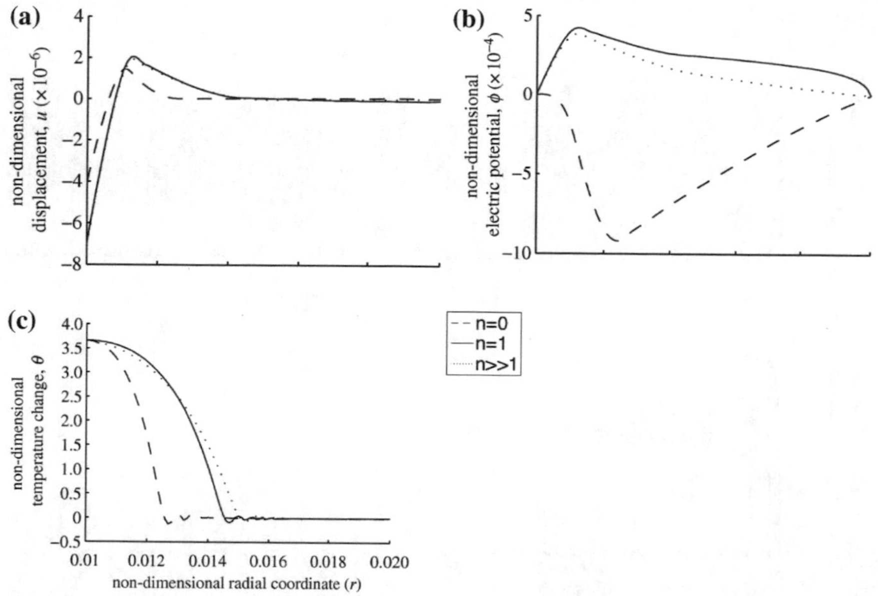

Fig. 4.11 The effect of different non-homogeneity index on the distribution of: **a** the displacement, **b** the electric potential, **c** temperature change, $t = 0.0011$, $\tau = 0.05$. Dashed line: $n = 0$ (cadmium selenide), continuous line: $n = 1$, dotted line: $n \gg 1$ (lead zirconate titanate). [Reproduced from [17] with permission from The Royal Society]

property, i.e. $n_\chi = n$. Three different non-homogeneity indices were used in the calculations, namely, $n = 0$ (Cadmium Selenide), $n = 1$ (linear variation), and $n \gg 1$ (Lead Zirconate Titanate).

The non-zero thermal relaxation time in Fig. 4.11 leads to non-Fourier heat conduction and consequently the presence of thermal wavefronts in the distributions. These wavefronts travel along the thickness of the cylinder from inner surface to outer surface, acting as messengers of the thermal shock. Intuitively, there exist undisturbed portions of the rod, located ahead of the wavefront, where the solutions remain at their initial values. As can be seen in Fig. 4.11a, c, an increase in n shortens the length of the unresponsive portion of the cylinder, although in Fig. 4.11b no wavefront is present. These contractions can be attributed to the fact that the velocity of a thermal wave, according to hyperbolic heat conduction, is equal to $\sqrt{\frac{K}{\rho c \tau}}$. When n is increased, the average of this term increases as well. It can also be noted that the change in the undisturbed portion of the displacement solution between the two different material indices in Fig. 4.11a is almost insignificant.

In Fig. 4.12a–c, the time history of the solutions at the midpoint of the cylinder are plotted for the same non-homogeneity indices seen in Fig. 4.11. It can be seen in Fig. 4.12a that the smallest amplitude of displacement occurs when $n = 0$, while

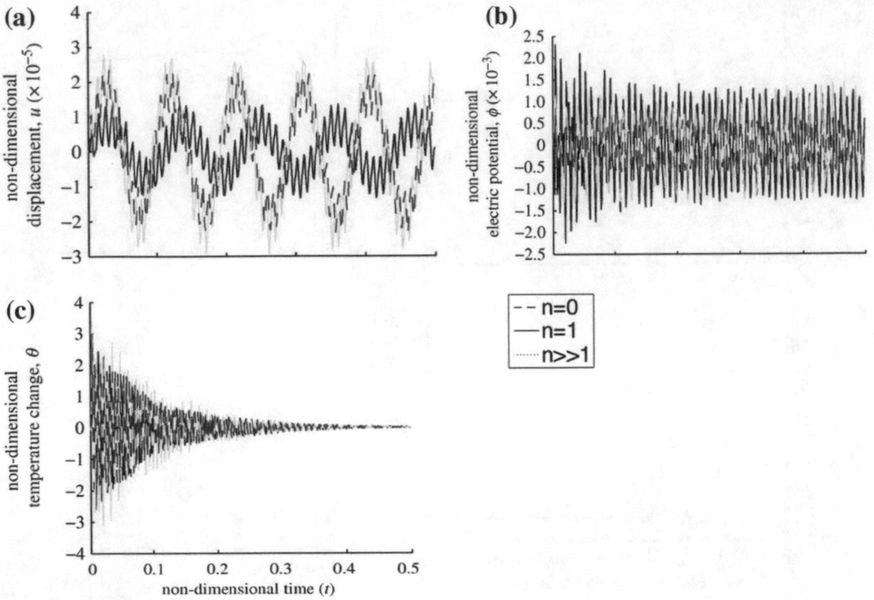

Fig. 4.12 The effect of the non-homogeneity index on the history of: **a** displacement, **b** electric potential, **c** temperature change, Middle point, $\tau = 0.05$. Dashed line: $n = 0$ (cadmium selenide), continuous line: $n = 1$, dotted line: $n \gg 1$ (lead zirconate titanate). [Reproduced from [17] with permission from The Royal Society]

this same index results in the largest amplitude of electric potential as seen in Fig. 4.12b. Furthermore, both the displacement and temperature change exhibit the largest fluctuations when $n = 1$, as depicted in Fig. 4.12a, c, respectively.

In the following two sets of figures, a non-homogeneity index of 0 is selected for analysis. Figure 4.13a–c depicts the effects of thermal relaxation time, τ, on the results. By definition, when $\tau = 0$ there exists no wavefront in the distributions of the solutions. As such, the thermal disturbance spreads throughout the cylinder's thickness immediately after the inner side is subjected to the thermal shock and no portion of the cylinder is undisturbed, as shown in Fig. 4.13a, c. In these same figures, it can also be noted that a larger thermal relaxation time leads to an expanded unaffected portion of the cylinder where the variables remain at their initial values.

Figure 4.14a–d displays the time histories of the solutions under conditions of varying thermal relaxation time. It can be seen that the fluctuations of all the variables are smallest when $\tau = 0$. In Fig. 4.14a, the amplitude of the displacement increases in a monotonic fashion as the thermal relaxation time becomes larger. Moreover, by increasing the thermal relaxation time, the fluctuations of temperature are maintained for a longer time period such that the time needed for the temperature difference of the cylinder to return to zero is extended. The time history of radial stress is shown in Fig. 4.14d for interest.

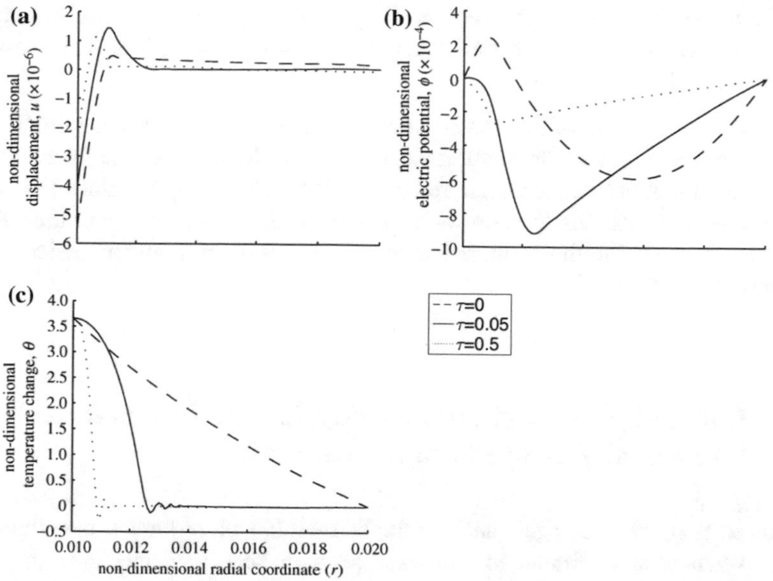

Fig. 4.13 The effect of the thermal relaxation time on the distribution of: **a** displacement, **b** electric potential, **c** temperature change, $t = 0.0011$, $n = 0$. Dashed line: $\tau = 0$, continuous line: $\tau = 0.05$, dotted line: $\tau = 0.5$. [Reproduced from [17] with permission from The Royal Society]

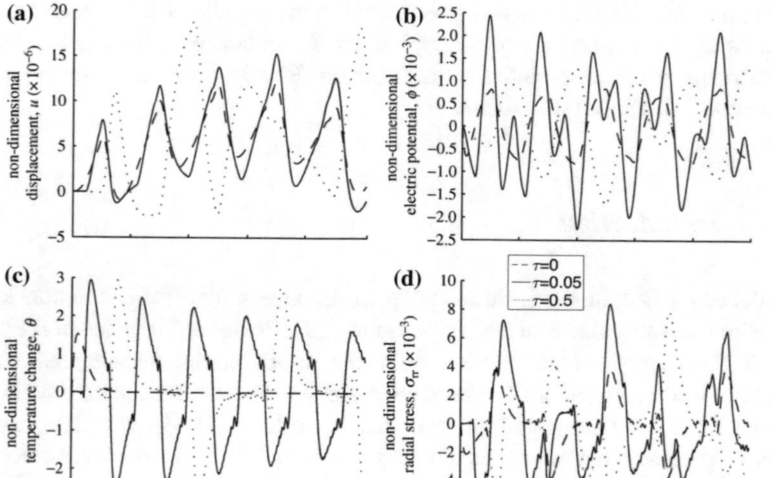

Fig. 4.14 The effect of the thermal relaxation time on the history of: **a** displacement, **b** electric potential, **c** temperature change, **d** radial stress, Middle point, $n = 0$. Dashed line: $\tau = 0$, continuous line: $\tau = 0.05$, dotted line: $\tau = 0.5$. [Reproduced from [17] with permission from The Royal Society]

The figures presented in this section all provide insight into the effects of varying both non-homogeneity and thermal relaxation time. In general, it is vital to note the following [17]:

1. Of the three values of the non-homogeneity index studied, the amplitude of the displacement reaches its smallest value at $n = 0$. In addition, the fluctuations of the displacement and temperature change reach their largest values at $n = 1$.
2. As thermal relaxation time grows larger, the undisturbed regions of the cylinder grow in the radial direction and when $\tau = 0$, all distributions experience minimal fluctuations.

4.4 Coupled Thermopiezoelectricity in One-Dimensional Functionally Graded Smart Materials

In this section, we once again analyze the behaviour of a rod under the effects of a moving heat source, similar to what was seen in Sect. 4.2, only now the rod is assumed to be functionally graded [3]. As such, its material properties, except specific and thermal relaxation time, are assumed to vary exponentially throughout its length. In this particular section, the governing equations of displacement, temperature and electric potential are stated in a general form than incorporates both coupled and uncoupled thermoelasticity theories. Within the coupled formulation, we can find that both classic and generalized thermoelasticity are considered. Once again using the Laplace transform, solutions are obtained in the Laplace domain and subsequently inverted into the time domain. Finally, a numerical example helps to illustrate the analyzed problem.

4.4.1 Introduction

Introduced by Biot in 1956, classic coupled thermoelasticity helped to consider the effects of the elastic terms in the heat equation [9]. However, since the heat equation used in this theory is parabolic in nature, the results predict an unrealistic infinite speed for heat propagation. Consequently, generalized theories were developed to account for the second sound effect of the wave [26, 27]. For instance, Lord and Shulman presented a new heat conduction law including a so-called relaxation time, accounting for the time required for acceleration of heat flow [10]. Following this, Green and Lindsay introduced another interpretation of generalized thermoelasticity using two relaxation times in the constitutive equations of stress and entropy [28].

In the past, many authors have produced works on the different theories of thermoelasticity. Tzou investigated the thermodynamic and mechanic nature of relaxation time in 1993 [29]. He found that this phenomenon can be interpreted as the phase-lag between the heat flux vector and the temperature gradient. In 2001, the thermally induced displacement of a rod was analyzed by Al-Huniti et al. based on the hyperbolic heat conduction model [30]. Using the Lord-Shulman theory of generalized thermoelasticity, a one-dimensional piezoelectric rod exposed to a sudden heat source was studied by He et al. The Laplace transform and the state-space approach allowed the solving of the governing partial differential equations [31]. In addition, Aouadi solved the problem of a coupled, two-dimensional, thermopiezoelectric, thick infinite plate by making use of the hybrid Laplace transform-finite element method. By considering both generalized and classical coupled thermoelasticity, the author was able to quantify the wave-like heat propagation in the plate [32].

Although previously investigated by Babaei and Chen [33], this section seeks a solution to the problem of a one-dimensional FGPM based on the Lord-Shulman theory of thermoelasticity. In particular, a more general formulation is sought out for understanding the coupled and uncoupled behaviour of the functionally graded piezoelectric medium at hand [3]. In the following section, the governing equations of the problem are given in general forms which include the generalized coupled (Lord-Shulman and Green-Lindsay), classical coupled, and classical uncoupled thermoelasticity theories. Solutions for three unknown fields, displacement, temperature and electric potential, are found in the Laplace domain for coupled thermoelasticity through the successive decoupling method. Then, a more straightforward approach is used for the analysis of the uncoupled theory. Finally, a numerical example provides results that are plotted for different non-homogeneity indices and for the different theories.

4.4.2 The Functionally Graded Rod Problem

Let us now consider a FGPM rod of length L that is aligned with the x-axis of a one-dimensional coordinate system as shown in Fig. 4.15. The left end of the rod is located at the origin. The rod's ends are fixed in space, thermally insulated and have zero voltage. A mobile heat source is located at the rod's left end at $t = 0$ and travels towards the right end at a constant speed υ.

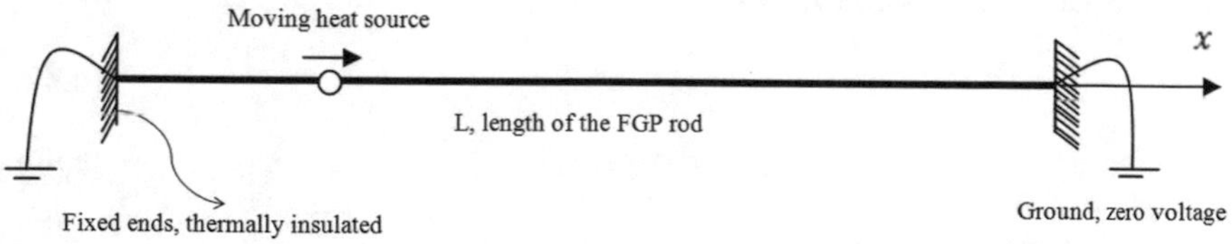

Fig. 4.15 FGP rod subjected to a moving heat source. [Reproduced from [3] with permission from World Scientific Publishing Co., Inc.]

4.4.2.1 Fundamental and Governing Equations

The governing equations are written below in a general form for the analysis according to the coupled and uncoupled thermoelastic theories of a piezoelectric rod [26, 34, 35].

$$\sigma_{ij} = c_{ijkl}\varepsilon_{kl} - \beta_{ij}(\theta + \nu\theta) - e_{ijk}E_k \tag{4.47a}$$

$$D_i = e_{ijk}\varepsilon_{jk} + p_i(\theta + \nu\theta) + \in_{ij} E_j \tag{4.47b}$$

$$\left(1 + n_0\tau\frac{\partial}{\partial t}\right)q_i = -K_{ij}\theta_{,j} \tag{4.47c}$$

The variables and constants in these equations have previously been defined in Sect. 4.2.2 and are omitted here. New additions include a constant and two relaxation times, represented by n_0, τ, and ν, respectively. Given Eq. (4.47), it is possible to obtain equations for the following thermoelasticity theories:

1. Classical coupled thermoelasticity, i.e. when $\tau = \nu = n_0 = 0$.
2. Lord-Shulman (L-S) generalized theory with one relaxation time, i.e. when $\nu = 0$, $n_0 = 1$, $\tau > 0$.
3. Green-Lindsay (G-L) generalized theory with two relaxation times, i.e. when $n_0 = 0$, $\nu > \tau > 0$.

For the classical coupled and L-S theories we can write:

$$\rho S = \beta_{ij}\varepsilon_{ij} + \frac{\rho C}{\theta_0}\theta + p_i E_i \tag{4.48a}$$

while, for the G-L theory, the entropy equation is given as:

$$\rho S = \beta\varepsilon_{ij} + \frac{\rho C}{\theta_0}\left(\theta + \tau\dot{\theta}\right) + p_i E_i \tag{4.48b}$$

where S, ρ, and C are, respectively, entropy per unit mass, density, and the specific heat. For thermopiezoelectricity in the absence of body force and volume charges, the set of governing equations, i.e. the equation of motion (a), Gaussian law (b), and energy equation (c), is as follows:

$$\sigma_{ij,j} = \rho\ddot{u}_i \tag{4.49a}$$

$$D_{i,i} = 0 \tag{4.49b}$$

$$\rho\left(\dot{S}\theta_0 - R\right) + q_{i,i} = 0 \tag{4.49c}$$

in which, a superposed dot represents a time derivative, while a comma subscript denotes partial differentiation with respect to the space variable x. In addition, the quasi-stationary electric field equation (a) and the linear strain-displacement relationship (b) are:

$$E_i = -\phi_{,i} \tag{4.50a}$$

$$\varepsilon_{ij} = \frac{1}{2}\left(u_{i,j} + u_{j,i}\right) \tag{4.50b}$$

where ϕ is the electric potential. For the current problem, the constitutive and governing equations depend purely on x and t. Knowing this, using Eq. (4.50) we can simplify Eqs. (4.47) and (4.49):

$$\sigma_{xx} = cu_{,x} - \beta(\theta + v\theta) - eE_1 \tag{4.51a}$$

$$D_x = eu_{,x} + p(\theta + v\theta) + \in E_1 \tag{4.51b}$$

$$\left(1 + n_0\tau\frac{\partial}{\partial t}\right)q_x = -K\theta_{,x} \tag{4.51c}$$

$$\sigma_{xx,x} = \rho\ddot{u} \tag{4.51d}$$

$$D_{x,x} = 0 \tag{4.51e}$$

$$\rho\left(\dot{S}\theta_0 - R\right) + q_{x,x} = 0 \tag{4.51f}$$

where u is the displacement in the x direction and the subscripts of material properties have been omitted for convenience. Once again, the rod is non-homogenous and all material properties are assumed to be functions of x, varying along the rod's length. The specific heat and the relaxation times are assumed constant for simplicity. Using Eqs. (4.48a), (4.48b) and (4.51), the governing differential equations for the aforementioned theories can be written as follows:

$$c_{,x}u_{,x} + cu_{,xx} - \rho u - \beta_{,x}\theta - \beta\theta_{,x} - \beta_{,x}v\theta - \beta v\theta_{,x} + e_{,x}\phi_{,x} + e\phi_{,xx} = 0 \tag{4.52a}$$

$$e_{,x}u_{,x} + eu_{,xx} + p_{,x}\theta + p\theta_{,x} + vp_{,x}\theta + vp\theta_{,x} - \varepsilon_{,x}\phi_{,x} - \varepsilon\phi_{,xx} = 0 \tag{4.52b}$$

$$\begin{aligned} &K_{,x}\theta_{,x} + K\theta_{,xx} + \rho(R + n_0\tau R) - \rho C(\theta + n_0\tau\theta) \\ &- \theta_0\beta\left(u_{,x} + n_0\tau u_{,x}\right) + \theta_0 p\left(\phi_{,x} + n_0\tau\phi_{,x}\right) = 0 \end{aligned} \tag{4.52c}$$

$$\begin{aligned} &K_{,x}\theta_{,x} + K\theta_{,xx} + \rho\left(R + n_0\tau\dot{R}\right) - \rho C\left(\dot{\theta} + \tau\ddot{\theta}\right) - n_0\tau\rho C\left(\ddot{\theta} + \tau\dddot{\theta}\right) \\ &- \theta_0\beta\left(\dot{u}_{,x} + n_0\tau\ddot{u}_{,x}\right) + \theta_0 p\left(\dot{\phi}_{,x} + n_0\tau\ddot{\phi}_{,x}\right) = 0 \end{aligned} \tag{4.52d}$$

The following non-dimensional parameters are introduced to streamline the solution procedure:

$$\eta^* = \frac{\rho_0 C}{K_0}, c^* = \sqrt{\frac{c_0}{\rho_0}}, (\bar{x}, \bar{u}) = c^*\eta^*(x, u), (\bar{t}, \bar{\tau}, \bar{v}) = c^{*2}\eta^*(t, \tau, v)$$
$$\bar{\theta} = \frac{\theta}{\theta_0}, \bar{\phi} = \frac{\in_0}{e_0 L}\phi, \bar{R} = \frac{R}{K_0\theta_0 c^{*2}\eta^{*2}}, \bar{\sigma}_{xx} = \frac{\sigma_{xx}}{c_0}, \bar{D}_x = \frac{D_x}{e_0}, l = Lc^*\eta^* \tag{4.53}$$

in which, K_0, c_0, ρ_0, e_0 and $\in_0$ are the thermal conductivity, elastic coefficients, density, piezoelectric and dielectric coefficients at $x = 0$, respectively. The material properties of the rod vary exponentially along the x-axis, following the relation shown below:

$$\chi = \chi_0 e^{\lambda x} \tag{4.54}$$

where χ is an arbitrary material property, and λ is an arbitrary non-homogeneity index. Finally, the moving thermal disturbance is defined in the following form:

$$\bar{R} = R_0\delta(\bar{x} - \upsilon\bar{t}) \tag{4.55}$$

where R_0 and υ are the intensity and velocity of the heat source while δ is the Dirac delta function.

4.4.3 Solution Procedures

4.4.3.1 Solution in Laplace Domain

Anytime a solution is obtained in the Laplace domain, a numerical Laplace inversion must be used to obtain the final solution in the time domain. The Laplace transform, defined in Sect. 4.3.3, is also used in this section. The initial conditions for the displacement, electric potential, and temperature change are assumed to be zero [3]. The non-dimensional governing equations in the Laplace domain are written below:

$$u_{,xx} + a_1 u_{,x} - s^2 u + a_2 c^*\eta^*\phi_{,xx} + a_1 a_2 c^*\eta^*\phi_{,x} - a_3 a_8\theta_{,x} - a_1 a_3 a_8\theta = 0 \tag{4.56a}$$

$$u_{,xx} + a_1 u_{,x} - Lc^*\eta^*\phi_{,xx} - a_1 Lc^*\eta^*\phi_{,x} + a_4 a_8\theta_{,x} + a_1 a_4 a_8\theta = 0 \tag{4.56b}$$

$$a_9\theta_{,xx} + a_1 a_9\theta_{,x} - a_{10}\theta - a_6 a_{10} u_{,x} + a_7 a_{10}\phi_{,x} = -\frac{a_5 a_{10}}{s} e^{-\frac{s}{\upsilon}x} \tag{4.56c}$$

$$a_9\theta_{,xx} + a_1 a_9 \theta_{,x} - a_{11}\theta - a_6 a_{10} u_{,x} + a_7 a_{10}\phi_{,x} = -\frac{a_5 a_{10}}{s} e^{-\frac{s}{v}x} \tag{4.56d}$$

where

$$\begin{aligned} a_1 &= \frac{\lambda}{c^*\eta^*}, a_2 = \frac{e_0^2 L}{\in_0 c_0}, a_3 = \frac{\beta_0\theta_0}{c_0}, a_4 = \frac{p_0\theta_0}{e_0}, a_5 = \frac{\rho_0 R_0}{v}, a_6 = \frac{\beta_0}{\rho C} \\ a_7 &= \frac{p_0 c^* e_0 L}{\in_0 K_0}, a_8 = (1+sv), a_9 = 1, a_{10} = s(1+n_0\tau s) \\ a_{11} &= s\left(1 + s\tau(1+n_0) + s^2\tau^2 n_0\right) \end{aligned} \tag{4.57}$$

For convenience, overbar and tilde signs have been omitted in the above equations.

4.4.3.2 Coupled Thermopiezoelectricity Analysis

In this section, the coupled thermopiezoelectrical response of the FGPM rod is analyzed based on the classical and generalized L-S theories. As such, in Eq. (4.56), only the energy equation based on the L-S theory is considered. The solution of the linear ordinary differential equations of Eq. (4.56) contains two components; the particular solution and the general solutions. The former can be written in the following format:

$$\left(u_p, \phi_p, \theta_p\right) = \left(P_u, P_\phi, P_\theta\right) e^{-\frac{s}{v}x} \tag{4.58}$$

Here, subscript p is used to denote the particular solution. By substituting Eq. (4.58) into Eq. (4.56), we can solve the algebraic equation and obtain:

$$\begin{bmatrix} \left(\frac{s}{v}\right)^2 - a_1\frac{s}{v} - s^2 & I\left(\left(\frac{s}{v}\right)^2 - a_1\frac{s}{v}\right) & -a_3 a_8\left(-\frac{s}{v} + a_1\right) \\ \left(\frac{s}{v}\right)^2 - a_1\frac{s}{v} & -J\left(\left(\frac{s}{v}\right)^2 - a_1\frac{s}{v}\right) & a_4\left(-a_8\frac{s}{v} + a_1 a_8\right) \\ a_6 a_{10}\frac{s}{v} & -a_7 a_{10}\frac{s}{v} & a_9\left(\left(\frac{s}{v}\right)^2 - a_1\frac{s}{v}\right) - a_{10} \end{bmatrix} \begin{Bmatrix} P_u \\ P_\phi \\ P_\theta \end{Bmatrix} = \begin{Bmatrix} 0 \\ 0 \\ -\frac{a_5 a_{10}}{s} \end{Bmatrix} \tag{4.59}$$

where, $I = a_2 c^*\eta^*$ and $J = Lc^*\eta^*$. To obtain the general solution, we successively eliminate ϕ and u in the governing equations, which results in an ordinary differential equation containing θ. Then we find $\theta_{g,x}$ from the third equation of the homogenous form of Eq. (4.56):

$$\phi_{g,x} = -\frac{a_9}{a_7 a_{10}}\theta_{g,xx} - \frac{a_1 a_9}{a_7 a_{10}}\theta_{g,x} + \frac{1}{a_7}\theta_g + \frac{a_6}{a_7} u_{g,x} \tag{4.60}$$

where subscript g is used to indicate the general solution. Using Eq. (4.60), the governing equations are reduced to:

$$A_1 u_{g,xx} + a_1 A_1 u_{g,x} + A_2 \theta_{g,xxx} + 2a_1 A_2 \theta_{g,xx} + A_3 \theta_{g,x} + a_1 A_4 \theta_g = 0 \quad (4.61a)$$

$$B_1 u_{g,xx} + a_1 B_1 u_{g,x} - s^2 u_g - B_2 \theta_{g,xxx} - 2a_1 B_2 \theta_{g,xx} + B_3 \theta_{g,x} + a_1 B_4 \theta_g = 0 \quad (4.61b)$$

where

$$A_1 = 1 - J\frac{a_6}{a_7}, A_2 = \frac{Ja_9}{a_7 a_{10}}, A_3 = \frac{J}{a_7}\left(\frac{a_1^2 a_9}{a_{10}} - 1\right) + a_4 a_8, A_4 = -\frac{J}{a_7} + a_4 a_8$$
$$B_1 = 1 + \frac{Ia_6}{a_7}, B_2 = \frac{Ia_9}{a_7 a_{10}}, B_3 = \frac{I}{a_7}\left(1 - \frac{a_1^2 a_9}{a_7 a_{10}}\right) - a_3 a_8, B_4 = \frac{I}{a_7} - a_3 a_8 \quad (4.62)$$

To obtain the following equation for u_g, we can multiply Eq. (4.61b) by $-\frac{A_1}{B_1}$ and add this result with (4.61a):

$$u_g = D_1 \theta_{g,xxx} + 2a_1 D_1 \theta_{g,x} + D_2 \theta_{g,x} + a_1 D_3 \theta_g \quad (4.63)$$

in which,

$$D_1 = -\frac{1}{s^2}\left(B_2 + \frac{B_1 A_2}{A_1}\right), D_2 = \frac{1}{s^2}\left(B_3 - \frac{B_1 A_3}{A_1}\right), D_3 = \frac{1}{s^2}\left(B_4 - \frac{B_1 A_4}{A_1}\right) \quad (4.64)$$

Substituting Eq. (4.63) and its derivatives into Eq. (4.61a) leads to the ordinary differential equation with constant coefficients below:

$$E_1 \theta_{g,xxxxx} + 3a_1 E_1 \theta_{g,xxxx} + E_2 \theta_{g,xxx} + a_1 E_3 \theta_{g,xx} + E_4 \theta_{g,x} + a_1 A_4 \theta_g = 0 \quad (4.65)$$

where

$$E_1 = A_1 D_1, E_2 = A_1\left(D_2 + 2a_1^2 D_1\right) + A_2, E_3 = A_1(D_2 + D_3) + 2A_2, E_4 = a_1^2 A_1 D_3 + A_3 \quad (4.66)$$

Solving Eq. (4.65) allows us to obtain the following characteristic equation:

$$E_1 \zeta^5 + 3a_1 E_1 \zeta^4 + E_2 \zeta^3 + a_1 E_3 \zeta^2 + E_4 \zeta + a_1 A_4 = 0 \quad (4.67)$$

Analytical methods have been proposed for solving quintic equations, such as using the Hermit-Kronecker method and the Mellin method [36, 37]. However, it can be noted that one of the characteristic roots of Eq. (4.67) is $\zeta_1 = -a_1$, and as such we can factor this term from Eq. (4.67). Therefore, it is possible to analytically solve the resulting fourth order algebraic equation [38] and the solution for the temperature can be written as follows:

$$\theta(x,s) = \theta_g + \theta_p = C_{\theta_i} e^{\zeta_i x} + P_\theta e^{-\frac{s}{v}x} \tag{4.68}$$

Substituting Eq. (4.68) into Eqs. (4.63) and (4.60) leads to the following equations for displacement and electric potential:

$$u(x,s) = C_{u_i} C_{\theta_i} e^{\zeta_i x} + P_u e^{-\frac{s}{v}x} \tag{4.69a}$$

$$\phi(x,s) = C_{\phi_i} C_{\theta_i} e^{\zeta_i x} + C' + P_\phi e^{-\frac{s}{v}x} \tag{4.69b}$$

where C_{θ_i} and C' are integration constants. C_{u_i} and C_{ϕ_i} are defined in the following forms:

$$C_{u_i} = D_1 \zeta_i^3 + 2a_1 D_1 \zeta_i^2 + D_2 \zeta_i + a_1 D_3 \tag{4.70a}$$

$$C_{\phi_i} = \frac{1}{a_7}\left(-\frac{a_9}{a_{10}}\zeta_i - \frac{a_1 a_9}{a_{10}} + \frac{1}{\zeta_i} + a_6 C_{u_i}\right) \tag{4.70b}$$

Applying the boundary condition $\left(u, \phi, \frac{\partial\theta}{\partial x}\right)\big|_{x=0,l} = (0,0,0)$ on the system results in the following system of equations which can be solved for the integration constants:

$$\begin{pmatrix} C_{u_1} & C_{u_2} & C_{u_3} & C_{u_4} & C_{u_5} & C_{u_6} \\ C_{u_1} e^{\zeta_1 l} & C_{u_2} e^{\zeta_2 l} & C_{u_3} e^{\zeta_3 l} & C_{u_4} e^{\zeta_4 l} & C_{u_5} e^{\zeta_5 l} & C_{u_6} e^{\zeta_6 l} \\ C_{\phi_1} & C_{\phi_2} & C_{\phi_3} & C_{\phi_4} & C_{\phi_5} & C_{\phi_6} \\ C_{\phi_1} e^{\zeta_1 l} & C_{\phi_2} e^{\zeta_2 l} & C_{\phi_3} e^{\zeta_3 l} & C_{\phi_4} e^{\zeta_4 l} & C_{\phi_5} e^{\zeta_5 l} & C_{\phi_6} e^{\zeta_6 l} \\ \zeta_1 & \zeta_2 & \zeta_3 & \zeta_4 & \zeta_5 & \zeta_6 \\ \zeta_1 e^{\zeta_1 l} & \zeta_2 e^{\zeta_2 l} & \zeta_3 e^{\zeta_3 l} & \zeta_4 e^{\zeta_4 l} & \zeta_5 e^{\zeta_5 l} & \zeta_6 e^{\zeta_6 l} \end{pmatrix} \begin{Bmatrix} C_{\theta_1} \\ C_{\theta_2} \\ C_{\theta_3} \\ C_{\theta_4} \\ C_{\theta_5} \\ C' \end{Bmatrix} = \begin{Bmatrix} -P_u \\ -P_u e^{-\frac{s}{v}l} \\ -P_\phi \\ -P_\phi e^{-\frac{s}{v}l} \\ P_\theta \frac{s}{v} \\ P_\theta \frac{s}{v} e^{-\frac{s}{v}l} \end{Bmatrix} \tag{4.71}$$

Moreover, the normalized stress and electric displacement can be found in the Laplace domain as shown below:

$$\sigma_x = \left\{ \begin{matrix} C_{\theta_i}\left[\zeta_i\left(C_{u_i} + \frac{a_2 J}{L} C_{\phi_i}\right) - a_3(1+vs)\right] e^{\zeta_i x} \\ -\left[\frac{s}{v}\left(P_u + \frac{a_2 J}{L} P_\phi\right) + a_3(1+vs) P_\theta\right] \end{matrix} \right\} e^{a_1 x} \tag{4.72a}$$

$$D_1 = \left\{ C_{\theta_i}\left[\zeta_i\left(C_{u_i} - J C_{\phi_i}\right) + a_4(1+vs)\right] e^{\zeta_i x} - \left[\frac{s}{v}\left(P_u - J P_\phi\right) - a_4(1+vs) P_\theta\right] \right\} e^{a_1 x} \tag{4.72b}$$

In the case where the strain rate or the time rate of change of thermal sources is relatively low, the displacement effects are ignored in the energy equation. Then we can solve the governing equations with a similar approach by using the following values:

$$a_6 = 0, a_8 = 1, a_9 = 1, a_{10} = s, a_{11} = s \tag{4.73}$$

4.4.3.3 Uncoupled Thermopiezoelectricity Analysis

In this section, the FGPM rod is studied on the basis of classical uncoupled thermopiezoelectricity. The classical uncoupled theory does not take into account the coupling effect of strain and electric potential on temperature. Despite this flaw, it remains accurate enough to successfully model many engineering applications, especially if the rate of strain and electric field and relatively small [39]. When this occurs, the governing equations from Eq. (4.56) are simplified as follows:

$$u_{,xx} + a_1 u_{,x} - s^2 u + a_2 c^* \eta^* \phi_{,xx} + a_1 a_2 c^* \eta^* \phi_{,x} - a_3 \theta_{,x} - a_1 a_3 \theta = 0 \tag{4.74a}$$

$$u_{,xx} + a_1 u_{,x} - Lc^* \eta^* \phi_{,xx} - a_1 L c^* \eta^* \phi_{,x} + a_4 \theta_{,x} + a_1 a_4 \theta = 0 \tag{4.74b}$$

$$\theta_{,xx} + a_1 \theta_{,x} - s\theta = -a_5 e^{-\frac{s}{v}x} \tag{4.74c}$$

Multiplying the second equation of Eq. (4.74) by $\frac{a_2}{L}$ and summing it to the first equation:

$$F_1 u_{,xx} + a_1 F_1 u_{,x} - s^2 u + F_2 \theta_{,x} + a_1 F_2 \theta = 0 \tag{4.75a}$$

$$\theta_{,xx} + a_1 \theta_{,x} - s\theta = -a_5 e^{-\frac{s}{v}x} \tag{4.75b}$$

where

$$F_1 = 1 + \frac{a_2}{L} \qquad F_2 = \frac{a_2 a_4}{L} - a_3 \tag{4.76}$$

It can be noted that the second equation of Eq. (4.75) is an ordinary differential equation. As such, the solution can be obtained as:

$$\theta = C_1 e^{r_1 x} + C_2 e^{r_2 x} + \theta_p \tag{4.77a}$$

$$\theta_p = -\frac{a_5}{\left(\frac{s}{v}\right)^2 - a_1\left(\frac{s}{v}\right) - s} e^{-\frac{s}{v}x} \tag{4.77b}$$

$$r_{1,2} = \frac{-a_1 \pm \sqrt{a_1^2 + 4s}}{2} \tag{4.77c}$$

Substituting the temperature relation from Eq. (4.77) into Eq. (4.75a) leads to:

$$\begin{aligned} F_1 u_{,xx} + a_1 F_1 u_{,x} - s^2 u = & -F_2 C_1 (r_1 + a_1) e^{r_1 x} - F_2 C_2 (r_2 + a_1) e^{r_2 x} \\ & - F_2 \theta_p \left(-\frac{s}{v} + a_1 \right) \end{aligned} \tag{4.78}$$

The solution of Eq. (4.78) contains two parts; the general solution and the particular solution:

$$u = C_3 e^{\xi_1 x} + C_4 e^{\xi_2 x} + u_{p_1} + u_{p_2} + u_{p_3} \tag{4.79a}$$

$$\xi_{1,2} = \frac{-a_1 F_1 \pm \sqrt{a_1^2 F_1^2 + 4 F_1 s^2}}{2 F_1} \tag{4.79b}$$

$$u_{p_1} = -\frac{F_2 C_1 (r_1 + a_1)}{F_1 r_1^2 + a_1 F_1 r_1 - s^2} e^{r_1 x} \tag{4.79c}$$

$$u_{p_2} = -\frac{F_2 C_1 (r_2 + a_1)}{F_1 r_2^2 + a_1 F_1 r_2 - s^2} e^{r_2 x} \tag{4.79d}$$

$$u_{p_3} = -\frac{F_2 C_{\theta_1} \left(-\frac{s}{v} + a_1 \right)}{F_1 \left(\frac{s}{v} \right)^2 - a_1 F_1 \left(\frac{s}{v} \right) - s^2} e^{-\frac{s}{v} x} \tag{4.79e}$$

Inserting the displacement and the temperature equations into Eq. (4.74b) results in the ordinary differential equation for electric potential written below:

$$\begin{aligned} Lc^* \eta^* \phi_{,xx} + a_1 Lc^* \eta^* \phi_{,x} = & \, C_1 a_4 (r_1 + a_1) e^{r_1 x} + C_2 a_4 (r_2 + a_1) e^{r_2 x} \\ & + \theta_p a_4 \left(\left(-\frac{s}{v} \right) + a_1 \right) + C_3 \left(\xi_1^2 + a_1 \xi_1 \right) e^{\xi_1 x} + C_4 \left(\xi_2^2 + a_1 \xi_2 \right) e^{\xi_2 x} \\ & + u_{p_1} \left(r_1^2 + a_1 r_1 \right) + u_{p_2} \left(r_2^2 + a_1 r_2 \right) + u_{p_3} \left(\left(\frac{s}{v} \right)^2 - a_1 \left(\frac{s}{v} \right) \right) \end{aligned} \tag{4.80}$$

By solving Eq. (4.80), we complete the solution procedure for the uncoupled thermoelasticity analysis:

$$\phi = C_5 e^{-a_1 x} + C_6 + \phi_{p_1} + \phi_{p_2} + \phi_{p_3} + \phi_{p_4} + \phi_{p_5} \tag{4.81a}$$

$$\phi_{p_1} - \frac{C_3 \left(\xi_1^2 + a_1 \xi_1 \right)}{Lc^* \eta^* \xi_1^2 + a_1 Lc^* \eta^* \xi_1} e^{\xi_1 x} \tag{4.81b}$$

$$\phi_{p_2} = \frac{C_4(\xi_2^2 + a_1\xi_2)}{Lc^*\eta^*\xi_2^2 + a_1 Lc^*\eta^*\xi_2} e^{\xi_2 x} \tag{4.81c}$$

$$\phi_{p_3} = \frac{u_{p_1}(r_1^2 + a_1 r_1) + C_1 a_4 (r_1 + a_1) e^{r_1 x}}{Lc^*\eta^* r_1^2 + a_1 Lc^*\eta^* r_1} \tag{4.81d}$$

$$\phi_{p_4} = \frac{u_{p_2}(r_2^2 + a_1 r_2) + C_2 a_4 (r_2 + a_1) e^{r_2 x}}{Lc^*\eta^* r_2^2 + a_1 Lc^*\eta^* r_2} \tag{4.81e}$$

$$\phi_{p_5} = \frac{u_{p_3}\left(\left(\frac{s}{v}\right)^2 - a_1\left(\frac{s}{v}\right)\right) + \theta_p a_4\left(-\frac{s}{v} + a_1\right)}{Lc^*\eta^*\left(\frac{s}{v}\right)^2 - a_1 Lc^*\eta^*\left(\frac{s}{v}\right)} \tag{4.81f}$$

4.4.3.4 Numerical Inversion of the Laplace Transform

Having obtained the coupled and uncoupled solutions in the Laplace domain, we now employ the so-called fast Laplace inverse transform [14]. Performing this numerical inversion will transform the results into the time domain for later analysis. The form of this inversion process have been previously stated in Sect. 4.2.3 and is omitted here. The only change concerns the values of the constraint parameters, which are stated below:

$$aT = \frac{100}{15}, L_n = 5, N = 900$$

4.4.4 Results and Discussion

The numerical example presented in this section will help to quantify the results of the solution procedure. Consider an FGPM rod with the properties listed in Table 4.2 [31], in which the left side is composed of Cadmium Selenide. The rod's initial temperature is taken to be the ambient temperature $\theta_0 = 293\,\mathrm{K}$ and the intensity of the mobile heat source is $R_0 = \frac{10}{\rho_0}$. The relaxation time incorporated in the L-S theory has a non-dimensional value of 0.05, and the velocity of the thermal disturbance is 0.5 [3].

For the first set of figures, the rod is assumed to be homogeneous and is analyzed based on the generalized coupled (L-S), classical coupled, and classical uncoupled theories. The distributions of temperature change, displacement, stress, and electric potential at non-dimensional time $t = 0.2$ in response to a moving heat source are displayed in Figs. 4.16, 4.17, 4.18 and 4.19 respectively. It is immediately noticed that the results for classical coupled and classical uncoupled are practically identical

Table 4.2 Material properties of the left end of the rod [3, 31]

Properties	Cadmium Selenide
c_0 (GPa)	74.1
$e_0\left(\frac{C}{m^2}\right)$	0.347
$\beta_0 \times 10^6\left(\frac{N}{Km^2}\right)$	0.621
$\in_0 \times 10^{-11}\left(\frac{C^2}{Nm^2}\right)$	9.03
$p_0 \times 10^{-6}\left(\frac{C}{Km^2}\right)$	−2.94
$\rho_0 \times 10^3\left(\frac{kg}{m^3}\right)$	7.60
$C_E \times 10^3\left(\frac{J}{kg\ K}\right)$	0.42
$K_0\left(\frac{W}{mK}\right)$	12.9

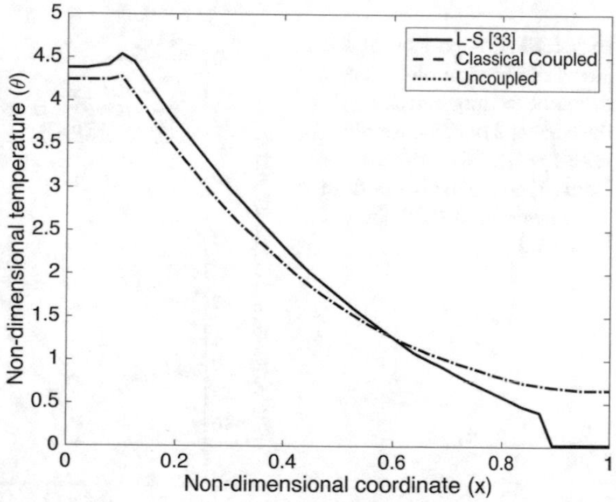

Fig. 4.16 Comparison of the temperature distribution based on different thermoelasticity theories at non-dimensional time $t = 0.2$. [Reproduced from [3] with permission from World Scientific Publishing Co., Inc.]

and are therefore indiscernible in the figures. Intuitively, the maximum temperature occurs at the location of the heat source at this time ($x = vt = 0.1$). On the other hand, the maxima of displacement, stress and electric potential all occur ahead of this point as shown in Figs. 4.17, 4.18 and 4.19. These results are in agreement with those found by Babei and Chen [33] analytically.

The effects of non-Fourier heat conduction can be seen in Figs. 4.16 and 4.18. For the classical coupled and classical uncoupled solutions, the temperature distribution is diffusive, and as such thermal wave characteristics are not observed due to the parabolic nature of Fourier heat conduction. Contrarily, in the generalized distributions, the presence of thermal wavefronts is obvious due to the finite thermal wave speed and hyperbolic heat conduction. Therefore, there are distinguishable undisturbed portions of the rod in the distributions of temperature and stress.

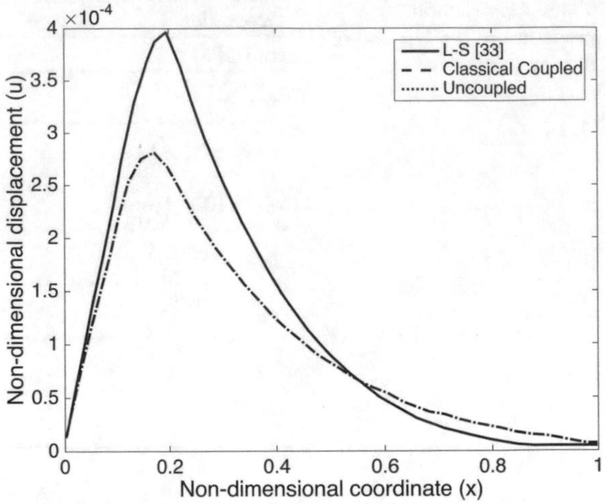

Fig. 4.17 Comparison of the displacement distribution based on different thermoelasticity theories at non-dimensional time $t = 0.2$. [Reproduced from [3] with permission from World Scientific Publishing Co., Inc.]

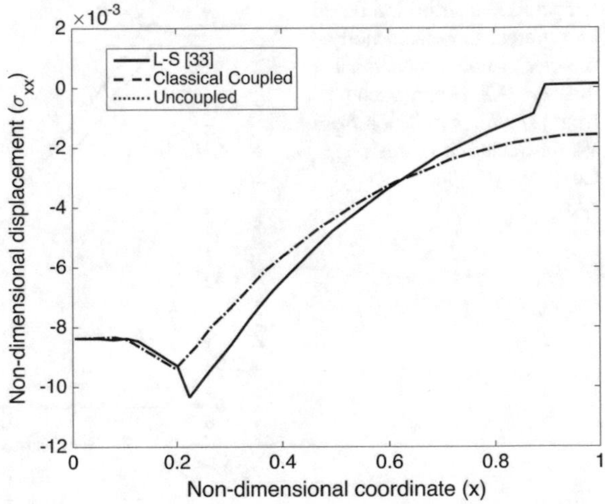

Fig. 4.18 Comparison of the stress distribution based on different thermoelasticity theories at non-dimensional time $t = 0.2$. [Reproduced from [3] with permission from World Scientific Publishing Co., Inc.]

The time history of the difference between the temperature solutions of the coupled and uncoupled theories is depicted in Fig. 4.20. The amplitude of fluctuation remains constant after the heat source leaves the rod at $t_{exit} = 2$.

Figures 4.21, 4.22 and 4.23 display the effect of the non-homogeneity index λ on the histories of displacement, temperature and electric displacement, respectively. This analysis is performed at a non-dimensional location of $x = 0.5$ based on the classical coupled thermoelasticity theory. The absolute mean value of fluctuation for the displacement distribution increases as the value of λ increases. It is shown in Fig. 4.22 that before the heat source exits the rod, the temperature at each location increases monotonically. After this point, the temperature will reach its constant value while exhibiting small fluctuations.

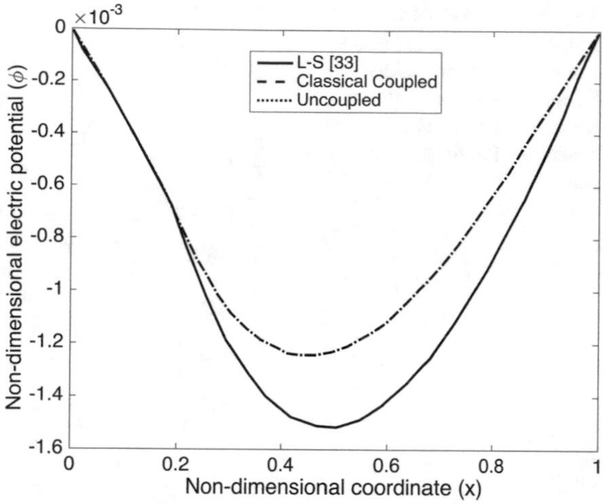

Fig. 4.19 Comparison of the electric potential distribution based on different thermoelasticity theories at non-dimensional time $t = 0.2$. [Reproduced from [3] with permission from World Scientific Publishing Co., Inc.]

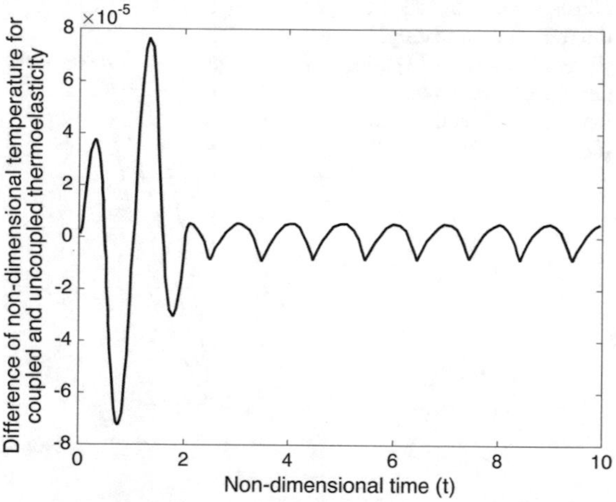

Fig. 4.20 Time history of the difference of the temperature distribution for coupled and uncoupled thermoelasticity at non-dimensional time $t = 0.2$. [Reproduced from [3] with permission from World Scientific Publishing Co., Inc.]

In Fig. 4.23 the history of electric displacement with different non-homogeneity indices is depicted. It is clear to see that the absolute value of electric displacement increases when λ becomes larger. The distribution smoothly increases until the thermal disturbance reaches the end of the rod, at which point it remains constant. This phenomenon is consistent with the results obtained by Babei and Chen for a homogeneous rod under L-S theory [33].

The effect of non-homogeneity index on the distribution of stress is depicted in Fig. 4.24. The results are again analyzed for classical coupled thermoelasticity at non-dimensional time $t = 0.2$. Before it reaches its maximum, the absolute value of stress decreases when λ increases. This relationship is completely reversed after the maximum of stress occurs. Once again, these findings can also be observed for the coupled thermoelasticity analysis based on L-S theory.

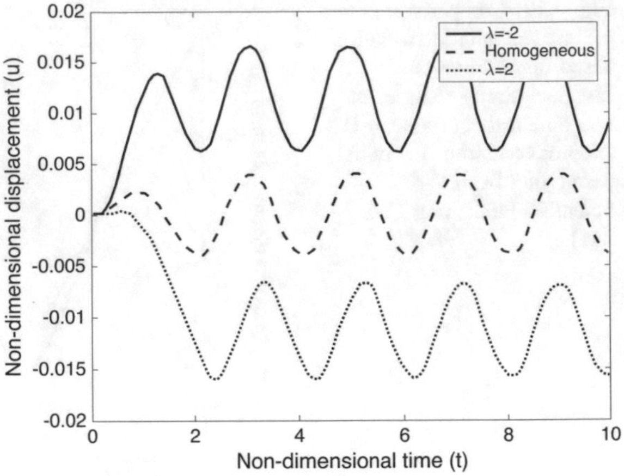

Fig. 4.21 Effect of the non-homogeneity index on the displacement history. [Reproduced from [3] with permission from World Scientific Publishing Co., Inc.]

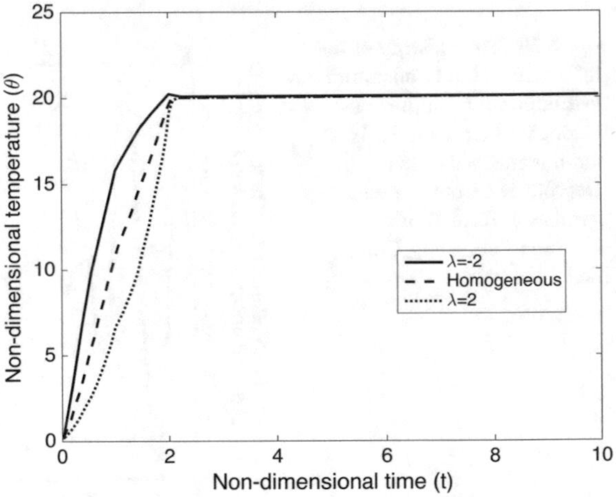

Fig. 4.22 Effect of the non-homogeneity index on the temperature history. [Reproduced from [3] with permission from World Scientific Publishing Co., Inc.]

In general, certain phenomena can be noted based on the results found in this section. In classical coupled thermoelasticity, there are no wave fronts in the distributions of temperature or stress, but they exist for the generalized L-S solutions. Additionally, the extrema of temperature and stress based on classical coupled and classical uncoupled thermoelasticity are lower compared to those based on the generalized theory. For any thermoelasticity theory discussed, an increase in λ results in an increase of the absolute value of electric displacement after the thermal disturbance has left the rod. For classical coupled thermoelasticity, an increase in λ diminishes the dynamic response of displacement, temperature and electric potential. Nonetheless, variations in non-homogeneity have no effect on the constant temperature reached once the heat source exits the rod [3].

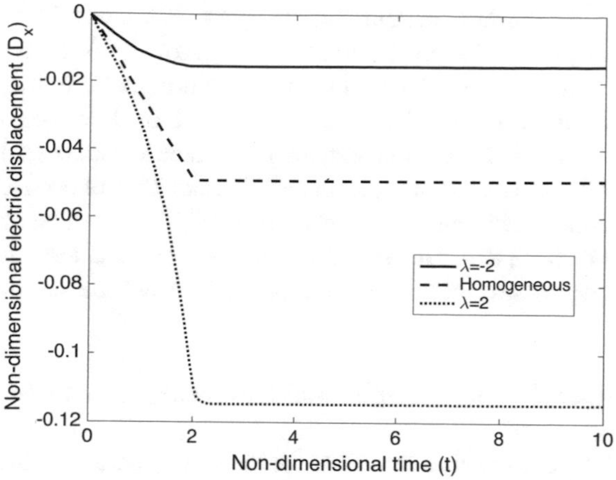

Fig. 4.23 Effect of the non-homogeneity index on the electric displacement history. [Reproduced from [3] with permission from World Scientific Publishing Co., Inc.]

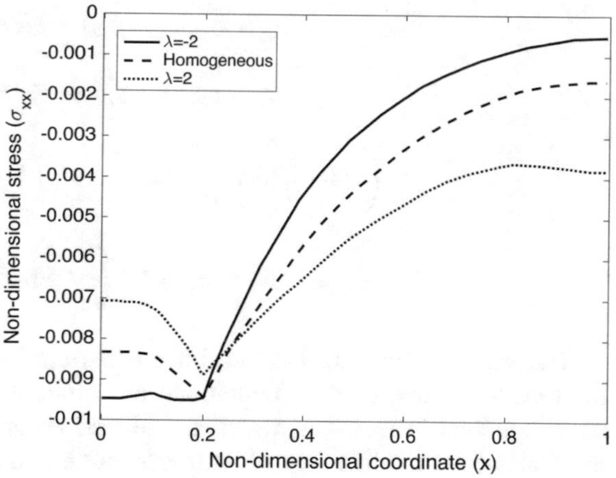

Fig. 4.24 Effect of the non-homogeneity index on the distribution of stress. [Reproduced from [3] with permission from World Scientific Publishing Co., Inc.]

4.4.5 Introduction of Dual Phase Lag Models

As previously mentioned, the work of Green and Lindsay (G-L) introduced a new version of generalized thermoelasticity which incorporated two relaxation times for the relations of stress and entropy [28, 40]. More recently, Chandrasekharaiah and Tzou (C-T) proposed another generalized theory that considers the dual phase lag of heat flux and temperature gradient [41, 42]. Essentially, this theory establishes the hypothesis that the temperature gradient or the heat flux may precede one another [40]. Given that it has had close agreements with a variety of experiments on both microscale and macroscale ranges, this new theory has gained prevalence [43, 44].

In the previous section, results were presented from the response of a thermopiezoelectric, one-dimensional, functionally graded rod on the basis of L-S generalized theory. However, in certain applications such as pulsed laser heating and ultra-fast heating sources, there is a delay in the response of the object with respect to the heat source. Furthermore, the delay of the heat flux may not be identical to the delay of the temperature gradient [40]. It is therefore very beneficial to consider the dual-phase-lag model in thermoelastic analysis in order to fully understand the behaviour of piezoelectric media. It is with this reasoning that the following section is presented. The same problem from the beginning of Sect. 4.4 is considered, only this time based on C-T thermoelasticity theory with two phase lags present [40]. The equations, unknowns, and solution procedures are identical in both cases, with a few exceptions that will be outlined.

4.4.5.1 Fundamental and Governing Equations

The constitutive equations for the C-T generalized thermoelasticity for piezoelectric materials are defined below [26, 34, 35]:

$$\sigma_{ij} = c_{ijkl}\varepsilon_{kl} - \beta_{ij}\theta - e_{ijk}E_k \tag{4.82a}$$

$$D_i = e_{ijk}\varepsilon_{jk} + p_i\theta + \in_{ij} E_j \tag{4.82b}$$

$$\left(1 + \tau\frac{\partial}{\partial t} + t_2^2\frac{\partial^2}{\partial t^2}\right)q_i = -K_{ij}\left(1 + t_1\frac{\partial}{\partial t}\right)\theta_{,j} \tag{4.82c}$$

$$\rho S = \beta_{ij}\varepsilon_{ij} + \frac{\rho C}{\theta_0}\theta + p_i E_i \tag{4.82d}$$

The variables contained in the above equations have been previously described in Sect. 4.3.2 and so their definitions are omitted here. The additions to these equations include t_1 and t_2, which are defined according to the following approximations of the modification of Fourier's law based on C-T theory [26, 45]:

$$t_1 = \tau_\theta > 0, \tau = \tau_q > 0, t_2^2 = 0, \tau_q > \tau_\theta > 0 \tag{4.83a}$$

$$t_1 = \tau_\theta > 0, \tau = \tau_q > 0, t_2^2 = \frac{1}{2}\tau_q^2 \tag{4.83b}$$

where τ_q is the heat flux phase-lag and τ_θ is the temperature gradient phase-lag. Giving the phase lags in this structure allows us to investigate the generalized thermoelasticity based on the Lord-Shulman (L-S) theory as well. From this point onward, the solution procedure used to solve these equations is identical to that found in Sect. 4.4.3 for the coupled theories. For this reason, the process is omitted for brevity and the results are directly shown below.

4.4.6 Results of Dual Phase Lag Model Analysis

The problem analyzed in this section is identical to that of Sect. 4.4.4. The only modification is the addition of the temperature gradient phase lag, which has a non-dimensional value of 0.04. Since the solution in this section was based on a general approximation of C-T thermoelasticity theory, it is possible to study the response of the rod based on L-S theory by setting t_1 and t_2 equal to zero. In this case, the problem truly is identical to the preceding one and so these results are used in this section for comparison purposes [12]. The analysis based on C-T theory, however, is performed for both $t_2 = 0$ and $t_2 = \frac{\tau_q}{\sqrt{2}}$ in order to investigate the effects of this parameter on the results.

In Figs. 4.25, 4.26, 4.27 and 4.28, the distributions of displacement, temperature change, stress, and electric potential are depicted based on two approximations of

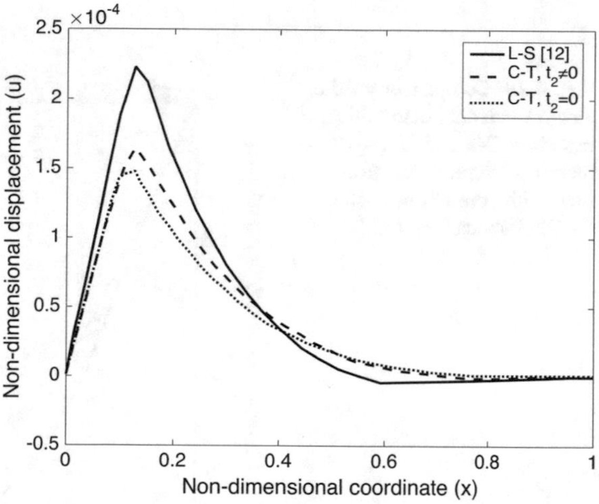

Fig. 4.25 Comparison of the displacement distribution based on C-T and L-S theories. [Reproduced from [40] with permission from SAGE Publications Ltd.]

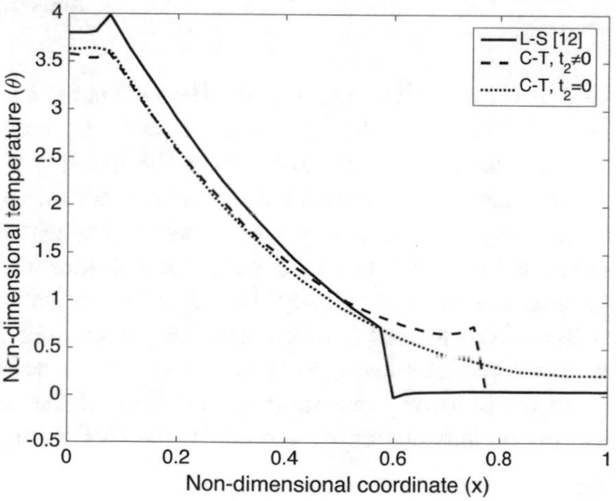

Fig. 4.26 Comparison of the temperature distribution based on C-T and L-S theories. [Reproduced from [40] with permission from SAGE Publications Ltd.]

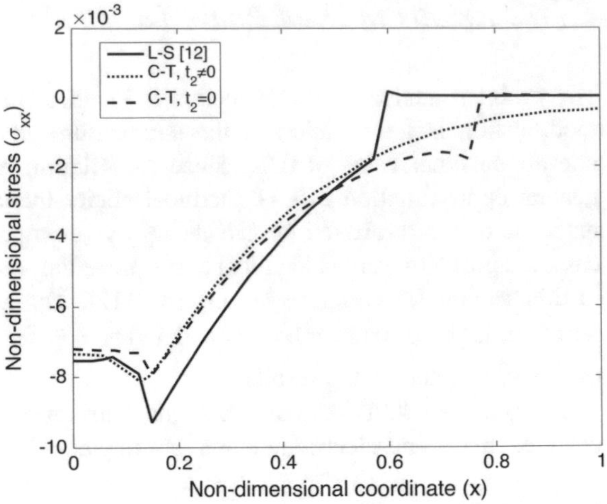

Fig. 4.27 Comparison of the stress distribution based on C-T and L-S theories. [Reproduced from [40] with permission from SAGE Publications Ltd.]

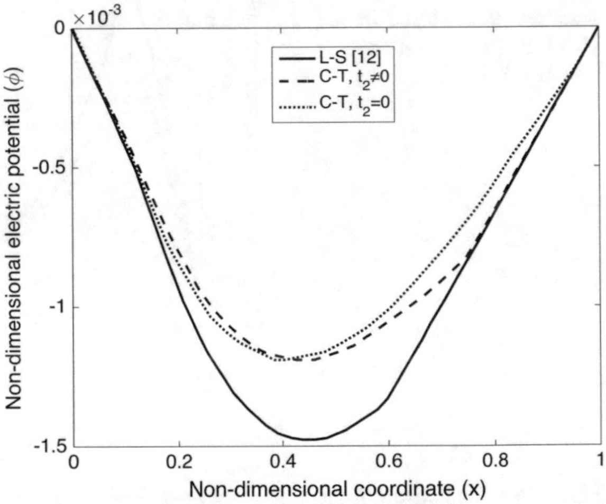

Fig. 4.28 Comparison of the electric potential distribution based on C-T and L-S theories. [Reproduced from [40] with permission from SAGE Publications Ltd.]

C-T theory as well as L-S theory. The results are shown for non-dimensional time $t = 0.1333$, thus the non-dimensional location of the heat source is $x = \upsilon t = 0.0667$. As seen previously, the maximum temperature in the rod occurs at this point for L-S theory, but the same is not true for C-T. Nonetheless, it will be seen in Fig. 4.29 that as the temperature gradient phase lag decreases, the temperature maximum tends to occur at the location of the thermal disturbance. Figures 4.26 and 4.27 clearly show that the thermal wavefront based on C-T theory with $t_2 \neq 0$ is located farther ahead than the wavefront based on the L-S theory, and furthermore that when $t_2 = 0$, no wavefront is observed.

In the following two figures, the effect of the temperature gradient phase lag $\tau_\theta = t_1$ on temperature is studied for the C-T theory with $t_2 \neq 0$ and $t_2 = 0$. The

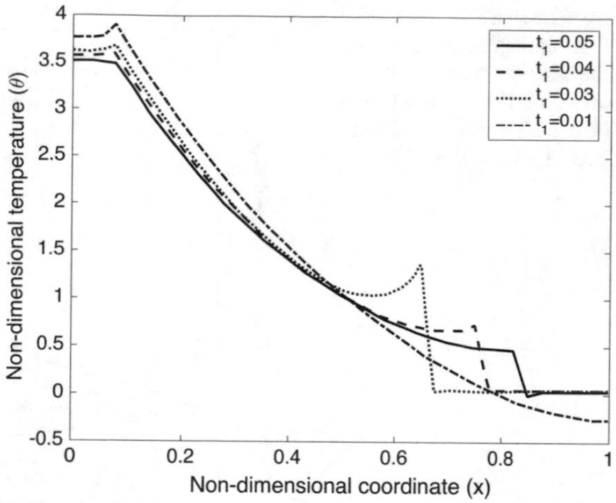

Fig. 4.29 Effect of the phase-lag of temperature gradient on the temperature distribution for C-T theory with $t_2 \neq 0$. [Reproduced from [40] with permission from SAGE Publications Ltd.]

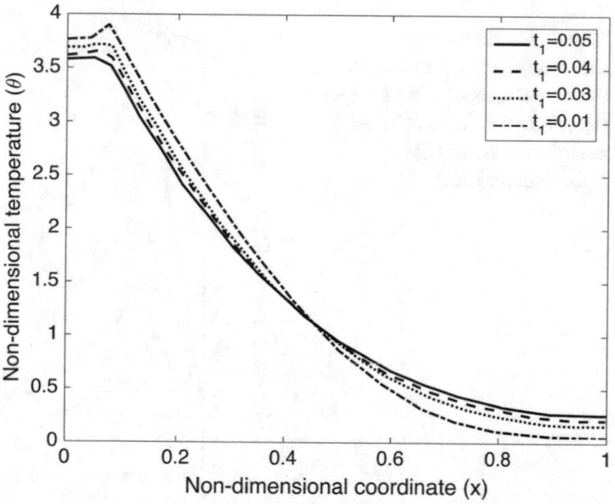

Fig. 4.30 Effect of the phase-lag of temperature gradient on the temperature distribution for C-T theory with $t_2 = 0$. [Reproduced from [40] with permission from SAGE Publications Ltd.]

material is still considered homogeneous at this point and the results are shown at the same non-dimensional time $t = 0.1333$. It can be concluded that as the phase lag increases, the wave propagation speed increases, thus forcing the wave fronts to move ahead farther. Moreover, the wavefronts weaken as τ_θ decreases and they eventually disappear at $t_1 = 0.01$, as seen in Fig. 4.29. In Fig. 4.30, no wavefronts are observed at all due to the heat flux phase lag being equal to zero. We can also conclude that a decrease in t_1 increases the absolute values of the extrema for the temperature distribution in the C-T theory whether or not $t_2 = 0$.

We will now study this problem on the basis of functionally graded media. As seen below in Fig. 4.31 for the elastic constant, the non-homogeneity index λ holds an exponential relationship with the material properties of the thermopiezo-electric rod.

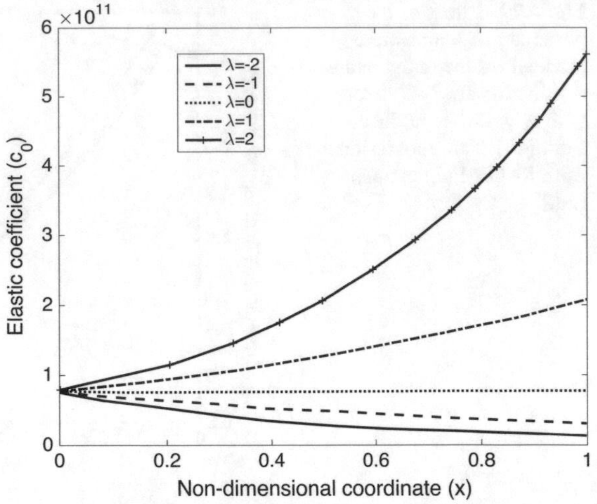

Fig. 4.31 Effect of non-homogeneity index on the distribution of elastic constant. [Reproduced from [40] with permission from SAGE Publications Ltd.]

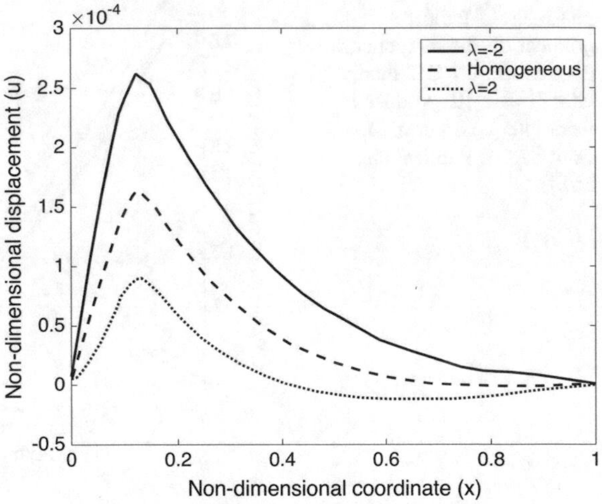

Fig. 4.32 Effect of non-homogeneity index on the displacement distribution. [Reproduced from [40] with permission from SAGE Publications Ltd.]

In Figs. 4.32, 4.33, 4.34 and 4.35, the distributions of displacement, temperature, electric potential, and stress are depicted with varying non-homogeneity indices for C-T theory with $\tau_\theta = t_1 = 0.04$ and $t_2 \neq 0$ at non-dimensional time $t = 0.1333$. An increase in λ lowers the absolute value of displacement, temperature and electric potential and also reduces the heights of the wavefronts seen in the distributions. The locations of the wavefronts, however, remains the same for varying values of non-homogeneity. Finally, when λ is increased, the absolute value of stress decreases before it reaches its maximum and increases after its maximum. For C-T theory with $t_2 = 0$, as well as L-S theory, similar results can be observed.

Through the dual phase lag results presented in this section, it is possible to study C-T theory with different values for phase lags, as well as generalized L-S

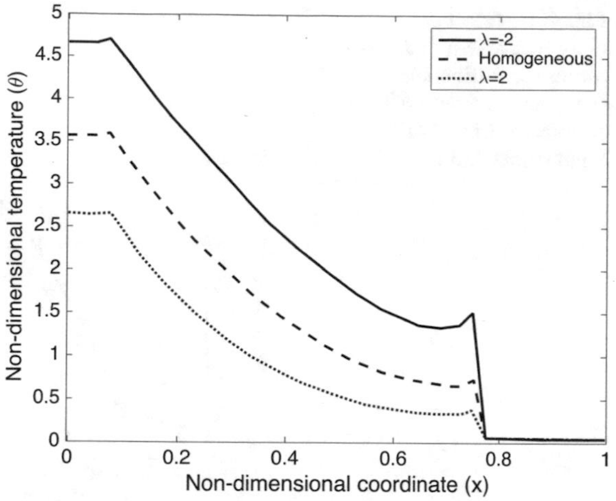

Fig. 4.33 Effect of non-homogeneity index on the temperature distribution. [Reproduced from [40] with permission from SAGE Publications Ltd.]

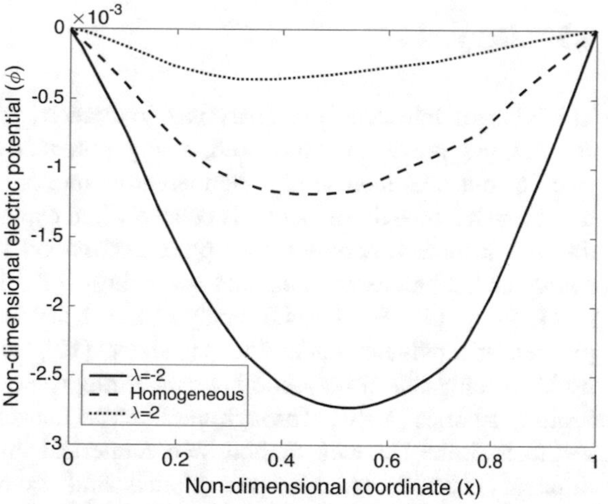

Fig. 4.34 Effect of non-homogeneity index on the electric potential distribution. [Reproduced from [40] with permission from SAGE Publications Ltd.]

theory. In the C-T theory, τ_q and τ_θ can be interpreted as two relaxation times, whereas in L-S theory we only account for one relaxation time [40]. The results found are reduced to those for coupled L-S theory when $t_1 = t_2 = 0$. However, using C-T theory with two phase lags τ_θ and τ_q allows the consideration of the fact that the heat flux vector may precede the temperature gradient or vice versa. In addition, non-equilibrium thermodynamic transitions and microscope effects of energy exchange in high-rate heating applications are significant setbacks which are addressed by the dual phase lag C-T theory [40]. This approach to thermopiezo-electric problems provides a multiphysical description of functionally graded materials on microscopic and macroscopic scales while including other more generalized thermoelasticity theories.

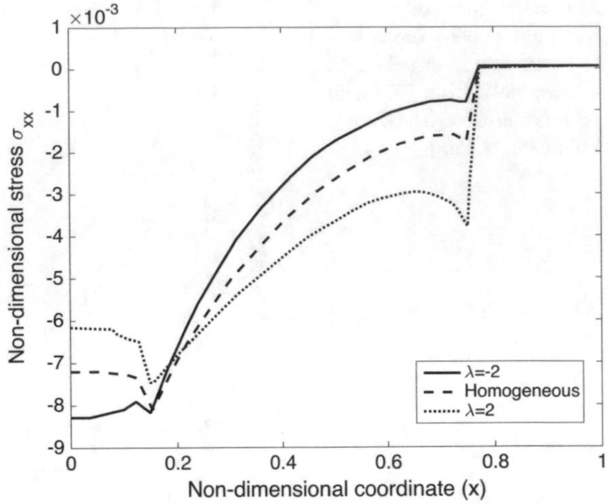

Fig. 4.35 Effect of non-homogeneity index on the stress distribution. [Reproduced from [40] with permission from SAGE Publications Ltd.]

4.5 Remarks

This chapter introduce the analysis of advanced smart materials under thermal stresses. As previously discussed, many potential applications of thermopiezoelectric structures include high-temperature conditions [2]. As such, it is important to understand how these materials behave when exposed to thermal disturbances. In the first section, a homogeneous piezoelectric rod is subjected to a moving heat source and solutions are obtained according to the Lord-Shulman theory of thermoelasticity [5]. Section 4.3 analyze the behaviour of a functionally graded piezoelectric cylinder under thermal shock [17], while Sect. 4.4 revisite the rod problem, only this time assuming non-homogeneity and three different thermoelasticity theories [3, 40]. Through the Laplace transform, successive elimination of variables, finite element method, and numerical inversion method, solutions for different material variables are plotted and examined. It is found that the non-homogeneity of the material has profound effects on the solutions, as did the governing theories chosen for analysis. The presence of thermal wavefronts in temperature and stress, as well as the overall behaviour of distributions depend heavily on whether the equations are coupled, uncoupled or generalized. Furthermore, the degree to which the material properties are varied change the amplitudes and shapes of the distributions drastically. These two factors represent intrinsic details that determine the useful role of advanced and smart materials in thermal applications.

References

1. Ray AK et al (2005) Bamboo—a functionally graded composite-correlation between microstructure and mechanical strength. J Mater Sci 40(19):5249–5253
2. Rubio WM et al (2010) Functionally graded piezoelectric material systems—a multiphysics perspective, in advanced computational materials modeling. Wiley-VCH Verlag GmbH & Co. KGaA, pp 301–339
3. Akbarzadeh A, Babaei M, Chen Z (2011) Thermopiezoelectric analysis of a functionally graded piezoelectric medium. Int J Appl Mech 3(01):47–68
4. Akbarzadeh A, Cui Y, Chen Z (2017) Thermal wave: from nonlocal continuum to molecular dynamics. RSC Adv 7(22):13623–13636
5. Babaei M, Chen Z (2008) Dynamic response of a thermopiezoelectric rod due to a moving heat source. Smart Mater Struct 18(2):025003
6. Parton V, Kudryavtsev B (1988) Electromagnetoelasticity, vol 2. Gordon and Breach Science Publishers, New York, p 90059-0
7. Vernotte P (1958) Les paradoxes de la théorie continue de léquation de la chaleur. C R Hebd Seances Acad Sci 246(22):3154–3155
8. Cattaneo C (1948) Sullaconduzione de calore. Atti del. Seminar 3:83–101
9. Biot MA (1956) Thermoelasticity and irreversible thermodynamics. J Appl Phys 27(3): 240–253
10. Lord HW, Shulman Y (1967) A generalized dynamical theory of thermoelasticity. J Mech Phys Solids 15(5):299–309
11. Chandrasekharaiah D (1988) A generalized linear thermoelasticity theory for piezoelectric media. Acta Mech 71(1):39–49
12. He T, Cao L, Li S (2007) Dynamic response of a piezoelectric rod with thermal relaxation. J Sound Vib 306(3):897–907
13. Boyce WE, DiPrima RC, Haines CW (1969) Elementary differential equations and boundary value problems, vol 9. Wiley, New York
14. Durbin F (1974) Numerical inversion of Laplace transforms: an efficient improvement to Dubner and Abate's method. Comput J 17(4):371–376
15. Wojnar R, Bytner S, Galka A (1999) Effective properties of elastic composites subject to thermal Fields. Therm Stresses, 5
16. Landolt HH et al (1980) Numerical data and functional relationships in science and technology: new series. Nuclear and particle physics. Elastic and charge exchange scattering of elementary particles. Nucleon nucleon and kaon nucleon scattering, vol 9. Springer, Berlin
17. Babaei M, Chen Z (2009) The transient coupled thermo-piezoelectric response of a functionally graded piezoelectric hollow cylinder to dynamic loadings. Proc R Soc London: Math Phys Eng Sci (The Royal Society)
18. Liu G et al (2003) Dispersion of waves and characteristic wave surfaces in functionally graded piezoelectric plates. J Sound Vib 268(1):131–147
19. Takagi K et al (2002) Design and fabrication of functionally graded PZT/Pt piezoelectric bimorph actuator. Sci Technol Adv Mater 3(2):217–224
20. Richard BH, Eslami MR (2008) Thermal stresses-advanced theory and applications. Springer, The Netherlands
21. Babaei MH, Akhras G (2011) Temperature-dependent response of radially polarized piezoceramic cylinders to harmonic loadings. J Intell Mater Syst Struct 22(7):645–654
22. Eslami M (2003) A first course in finite element analysis. Amirkabir University Press, Tehran
23. Tsamasphyros G, Song Z (2006) The general solution for a finite thermopiezoelectric plate containing a hole and a crack. Arch Appl Mech 76(1):1–17
24. Liu W et al (2003) Noise and specific detectivity of pyroelectric detectors using lead titanate zirconate (PZT) thin films. Microelectron Eng 66(1):785–791
25. Ootao Y, Akai T, Tanigawa Y (2008) Transient piezothermoelastic analysis for a functionally graded thermopiezoelectric hollow cylinder. J Therm Stresses 31(10):935–955

26. Hetnarski RB, Ignaczak J (1999) Generalized thermoelasticity. J Therm Stresses 22(4–5): 451–476
27. Hetnarski R, Ignaczak J (2000) Nonclassical dynamical thermoelasticity. Int J Solids Struct 37(1):215–224
28. Green A, Lindsay K (1972) Thermoelasticity. J Elast 2(1):1–7
29. Tzou DY (1993) An engineering assessment to the relaxation time in thermal wave propagation. Int J Heat Mass Transf 36(7):1845–1851
30. Al-Huniti NS, Al-Nimr M, Naji M (2001) Dynamic response of a rod due to a moving heat source under the hyperbolic heat conduction model. J Sound Vib 242(4):629–640
31. He T, Tian X, Shen Y (2002) State space approach to one-dimensional thermal shock problem for a semi-infinite piezoelectric rod. Int J Eng Sci 40(10):1081–1097
32. Aouadi M (2007) Generalized thermoelastic-piezoelectric problem by hybrid Laplace transform-finite element method. Int J Comput Methods Eng Sci Mech 8(3):137–147
33. Babaei M, Chen Z (2010) Transient thermopiezoelectric response of a one-dimensional functionally graded piezoelectric medium to a moving heat source. Arch Appl Mech 80(7):803–813
34. El-Karamany AS, Ezzat MA (2005) Propagation of discontinuities in thermopiezoelectric rod. J Therm Stresses 28(10):997–1030
35. Mindlin R (1961) On the equations of motion of piezoelectric crystals. Prob Continuum Mech, 282–290
36. Davis S (2009) A finite algorithm for the solution to an algebraic equation. Int J Algebra 3(10):449–460
37. Mikhalkin EN (2009) Solution of fifth-degree equations. Russ Math (Iz VUZ) 53(6):15–23
38. Tignol J (1987) Theory of algebraic equations. Wiley, New York
39. Hetnarski RB, Eslami MR, Gladwell G (2009) Thermal stresses: advanced theory and applications, vol 41. Springer, Berlin
40. Akbarzadeh A, Babaei M, Chen Z (2011) Coupled thermopiezoelectric behaviour of a one-dimensional functionally graded piezoelectric medium based on C-T theory. Proc Inst Mech Eng, Part C: J Mech Eng Sci 225(11):2537–2551
41. Chandrasekharaiah D (1998) Hyperbolic thermoelasticity: a review of recent literature. Appl Mech Rev 51:705–730
42. Tzou DY (1995) The generalized lagging response in small-scale and high-rate heating. Int J Heat Mass Transf 38(17):3231–3240
43. Antaki PJ (1998) Solution for non-Fourier dual phase lag heat conduction in a semiinfinite slab with surface heat flux. Int J Heat Mass Transf 41(14):2253–2258
44. Tzou DY (1995) Experimental support for the lagging behavior in heat propagation. J Thermophys Heat Transfer 9(4):686–693
45. Tzou D (1997) Macro- to micro-scale heat transfer: the lagging behavior. Taylor & Francis, Washington, DC

Chapter 5
Thermal Fracture of Advanced Materials Based on Fourier Heat Conduction

5.1 Introduction

In this chapter, we introduce a so-called extended displacement discontinuity approach to deal with three-dimensional (3D) thermoelastic plane crack problems in advanced materials. The method can be used to treat general 3D thermoelastic crack problems in advanced materials, including 3D interface crack problems. As the interface crack problem is more general and can be reduced to an embedded crack problem in a single material, we will directly introduce the methodology in the first section for a 3D interface plane crack based on general thermoelasticity. Then, we show that the methodology can be extended readily for advanced smart materials in the subsequent sections, such as piezoelectric materials, electromagnetic materials, and quasi-crystals. To illustrate the application of the method, a boundary integral approach based on the analytical results is introduced to deal with arbitrarily shaped, 3D cracks in advanced materials. Some numerical results are presented to illustrate the interaction among different physical fields.

5.2 Extended Displacement Discontinuity Method and Fundamental Solutions for Thermoelastic Crack Problems

Due to the increasing use of composite materials in thermomechanical environments, the study on the fracture behavior of interfacial cracks is of great importance in engineering and has attracted much attention. As the failure of composite laminates is dominated by the development of interface cracks, understanding the interface fracture behavior is of great importance. Sih [1] (1962) pointed out that in homogeneous elastic media, the thermal stress near the crack tip has the classical singularity $r^{-1/2}$ as mechanical stresses. Chen and Ting [2] pointed out that the

Z. T. Chen and A. H. Akbarzadeh, *Advanced Thermal Stress Analysis of Smart Materials and Structures*, Structural Integrity 10,
https://doi.org/10.1007/978-3-030-25201-4_5

temperature is proportional to $r^{1/2}$, while temperature gradient and stresses to $r^{-1/2}$ near the crack tip for an insulated crack. The discontinuities of the material properties and geometries leads to a much complex stress field around the interface crack tip with an oscillatory singularity, $r^{-1/2+\varepsilon,}$ with ε being the bi-material constant even under uniaxial tensile loading about the crack. For two dimensional (2D) problems, Brown and Erdogan [3] initially studied an insulated, Griffith interfacial crack under uniform heat flow and obtained the stress fields. Herrmann et al. [4] compared the experimental results with numerical results of thermal cracking in dissimilar materials. Herrmann and Grebner [5] studied a curved, thermal crack problem in a brittle, two-phase, compound material and built a closed form solution for the stress field. Martin-Moran et al. [6] and Barber and Comninou [7] studied a penny-shaped interface crack subjected to a heat flow with either perfect or imperfect contact and compared the difference between the two contact conditions. Later, Takakuda et al. [8] used the complex function method to solve an external interface crack subjected to a uniform temperature change or heat flow, and obtained the distributions of displacements and stresses on the interface. Similar work on interface crack problems can aslo be found in [9–16]. Contact crack faces of interface cracks in thermomechanical analysis may reflect actual crack face boundary conditions. Thermal stress analysis of interface cracks based on contact zone models can be found in [17–20]. Ratnesh and Chandra [21] employed the weight function method to analyze a 2D interface and found that the general expression of the stress field for the interface crack is in the same form as that of the homogeneous one. Pant et al. [22] extended the element free, Galerkin method and employed jump function to solve interfacial crack problems in bi-materials. Khandelwal and Chandra [23] utilized body analogy method to analyze an interfacial crack subjected to thermal loads and obtained the analytical solution by computing the thermal weight function, with which the stress intensity factors are computed as well. Ma et al. [24] studied the Zener-Stroh model of an interface crack subjected to a uniform temperature shift, and evaluated the interface defect tolerant size, which can be used to assess the interface integrity and reliability under thermal loading.

For 3D cases, Bregman and Kassir [25] employed the Muskhelishvili's method [26] to study a penny-shaped, interface crack subjected to a uniform heat flow and got the stress intensity factors and energy release rate. Andrzej and Stanislaw [27] used the potential theory method to study a plane crack on an interface in a microperiodic, two-layered composite under a uniform, vertical heat flow. Johnson and Qu [28] extended the interaction integral method to analyze curvilinear cracks in a bimaterial interface under a non-uniform, temperature distribution and obtained the induced stress intensity factors. Nomura et al. [29] developed a numerical method using a path-independent, H-integral to analyze the singular stress field of a 3D interface corner between anisotropic bimaterials subjected to thermal stresses. Guo et al. [30] investigated a plane crack problem of inhomogeneous materials with interfaces subjected to thermal loading using a modified, interaction energy integral method, and obtained the thermal stress intensity factors. Li et al. [31] used the weight function method to study a 3D, interface crack in a bi-material under combined, thermomechanical loading.

The displacement discontinuity boundary integral equation method proves to be very efficient in solving crack problems as it grasps the intrinsic characteristics of crack problems that physical fields are discontinuous across crack faces [32]. This method has also been extended to solve interface crack problems in elastic media [33–35], piezoelectric media [36] and magnetoelectroelastic media [37, 38].

Motivated by the current research on interface crack problems, we developed the displacement and temperature discontinuity, boundary hyper-singular, integral-differential equation method for interface cracks in dissimilar, isotropic, thermal elastic bi-materials [39].

The displacement discontinuity method was first proposed by Tang et al. [33] to deal with planar crack problems in a three dimensional (3D) solid. It provides an efficient way to approach the complicated, 3D crack problems via a simple, numerical algorithm.

In the absence of body forces, the governing equations for a 3D, homogeneous thermoelastic medium in a steady state are [1]

$$\sigma_{ij,j} = 0 \tag{5.1a}$$

$$h_{i,i} = 0 \tag{5.1b}$$

where σ_{ij} and h_i are the stress and heat flux, respectively, $i,j = 1,2,3$ or $i,j = x,y,z$, and the index i or j after the comma denotes differentiation with respect to the coordinate.

In the Cartesian coordinates (x, y, z) and cylindrical coordinates (r, ϕ, z), the constitutive equations are expressed, respectively, in the form

$$\begin{aligned}
\sigma_x &= 2\mu\frac{\partial u}{\partial x} + \lambda\left(\frac{\partial u}{\partial x} + \frac{\partial v}{\partial y} + \frac{\partial w}{\partial z}\right) - \frac{E}{1-2v}\alpha\theta,\\
\sigma_y &= 2\mu\frac{\partial v}{\partial y} + \lambda\left(\frac{\partial u}{\partial x} + \frac{\partial v}{\partial y} + \frac{\partial w}{\partial z}\right) - \frac{E}{1-2v}\alpha\theta,\\
\sigma_z &= 2\mu\frac{\partial w}{\partial z} + \lambda\left(\frac{\partial u}{\partial x} + \frac{\partial v}{\partial y} + \frac{\partial w}{\partial z}\right) - \frac{E}{1-2v}\alpha\theta,\\
\tau_{xy} &= \mu\left(\frac{\partial u}{\partial y} + \frac{\partial v}{\partial x}\right), \tau_{yz} = \mu\left(\frac{\partial w}{\partial y} + \frac{\partial v}{\partial z}\right), \tau_{zx} = \mu\left(\frac{\partial w}{\partial x} + \frac{\partial u}{\partial z}\right),\\
h_x &= -\beta\frac{\partial \theta}{\partial x}, h_y = -\beta\frac{\partial \theta}{\partial y}, h_z = -\beta\frac{\partial \theta}{\partial z},
\end{aligned} \tag{5.2a}$$

$$
\begin{aligned}
\sigma_r &= (\lambda+2\mu)\frac{\partial u_r}{\partial r} + \lambda\left(\frac{1}{r}\frac{\partial u_\phi}{\partial \phi} + \frac{u_r}{r}\right) + \lambda\frac{\partial w}{\partial z} - \frac{E}{1-2v}\alpha\theta,\\
\sigma_\phi &= \lambda\frac{\partial u_r}{\partial r} + (\lambda+2\mu)\left(\frac{1}{r}\frac{\partial u_\phi}{\partial \phi} + \frac{u_r}{r}\right) + \lambda\frac{\partial w}{\partial z} - \frac{E}{1-2v}\alpha\theta,\\
\sigma_z &= \lambda\frac{\partial u_r}{\partial r} + (\lambda+2\mu)\frac{\partial w}{\partial z} + \lambda\left(\frac{1}{r}\frac{\partial u_\phi}{\partial \phi} + \frac{u_r}{r}\right) - \frac{E}{1-2v}\alpha\theta,\\
\sigma_{r\phi} &= \mu\left(\frac{1}{r}\frac{\partial u_r}{\partial \phi} + \frac{\partial u_\phi}{\partial r} - \frac{u_\phi}{r}\right), \sigma_{\phi z} = \mu\left(\frac{1}{r}\frac{\partial w}{\partial \phi} + \frac{\partial u_\phi}{\partial z}\right),\\
\sigma_{zr} &= \mu\left(\frac{\partial w}{\partial r} + \frac{\partial u_r}{\partial z}\right),
\end{aligned}
\tag{5.2b}
$$

where u, v, and w are displacements, σ_{ij} is stress, and θ is temperature change (hereafter it is referred to as "temperature" for simplicity) with $\theta = 0$ corresponding to the free stress state; E, v, α, and β are the elastic modulus, Poisson's ratio, coefficient of linear thermal expansion, and coefficient of thermal conductivity, respectively, and λ and μ are the Lame constants, which are expressed in the following forms:

$$
\lambda = \frac{Ev}{(1+v)(1-2v)}, \quad \mu = \frac{E}{2(1+v)} \tag{5.3}
$$

5.2.1 *Fundamental Solutions for Unit Point Loading on a Penny-Shaped Interface Crack*

In order to build the solution for general loading and crack geometry, solutions for displacement and temperature discontinuities can be constructed first as fundamental solutions, which can then be used as Green's functions for general loading and interface crack geometry.

Consider a penny-shaped crack with radius a lying at the interface of two bonded dissimilar materials as illustrated in Fig. 5.1. The crack is located in the plane xoy, and the two bonded solids are assumed to occupy the upper and lower half-space, respectively. Then the relation between the Cartesian and cylindrical coordinates can be expressed as

$$
\begin{cases}
x = r\cos\phi,\\
y = r\sin\phi,\\
R^2 = r^2 + z^2 = x^2 + y^2 + z^2.
\end{cases}
\tag{5.4}
$$

According to Zhao et al. [40] and Zhao and Liu [41], if the radius of the crack a approaches zero, one can obtain the fundamental solutions corresponding to a

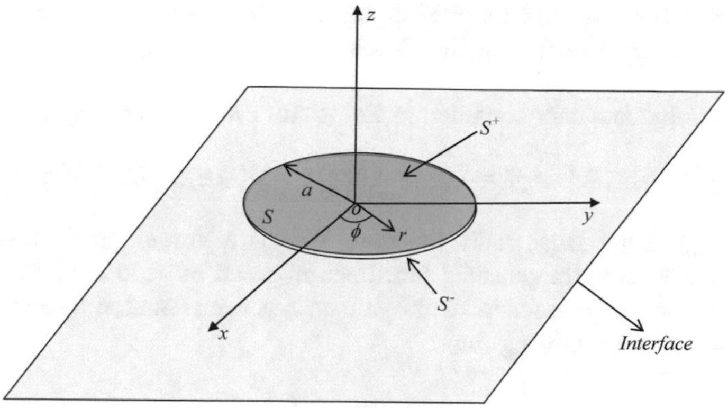

Fig. 5.1 A penny-shaped crack with radius of a lying in the interface plane [39]

unit, concentrated temperature discontinuity and displacement discontinuity satisfying the governing equations of thermoelasticity and the following conditions

$$\lim_{a\to 0}\int_S \{\|u\|,\|v\|,\|w\|,\|\theta\|\}dS=\{1,0,0,0\}, \tag{5.5a}$$

$$\lim_{a\to 0}\int_S \{\|u\|,\|v\|,\|w\|,\|\theta\|\}dS=\{0,1,0,0\}, \tag{5.5b}$$

$$\lim_{a\to 0}\int_S \{\|u\|,\|v\|,\|w\|,\|\theta\|\}dS=\{0,0,1,0\}, \tag{5.5c}$$

$$\lim_{a\to 0}\int_S \{\|u\|,\|v\|,\|w\|,\|\theta\|\}dS=\{0,0,0,1\}. \tag{5.5d}$$

where $\|u\|, \|v\|, \|w\|, \text{and}\|\theta\|$ are the elastic displacement and temperature discontinuities across the crack face. In the following sections, we present the fundamental solutions for each unit point loading case, which would be readily used as weight functions for general loading cases of arbitrary crack shapes. It is worth mentioning that as the point solutions are independent of the original shape of the penny-shaped crack, these solutions can be used to build the solutions for any interface crack of arbitrary shape under various loading conditions. We will show a few examples later, where fundamental solutions have been extended and used to build analytical solutions for interface crack problems in piezoelectric, piezoelectromagnetic, and quasi-crystalline materials.

5.2.1.1 Solution for Unit-Point Displacement Discontinuity in the Z-Direction of the Crack

In this case, the boundary condition in Eq. (5.5c) can be rewritten as:

$$\|u(\xi,\eta)\| = 0, \quad \|v(\xi,\eta)\| = 0, \quad \|w(\xi,\eta)\| = \delta(\xi,\eta), \quad \|\theta(\xi,\eta)\| = 0. \tag{5.6}$$

where $\delta(\xi,\eta)$ is the Dirac delta function. This is a non-torsional axisymmetric problem. Introducing the potential functions proposed by Hou et al. [42], and the relatively completed non-torsional axisymmetric general solution around the z-axis is expressed in the following form:

$$2\mu u_r = \frac{\partial \psi_1}{\partial r} + z\frac{\partial \psi_2}{\partial r}, \tag{5.7a}$$

$$2\mu w = \frac{\partial \psi_1}{\partial z} - (3-4v)\psi_2 + z\frac{\partial \psi_2}{\partial z} + 4(1-v)\psi_3, \tag{5.7b}$$

$$\sigma_z = \frac{\partial^2 \psi_1}{\partial z^2} - 2(1-v)\frac{\partial \psi_2}{\partial z} + z\frac{\partial^2 \psi_2}{\partial z^2} + 2(1-v)\frac{\partial \psi_3}{\partial z}, \tag{5.7c}$$

$$\sigma_{zr} = \frac{\partial^2 \psi_1}{\partial r \partial z} - (1-2v)\frac{\partial \psi_2}{\partial r} + z\frac{\partial^2 \psi_2}{\partial r \partial z} + 2(1-v)\frac{\partial \psi_3}{\partial r}, \tag{5.7d}$$

$$\frac{2\mu}{\Gamma}\theta = \frac{\partial \psi_3}{\partial z}, \tag{5.7e}$$

$$h_z = -\frac{\beta\Gamma}{2\mu}\frac{\partial^2 \psi_3}{\partial z^2}, \tag{5.7f}$$

where $\Gamma = \frac{2(1-v)}{\alpha(1+v)}$.

Using the zero-order Hankel transformation technique, the potential functions can be expressed as

$$\psi_1^+ = \int_0^\infty \xi A_1 e^{-\xi z} J_0(\xi r) d\xi, \tag{5.8a}$$

$$\psi_1^- = \int_0^\infty \xi A_2 e^{\xi z} J_0(\xi r) d\xi, \tag{5.8b}$$

$$\psi_2^+ = \int_0^\infty \xi B_1 e^{-\xi z} J_0(\xi r) d\xi, \tag{5.8c}$$

$$\psi_2^- = \int_0^\infty \xi B_2 e^{\xi z} J_0(\xi r) d\xi, \tag{5.8d}$$

$$\psi_3^+ = \int_0^\infty \xi C_1 e^{-\xi z} J_0(\xi r) d\xi, \tag{5.8e}$$

$$\psi_3^- = \int_0^\infty \xi C_2 e^{\xi z} J_0(\xi r) d\xi, \tag{5.8f}$$

where the superscript "+" and "–" denote the upper and lower domain, respectively. When the unit displacement discontinuity in z-direction is applied on the interfacial crack, the corresponding boundary conditions are given in the cylindrical coordinate system as

$$\begin{cases} u_r^+(r,0) - u_r^-(r,0) = 0, \\ w^+(r,0) - w^-(r,0) = \delta(r), \\ \theta^+(r,0) - \theta^-(r,0) = 0, \end{cases} \quad \begin{cases} \sigma_z^+(r,0) - \sigma_z^-(r,0) = 0, \\ \sigma_{zr}^+(r,0) - \sigma_{zr}^-(r,0) = 0, \\ h_z^+(r,0) - h_z^-(r,0) = 0. \end{cases} \tag{5.9}$$

where $0 \le r \le a$ belongs to the crack region. After inserting Eq. (5.8) into Eq. (5.9), one can obtain

$$\begin{cases} \frac{A_1}{\mu_1} = \frac{A_2}{\mu_2}, \\ \frac{1}{2\mu_1}[-\xi A_1 - (3-4\nu_1)B_1 + 4(1-\nu_1)C_1] \\ -\frac{1}{2\mu_2}[\xi A_2 - (3-4\nu_2)B_2 + 4(1-\nu_2)C_2] = \frac{1}{2\pi}, \\ -\frac{\Gamma_1}{\mu_1}C_1 = \frac{\Gamma_2}{\mu_2}C_2, \end{cases} \tag{5.10a}$$

$$\begin{cases} \xi A_1 + 2(1-\nu_1)B_1 - 2(1-\nu_1)C_1 = \xi A_2 - 2(1-\nu_2)B_2 + 2(1-\nu_2)C_2, \\ \xi A_1 + (1-2\nu_1)B_1 - 2(1-\nu_1)C_1 = -\xi A_2 + (1-2\nu_2)B_2 - 2(1-\nu_2)C_2, \\ \frac{\Gamma_1\beta_1}{\mu_1}C_1 = \frac{\Gamma_2\beta_2}{\mu_2}C_2, \end{cases} \tag{5.10b}$$

where the subscripts "1" and "2" denote the upper and lower domains, respectively. Solving these six equations one can get

$$C_1 = C_2 = 0 \tag{5.11a}$$

$$A_1 = -\frac{\mu_1(2 - 3\nu_1 - 3\nu_2 + 4\nu_1\nu_2)}{(3\mu_2 + \mu_1 - 4\mu_2\nu_1)(-\mu_2 - 3\mu_1 + 4\mu_1\nu_2)}\frac{2}{\pi}\frac{1}{\xi} = A_1^*\frac{1}{\xi} \tag{5.11b}$$

$$A_2 = -\frac{\mu_2(2 - 3\nu_1 - 3\nu_2 + 4\nu_1\nu_2)}{(-3\mu_2 - \mu_1 + 4\mu_2\nu_1)(\mu_2 + 3\mu_1 - 4\mu_1\nu_2)}\frac{2}{\pi}\frac{1}{\xi} = A_2^*\frac{1}{\xi} \tag{5.11c}$$

$$B_1 = \frac{\mu_1\mu_2}{2(-\mu_1 - 3\mu_2 + 4\mu_2\nu_1)}\frac{2}{\pi} = -\frac{\mu_1\mu_2}{\pi(\mu_1 + k_1\mu_2)} \tag{5.11d}$$

$$B_2 = \frac{\mu_1\mu_2}{2(\mu_2 + 3\mu_1 - 4\mu_1\nu_2)}\frac{2}{\pi} = \frac{\mu_1\mu_2}{\pi(\mu_2 + \mu_1 k_2)} \tag{5.11e}$$

where $k_\alpha = 3 - 4\nu_\alpha$, A_1^*, A_2^*, B_1 and B_2 are all constants, and the potential functions are obtained as:

$$\psi_1^+ = A_1^*\frac{1}{R} \tag{5.12a}$$

$$\psi_1^- = A_2^*\frac{1}{R} \tag{5.12b}$$

$$\psi_2^+ = B_1\frac{z}{R^3} \tag{5.12c}$$

$$\psi_2^- = -B_2\frac{z}{R^3} \tag{5.12d}$$

$$\psi_3^+ = 0 \tag{5.12e}$$

$$\psi_3^- = 0 \tag{5.12f}$$

where $R = \sqrt{r^2 + z^2}$. Then the corresponding displacements and stresses are obtained, for example, the stresses of an arbitrary point in the upper domain are obtained as:

$$\sigma_z^+ = \frac{\mu_1\mu_2}{2\pi(\mu_1 + k_1\mu_2)}\left[(1 + \eta_2)\frac{1}{R^3} + 3(5 - \eta_2)\frac{z^2}{R^5} - 30\frac{z^4}{R^7}\right] \tag{5.13a}$$

$$\sigma_{zr}^+ = \frac{\mu_1\mu_2}{2\pi(\mu_1 + k_1\mu_2)}\left[3(3 - \eta_2)\frac{rz}{R^5} - 30\frac{rz^3}{R^7}\right] \tag{5.13b}$$

and the forms of the stresses in the Cartesian coordinates are:

$$\sigma_z^+ = \frac{\mu_1\mu_2}{2\pi(\mu_1 + k_1\mu_2)}\left[(1+\eta_2)\frac{1}{R^3} + 3(5-\eta_2)\frac{z^2}{R^5} - 30\frac{z^4}{R^7}\right] \tag{5.14a}$$

$$\sigma_{zx}^+ = \frac{\mu_1\mu_2}{2\pi(\mu_1 + k_1\mu_2)}\left[3(3-\eta_2)\frac{xz}{R^5} - 30\frac{xz^3}{R^7}\right] \tag{5.14b}$$

$$\sigma_{yz}^+ = \frac{\mu_1\mu_2}{2\pi(\mu_1 + k_1\mu_2)}\left[3(3-\eta_2)\frac{yz}{R^5} - 30\frac{yz^3}{R^7}\right] \tag{5.14c}$$

where $\eta_2 = \frac{\mu_1 + \mu_2 k_1}{\mu_2 + \mu_1 k_2}$.

5.2.1.2 Unit Point Temperature Discontinuity of the Crack

The boundary condition in Eq. (5.5d) can be rewritten as

$$\|u(\xi,\eta)\| = 0, \quad \|v(\xi,\eta)\| = 0, \quad \|w(\xi,\eta)\| = 0, \quad \|\theta(\xi,\eta)\| = \delta(\xi,\eta). \tag{5.15}$$

It is also a non-torsional axisymmetric problem. We can use the same form of potential functions in Eq. (5.7) and adopt the same zero-order Hankel transformation in Eq. (5.8). When the unit temperature discontinuity is applied on the interface crack, the corresponding boundary conditions are as follows

$$\begin{cases} u_r^+(r,0) - u_r^-(r,0) = 0 \\ w^+(r,0) - w^-(r,0) = 0 \\ \theta^+(r,0) - \theta^-(r,0) = \delta(r) \end{cases} \quad \begin{cases} \sigma_z^+(r,0) - \sigma_z^-(r,0) = 0 \\ \sigma_{zr}^+(r,0) - \sigma_{zr}^-(r,0) = 0 \\ h_z^+(r,0) - h_z^-(r,0) = 0 \end{cases} \tag{5.16}$$

thus we can obtain

$$\begin{cases} \frac{A_1}{\mu_1} = \frac{A_2}{\mu_2} \\ \frac{1}{\mu_1}[-\xi A_1 - (3-4\nu_1)B_1 + 4(1-\nu_1)C_1] = \frac{1}{\mu_2}[\xi A_2 - (3-4\nu_2)B_2 + 4(1-\nu_2)C_2] \\ -\frac{\Gamma_1}{2\mu_1}C_1 - \frac{\Gamma_2}{2\mu_2}C_2 = \frac{1}{2\pi}\frac{1}{\xi} \end{cases} \tag{5.17a}$$

$$\begin{cases} \xi A_1 + 2(1-\nu_1)B_1 - 2(1-\nu_1)C_1 = \xi\Lambda_2 - 2(1-\nu_2)B_2 + 2(1-\nu_2)C_2 \\ \xi A_1 + (1-2\nu_1)B_1 - 2(1-\nu_1)C_1 = -\xi A_2 + (1-2\nu_2)B_2 - 2(1-\nu_2)C_2 \\ \frac{\Gamma_1\beta_1}{\mu_1}C_1 = \frac{\Gamma_2\beta_2}{\mu_2}C_2 \end{cases} \tag{5.17b}$$

Solving these six equations, we can get

$$A_1 = \frac{(v_1-1)\Gamma_2\beta_2\mu_1(\mu_2+k_2\mu_1)-(v_2-1)\Gamma_1\beta_1\mu_2(\mu_1+k_1\mu_2)}{\pi\Gamma_1\Gamma_2(\beta_1+\beta_2)(\mu_1+k_1\mu_2)(\mu_2+k_2\mu_1)}\frac{2\mu_1}{\xi^2} = A_1^*\frac{1}{\xi^2} \tag{5.18a}$$

$$A_2 = \frac{(v_1-1)\Gamma_2\beta_2\mu_1(\mu_2+k_2\mu_1)-(v_2-1)\Gamma_1\beta_1\mu_2(\mu_1+k_1\mu_2)}{\pi\Gamma_1\Gamma_2(\beta_1+\beta_2)(\mu_1+k_1\mu_2)(\mu_2+k_2\mu_1)}\frac{2\mu_2}{\xi^2} = A_2^*\frac{1}{\xi^2} \tag{5.18b}$$

$$B_1 = \frac{4\mu_1\mu_2(v_1-1)\beta_2}{\pi\Gamma_1(\beta_1+\beta_2)(\mu_1+k_1\mu_2)}\frac{1}{\xi} = B_1^*\frac{1}{\xi} \tag{5.18c}$$

$$B_2 = \frac{4\mu_1\mu_2(v_2-1)\beta_1}{\pi\Gamma_2(\beta_1+\beta_2)(\mu_2+k_2\mu_1)}\frac{1}{\xi} = B_2^*\frac{1}{\xi} \tag{5.18d}$$

$$C_1 = -\frac{\mu_1\beta_2}{\pi\Gamma_1(\beta_1+\beta_2)}\frac{1}{\xi} = C_1^*\frac{1}{\xi} \tag{5.18e}$$

$$C_2 = -\frac{\mu_2\beta_1}{\pi\Gamma_2(\beta_1+\beta_2)}\frac{1}{\xi} = C_2^*\frac{1}{\xi} \tag{5.18f}$$

where A_1^*, A_2^*, B_1^*, B_2^*, C_1^* and C_2^* are constants, and the potential functions are determined as:

$$\psi_1^+ = -A_1^*\ln(R+z) \tag{5.19a}$$

$$\psi_1^- = -A_2^*\ln(R-z) \tag{5.19b}$$

$$\psi_2^+ = B_1^*\frac{1}{R} \tag{5.19c}$$

$$\psi_2^- = B_2^*\frac{1}{R} \tag{5.19d}$$

$$\psi_3^+ = C_1^*\frac{1}{R} \tag{5.19e}$$

$$\psi_3^- = C_2^*\frac{1}{R} \tag{5.19f}$$

Furthermore, the corresponding displacements, stresses, temperature and heat flux caused by the unit temperature discontinuity can be obtained. For example, the stresses and heat flux of an arbitrary point in the upper domain are given as:

$$\sigma_z^+ = \frac{\mu_1\mu_2}{2\pi(\mu_1+k_1\mu_2)}\left\{\frac{2[\alpha_2(1+v_2)\beta_1\eta_2+\alpha_1(1+v_1)\beta_2]}{\beta_1+\beta_2}\frac{z}{R^3}-\frac{12\alpha_1(1+v_1)\beta_2}{\beta_1+\beta_2}\frac{z^3}{R^5}\right\} \tag{5.20a}$$

$$\sigma_{zr}^+ = \frac{\mu_1\mu_2}{2\pi(\mu_1+k_1\mu_2)}\left\{\frac{2[\alpha_2(1+v_2)\beta_1\eta_2+\alpha_1(1+v_1)\beta_2]}{\beta_1+\beta_2}\frac{r}{R^3}-\frac{12\alpha_1(1+v_1)\beta_2}{\beta_1+\beta_2}\frac{rz^2}{R^5}\right\} \tag{5.20b}$$

$$h_z^+ = -\frac{\beta_1\beta_2}{2\pi(\beta_1+\beta_2)}\left(\frac{1}{R^3}-\frac{3z^2}{R^5}\right) \tag{5.20c}$$

In terms of the Cartesian coordinates, they are:

$$\sigma_z^+ = \frac{\mu_1\mu_2}{2\pi(\mu_1+k_1\mu_2)}\left\{\frac{2[\alpha_2(1+v_2)\beta_1\eta_2+\alpha_1(1+v_1)\beta_2]}{\beta_1+\beta_2}\frac{z}{R^3}-\frac{12\alpha_1(1+v_1)\beta_2}{\beta_1+\beta_2}\frac{z^3}{R^5}\right\} \tag{5.21a}$$

$$\sigma_{zx}^+ = \frac{\mu_1\mu_2}{2\pi(\mu_1+k_1\mu_2)}\left\{\frac{2[\alpha_2(1+v_2)\beta_1\eta_2+\alpha_1(1+v_1)\beta_2]}{\beta_1+\beta_2}\frac{x}{R^3}-\frac{12\alpha_1(1+v_1)\beta_2}{\beta_1+\beta_2}\frac{xz^2}{R^5}\right\} \tag{5.21b}$$

$$\sigma_{yz}^+ = \frac{\mu_1\mu_2}{2\pi(\mu_1+k_1\mu_2)}\left\{\frac{2[\alpha_2(1+v_2)\beta_1\eta_2+\alpha_1(1+v_1)\beta_2]}{\beta_1+\beta_2}\frac{y}{R^3}-\frac{12\alpha_1(1+v_1)\beta_2}{\beta_1+\beta_2}\frac{yz^2}{R^5}\right\} \tag{5.21c}$$

$$h_z^+ = -\frac{\beta_1\beta_2}{2\pi(\beta_1+\beta_2)}\left(\frac{1}{R^3}-\frac{3z^2}{R^5}\right) \tag{5.21d}$$

5.2.1.3 Unit-Point Displacement Discontinuity in the y-Direction of the Crack

The boundary condition in Eq. (5.5b) can be rewritten as:

$$\|u(\xi,\eta)\| = 0, \quad \|v(\xi,\eta)\| = \delta(\xi,\eta), \quad \|w(\xi,\eta)\| = 0, \quad \|\theta(\xi,\eta)\| = 0. \tag{5.22}$$

Obviously this problem is no longer non-torsional axisymmetric, which requires a complete, general solution. The relatively completed, general solution around the z-axis is expressed in the following form [42]:

$$2\mu u_r = -\frac{\partial\psi_0}{r\partial\phi} + \frac{\partial\psi_1}{\partial r} + z\frac{\partial\psi_2}{\partial r}, \tag{5.23a}$$

$$2\mu u_\phi = \frac{\partial\psi_0}{\partial r} + \frac{\partial\psi_1}{r\partial\phi} + z\frac{\partial\psi_2}{r\partial\phi}, \tag{5.23b}$$

$$2\mu w = \frac{\partial\psi_1}{\partial z} - (3-4v)\psi_2 + z\frac{\partial\psi_2}{\partial z} + 4(1-v)\psi_3, \tag{5.23c}$$

$$\sigma_z = \frac{\partial^2\psi_1}{\partial z^2} - 2(1-v)\frac{\partial\psi_2}{\partial z} + z\frac{\partial^2\psi_2}{\partial z^2} + 2(1-v)\frac{\partial\psi_3}{\partial z}, \tag{5.23d}$$

$$\sigma_{zr} = -\frac{1}{2r}\frac{\partial^2\psi_0}{\partial\phi\partial z} + \frac{\partial^2\psi_1}{\partial r\partial z} - (1-2v)\frac{\partial\psi_2}{\partial r} + z\frac{\partial^2\psi_2}{\partial r\partial z} + 2(1-v)\frac{\partial\psi_3}{\partial r}, \tag{5.23e}$$

$$\sigma_{z\phi} = \frac{1}{2}\frac{\partial^2\psi_0}{\partial r\partial z} + \frac{1}{r}\frac{\partial^2\psi_1}{\partial\phi\partial z} - (1-2v)\frac{1}{r}\frac{\partial\psi_2}{\partial\phi} + \frac{z}{r}\frac{\partial^2\psi_2}{\partial\phi\partial z} + 2(1-v)\frac{1}{r}\frac{\partial\psi_3}{\partial\phi}, \tag{5.23f}$$

$$\frac{2\mu}{\Gamma}\theta = \frac{\partial\psi_3}{\partial z}, \tag{5.23g}$$

$$h_z = -\frac{\beta\Gamma}{2\mu}\frac{\partial^2\psi_3}{\partial z^2}, \tag{5.23h}$$

When the unit displacement discontinuity in y-direction is applied on the interfacial crack, the corresponding boundary conditions are:

$$\begin{cases} u_r^+(r,0) - u_r^-(r,0) = \delta(r)\sin\phi, \\ u_\phi^+(r,0) - u_\phi^-(r,0) = \delta(r)\cos\phi, \\ w^+(r,0) - w^-(r,0) = 0, \\ \theta^+(r,0) - \theta^-(r,0) = 0, \end{cases} \quad \begin{cases} \sigma_z^+(r,0) - \sigma_z^-(r,0) = 0, \\ \sigma_{zr}^+(r,0) - \sigma_{zr}^-(r,0) = 0, \\ \sigma_{z\phi}^+(r,0) - \sigma_{z\phi}^-(r,0) = 0, \\ h_z^+(r,0) - h_z^-(r,0) = 0. \end{cases} \tag{5.24}$$

According to the boundary conditions, the first-order Hankel transformation is introduced, and the corresponding potential functions are:

$$\psi_1^+ = \int_0^\infty \xi A_1 e^{-\xi z} J_1(\xi r) d\xi \cdot \sin\phi, \tag{5.25a}$$

$$\psi_1^- = \int_0^\infty \xi A_2 e^{\xi z} J_1(\xi r) d\xi \cdot \sin\phi, \tag{5.25b}$$

$$\psi_2^+ = \int_0^\infty \xi B_1 \mathrm{e}^{-\xi z} J_1(\xi r) d\xi \cdot \sin\phi, \tag{5.25c}$$

$$\psi_2^- = \int_0^\infty \xi B_2 \mathrm{e}^{\xi z} J_1(\xi r) d\xi \cdot \sin\phi, \tag{5.25d}$$

$$\psi_3^+ = \int_0^\infty \xi C_1 \mathrm{e}^{-\xi z} J_1(\xi r) d\xi \cdot \sin\phi, \tag{5.25e}$$

$$\psi_3^- = \int_0^\infty \xi C_2 \mathrm{e}^{\xi z} J_1(\xi r) d\xi \cdot \sin\phi, \tag{5.25f}$$

$$\psi_0^+ = \int_0^\infty \xi D_1 \mathrm{e}^{-\xi z} J_1(\xi r) d\xi \cdot \cos\phi, \tag{5.25g}$$

$$\psi_0^- = \int_0^\infty \xi D_2 \mathrm{e}^{\xi z} J_1(\xi r) d\xi \cdot \cos\phi, \tag{5.25h}$$

where superscripts "+" and "–" denote the upper and lower domains, respectively. Substituting Eq. (5.25) into the boundary conditions in Eq. (5.24) yields:

$$\begin{cases} \frac{D_1 - A_1}{\mu_1} - \frac{D_2 - A_2}{\mu_2} = 0 \\ \frac{D_1 + A_1}{2\mu_1} - \frac{D_2 + A_2}{2\mu_2} = \frac{1}{\pi\xi} \\ \frac{1}{\mu_1}[-\xi A_1 - (3 - 4\nu_1)B_1 + 4(1 - \nu_1)C_1] = \frac{1}{\mu_2}[\xi A_2 - (3 - 4\nu_2)B_2 + 4(1 - \nu_2)C_2] \\ -\frac{\Gamma_1}{\mu_1} C_1 = \frac{\Gamma_2}{\mu_2} C_2 \end{cases} \tag{5.26a}$$

$$\begin{cases} \xi A_1 + 2(1 - \nu_1)B_1 - 2(1 - \nu_1)C_1 = \xi A_2 - 2(1 - \nu_2)B_2 + 2(1 - \nu_2)C_2, \\ -\xi(D_1 + 2A_1) - 2(1 - 2\nu_1)B_1 + 4(1 - \nu_1)C_1 = \xi(D_2 + 2A_2) - 2(1 - 2\nu_2)B_2 + 4(1 - \nu_2)C_2, \\ -\xi(D_1 + 2A_1) + 2(1 - 2\nu_1)B_1 - 4(1 - \nu_1)C_1 = \xi(D_2 - 2A_2) + 2(1 - 2\nu_2)B_2 - 4(1 - \nu_2)C_2, \\ \frac{\Gamma_1\beta_1}{\mu_1} C_1 = \frac{\Gamma_2\beta_2}{\mu_2} C_2, \end{cases} \tag{5.26b}$$

where subscripts "1" and "2" denote the upper and lower domains, respectively. Solving Eq. (5.26), one can get

$$A_1 = -\frac{\mu_2(-3+4v_1)+\mu_1(-5+6v_1+6v_2-8v_1v_2)}{(\mu_1+\mu_2k_1)(\mu_2+\mu_1k_2)}\frac{1}{\pi\xi} = A_1^*\frac{1}{\xi}, \tag{5.27a}$$

$$A_2 = -\frac{\mu_1(3-4v_2)+\mu_2(5-6v_2-6v_2+8v_1v_2)}{(\mu_1+\mu_2k_1)(\mu_2+\mu_1k_2)}\frac{1}{\pi\xi} = A_2^*\frac{1}{\xi}, \tag{5.27b}$$

$$B_1 = \frac{\mu_1\mu_2}{(-\mu_2-3\mu_1+4\mu_1v_2)}\frac{1}{\pi} = -\frac{\mu_1\mu_2}{\pi(\mu_2+\mu_1k_2)}, \tag{5.27c}$$

$$B_2 = \frac{\mu_1\mu_2}{(-\mu_1-3\mu_2+4\mu_2v_1)}\frac{1}{\pi} = -\frac{\mu_1\mu_2}{\pi(\mu_1+\mu_2k_1)}, \tag{5.27d}$$

$$C_1 = C_2 = 0, \tag{5.27e}$$

$$D_1 = -\frac{\mu_1\mu_2}{\mu_1+\mu_2}\cdot\frac{1}{\pi\xi} = D_1^*\frac{1}{\xi}, \tag{5.27f}$$

$$D_2 = \frac{\mu_1\mu_2}{\mu_1+\mu_2}\cdot\frac{1}{\pi\xi} = D_2^*\frac{1}{\xi}, \tag{5.27g}$$

where $k_\alpha = 3-4v_\alpha$, A_1^*, A_2^*, D_1^*, D_2^*, B_1 and B_2 are all constants, and the potential functions are obtained as:

$$\psi_0^+ = D_1^*\frac{1}{r}\left(1-\frac{z}{R}\right)\cos\phi, \tag{5.28a}$$

$$\psi_0^- = D_2^*\frac{1}{r}\left(1+\frac{z}{R}\right)\cos\phi, \tag{5.28b}$$

$$\psi_1^+ = A_1^*\frac{1}{r}\left(1-\frac{z}{R}\right)\sin\phi, \tag{5.28c}$$

$$\psi_1^- = A_2^*\frac{1}{r}\left(1+\frac{z}{R}\right)\sin\phi, \tag{5.28d}$$

$$\psi_2^+ = B_1\frac{r}{R^3}\sin\phi, \tag{5.28e}$$

$$\psi_2^- = -B_2\frac{r}{R^3}\sin\phi, \tag{5.28f}$$

$$\psi_3^+ = 0, \tag{5.28g}$$

$$\psi_3^- = 0, \tag{5.28h}$$

where $R = \sqrt{r^2 + z^2}$. Then, the corresponding displacements and stresses are all obtained. For example, the stresses of an arbitrary point in the upper domain are obtained in the Cartesian coordinates:

$$\sigma_z^+ = \frac{\mu_1\mu_2}{2\pi(\mu_1 + k_1\mu_2)}\left\{\frac{3(1+\eta_2)yz}{R^5} - \frac{30yz^3}{R^7}\right\}, \tag{5.29a}$$

$$\sigma_{zx}^+ = \frac{\mu_1\mu_2}{2\pi(\mu_1 + k_1\mu_2)}\left\{\frac{3(1-\eta_1+\eta_2)xy}{R^5} - \frac{30xyz^2}{R^7}\right\}, \tag{5.29b}$$

$$\sigma_{yz}^+ = \frac{\mu_1\mu_2}{2\pi(\mu_1 + k_1\mu_2)}\left\{\frac{2\eta_1 - \eta_2 - 1}{R^3} + 3\frac{(1-\eta_1+\eta_2)y^2 + (2-\eta_2)z^2}{R^5} - 30\frac{y^2z^2}{R^7}\right\} \tag{5.29c}$$

5.2.1.4 Unit Point Displacement Discontinuity in the x-Direction of the Crack

Similar to Sect. 5.2.1.3, as the problem is symmetric about x and y-axis, under unit point displacement discontinuity in the x-direction, the stresses of an arbitrary point in the upper domain are obtained in the Cartesian coordinates

$$\sigma_z^+ = \frac{\mu_1\mu_2}{2\pi(\mu_1 + k_1\mu_2)}\left\{\frac{3(1+\eta_2)xz}{R^5} - \frac{30xz^3}{R^7}\right\}, \tag{5.30a}$$

$$\sigma_{zx}^+ = \frac{\mu_1\mu_2}{2\pi(\mu_1 + k_1\mu_2)}\left\{\frac{2\eta_1 - \eta_2 - 1}{R^3} + 3\frac{(1-\eta_1+\eta_2)x^2 + (2-\eta_2)z^2}{R^5} - 30\frac{x^2z^2}{R^7}\right\}, \tag{5.30b}$$

$$\sigma_{yz}^+ = \frac{\mu_1\mu_2}{2\pi(\mu_1 + k_1\mu_2)}\left\{\frac{3(1-\eta_1+\eta_2)xy}{R^5} - \frac{30xyz^2}{R^7}\right\} \tag{5.30c}$$

5.2.2 *Boundary Integral-Differential Equations for Interfacial Cracks*

Consider a three-dimensional, isotropic, thermoelastic bimaterial with the two materials denoted as 1 and 2, respectively. A Cartesian coordinate system is set up with the *xoy* plane lying in the interface, and an arbitrarily shaped, planar crack S is located on the interface, as shown in Fig. 5.2.

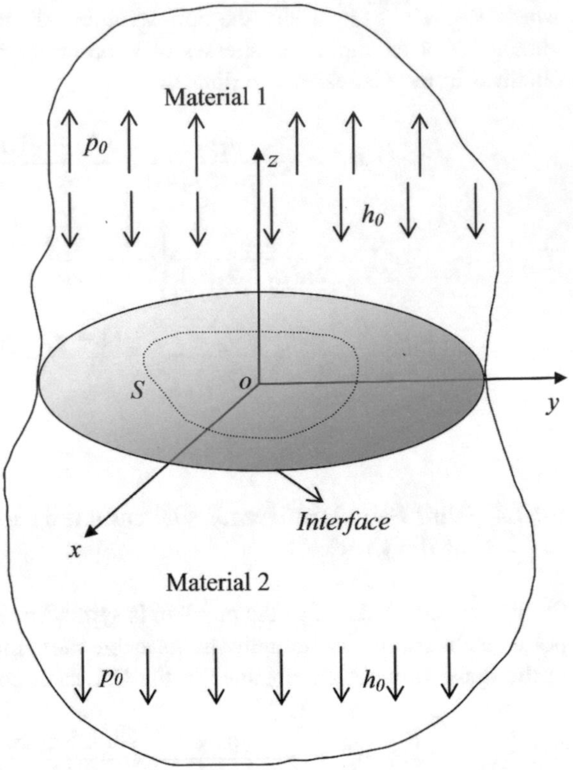

Fig. 5.2 A planar interface crack S of arbitrary shape in a bi-material system under thermomechanical loading [39]

With the aid of the fundamental solutions obtained above, one can get the fundamental solutions for the arbitrarily shaped, interface crack S. The corresponding stresses and heat flux at an arbitrary internal point in domain 1 can be expressed in terms of the displacement and temperature discontinuities across crack faces as follows:

$$\sigma_{zx}^{+} = \frac{\mu_1\mu_2}{2\pi(\mu_1+k_1\mu_2)}\int\limits_{S^+}\left[\frac{2\eta_1-\eta_2-1}{R^3}+3(2-\eta_2)\frac{z^2}{R^5}+\left(3\frac{1-\eta_1+\eta_2}{R^5}-30\frac{z^2}{R^7}\right)(\xi-x)^2\right]\|u\|$$
$$+\left(3\frac{1-\eta_1+\eta_2}{R^5}-30\frac{z^2}{R^7}\right)(\xi-x)(\eta-y)\|v\|+\left(30\frac{z^3}{R^7}-3\frac{(3-\eta_1)z}{R^5}\right)(\xi-x)\|w\|$$
$$+\left[\frac{12\alpha_1(1+v_1)\beta_2}{\beta_1+\beta_2}\frac{z^2}{R^5}-\frac{\alpha_1(1+v_1)\beta_2+\alpha_2(1+v_2)\beta_1\eta_2}{(\beta_1+\beta_2)/2}\frac{1}{R^3}\right](\xi-x)\|\theta\|\Big\}dS \tag{5.31a}$$

$$
\begin{aligned}
\sigma_{yz}^{+} = & \frac{\mu_1\mu_2}{2\pi(\mu_1+k_1\mu_2)}\int\limits_{S^+}\left\{\left(3\frac{1-\eta_1+\eta_2}{R^5}-30\frac{z^2}{R^7}\right)(\xi-x)(\eta-y)\|u\|\right. \\
& +\left[\frac{2\eta_1-\eta_2-1}{R^3}+3(2-\eta_2)\frac{z^2}{R^5}+\left(3\frac{1-\eta_1+\eta_2}{R^5}-30\frac{z^2}{R^7}\right)(\eta-y)^2\right]\|v\| \\
& +\left(30\frac{z^3}{R^7}-3\frac{(3-\eta_1)z}{R^5}\right)(\eta-y)\|w\| \\
& \left.+\left[\frac{12\alpha_1(1+v_1)\beta_2}{\beta_1+\beta_2}\frac{z^2}{R^5}-\frac{\alpha_1(1+v_1)\beta_2+\alpha_2(1+v_2)\beta_1\eta_2}{(\beta_1+\beta_2)/2}\frac{1}{R^3}\right](\eta-y)\|\theta\|\right\}dS
\end{aligned}
\tag{5.31b}
$$

$$
\begin{aligned}
\sigma_{z}^{+} = & \frac{\mu_1\mu_2}{2\pi(\mu_1+k_1\mu_2)}\int\limits_{S^+}\left\{\left[30\frac{z^3}{R^7}-3\frac{(1+\eta_2)z}{R^5}\right](\xi-x)\|u\|\right. \\
& +\left[30\frac{z^3}{R^7}-3\frac{(1+\eta_2)z}{R^5}\right](\eta-y)\|v\| \\
& +\left[\frac{1+\eta_2}{R^3}+\frac{3(5-\eta_2)z^2}{R^5}-30\frac{z^4}{R^7}\right]\|w\| \\
& \left.+\left[\frac{\alpha_1(1+v_1)\beta_2+\alpha_2(1+v_2)\beta_1\eta_2}{(\beta_1+\beta_2)/2}\frac{z}{R^3}-\frac{12\alpha_1(1+v_1)\beta_2}{\beta_1+\beta_2}\frac{z^2}{R^5}\right]\|\theta\|\right\}dS
\end{aligned}
\tag{5.31c}
$$

$$
h_z^{+} = -\frac{\beta_1\beta_2}{2\pi(\beta_1+\beta_2)}\int_{S^+}\left(\frac{1}{R^3}-\frac{3z^2}{R^5}\right)\|\theta\|dS
\tag{5.31d}
$$

where $R=\sqrt{(\xi-x)^2+(\eta-y)^2+z^2}$.

5.2.2.1 Hypersingular Integral-Differential Equations

As the internal point approaches the crack face, the integrals in Eq. (5.34) will become hypersingular. These integrals must be evaluated using finite-part integrals, and the following formulas are applied [33–35]:

$$
I_1 = \lim_{z\to 0}\int\limits_{S^+}\|u(\xi,\eta)\|\frac{(\xi-x)z}{R^5}d\xi d\eta = \frac{2\pi}{3}\frac{\partial\|u(x,y)\|}{\partial x}
\tag{5.32a}
$$

$$
I_2 = \lim_{z\to 0}\int\limits_{S^+}\|w(\xi,\eta)\|\frac{z^2}{R^5}d\xi d\eta = 0
\tag{5.32b}
$$

After some algebraic manipulations, one can obtain the boundary integral-differential equations for an arbitrarily shaped, interfacial crack:

$$\begin{aligned}
&\int_{S^+}\left\{\left[\frac{2\eta_1-\eta_2-1}{r^3}+\frac{3(1-\eta_1+\eta_2)\cos^2\phi}{r^3}\right]\|u(\xi,\eta)\|\right.\\
&\quad+\frac{3(1-\eta_1+\eta_2)\sin\phi\cos\phi}{r^3}\|v(\xi,\eta)\|\\
&\left.-\frac{\alpha_1(1+v_1)\beta_2+\alpha_2(1+v_2)\beta_1\eta_2}{(\beta_1+\beta_2)/2}\frac{\cos\phi}{r^2}\|\theta(\xi,\eta)\|\right\}dS\\
&+2\pi(\eta_2-1)\frac{\partial\|w(x,y)\|}{\partial x}=-\frac{2\pi(\mu_1+k_1\mu_2)}{\mu_1\mu_2}p_x(x,y)
\end{aligned}\tag{5.33a}$$

$$\begin{aligned}
&\int_{S^+}\left\{\left[\frac{2\eta_1-\eta_2-1}{r^3}+\frac{3(1-\eta_1+\eta_2)\sin^2\phi}{r^3}\right]\|v(\xi,\eta)\|\right.\\
&\quad+\frac{3(1-\eta_1+\eta_2)\sin\phi\cos\phi}{r^3}\|u(\xi,\eta)\|\\
&\left.-\frac{\alpha_1(1+v_1)\beta_2+\alpha_2(1+v_2)\beta_1\eta_2}{(\beta_1+\beta_2)/2}\frac{\sin\phi}{r^2}\|\theta(\xi,\eta)\|\right\}dS\\
&+2\pi(\eta_2-1)\frac{\partial\|w(x,y)\|}{\partial y}=-\frac{2\pi(\mu_1+k_1\mu_2)}{\mu_1\mu_2}p_y(x,y)
\end{aligned}\tag{5.33b}$$

$$\begin{aligned}
&\int_{S^+}\frac{1+\eta_2}{r^3}\|w(\xi,\eta)\|dS+2\pi(1-\eta_2)\left(\frac{\partial\|u(x,y)\|}{\partial x}+\frac{\partial\|v(x,y)\|}{\partial y}\right)\\
&=-\frac{2\pi(\mu_1+k_1\mu_2)}{\mu_1\mu_2}p_z(x,y)
\end{aligned}\tag{5.33c}$$

$$-\frac{\beta_1\beta_2}{2\pi(\beta_1+\beta_2)}\int_{S^+}\frac{1}{r^3}\|\theta(\xi,\eta)\|dS=-h_z(x,y)\tag{5.33d}$$

where

$$r=\sqrt{(\xi-x)^2+(\eta-y)^2},\cos\phi=(\xi-x)/r,\sin\phi=(\eta-y)/r.$$

5.2.2.2 Singular Behavior Near the Interface Crack Front

The singular behavior of interfacial crack in a three-dimensional, two-phase elastic medium were analyzed by Tang et al. [33]. Following the similar procedure, the interfacial crack in a three-dimensional, two-phase thermoelastic medium will be

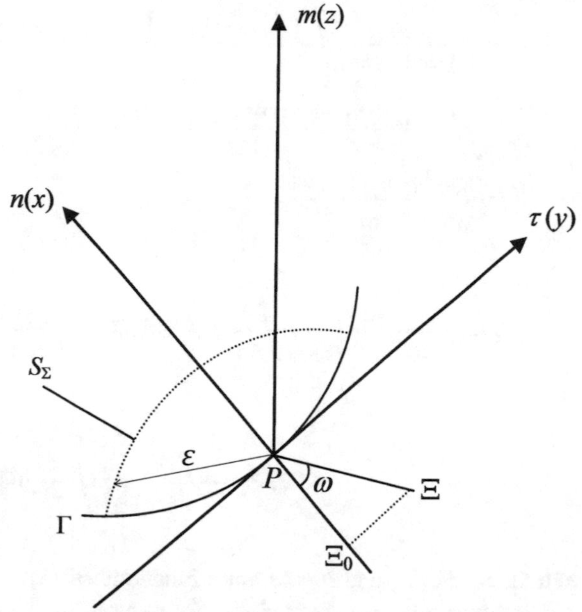

Fig. 5.3 The local intrinsic coordinate system at the interface crack front [39]

discussed. An arbitrary point P is chosen on the crack edge Γ of crack S. The edge Γ is smooth at point P. Without loss of generality, the Cartesian coordinate system $oxyz$ is oriented so that the x-direction and y-direction are normal and tangent to Γ, respectively, and a local orthogonal, intrinsic coordinate system $(n, \tau, m) = (x, y, z)$ at point P along the smooth periphery of the planar crack is set up. A given small circular area Σ contained in plane S with its radius defined as ε is centered at point P, as shown in Fig. 5.3.

The hypersingular parts of Eq. (5.33) should be finite in Σ for finite prescribed mechanical and heat flux loadings:

$$F_x = -\frac{\mu_1\mu_2}{2\pi(\mu_1 + k_1\mu_2)}\int_{\Sigma}\left\{\left[\frac{2\eta_1 - \eta_2 - 1}{r^3} + \frac{3(1-\eta_1+\eta_2)\cos^2\phi}{r^3}\right]\|u\| + \frac{3(1-\eta_1+\eta_2)\sin\phi\cos\phi}{r^5}\|v\| - \frac{\alpha_1(1+v_1)\beta_2 + \alpha_2(1+v_2)\beta_1\eta_2}{(\beta_1+\beta_2)/2}\frac{\cos\phi}{r^2}\|\theta\|\right\}dS - \frac{\mu_1\mu_2(\eta_2-1)}{(\mu_1+k_1\mu_2)}\frac{\partial\|w\|}{\partial x} \tag{5.34a}$$

$$F_y = -\frac{\mu_1\mu_2}{2\pi(\mu_1+k_1\mu_2)}\int\limits_{\Sigma}\left\{\left[\frac{2\eta_1-\eta_2-1}{r^3}+\frac{3(1-\eta_1+\eta_2)\sin^2\phi}{r^3}\right]\|v\|\right.$$
$$\left.+\frac{3(1-\eta_1+\eta_2)\sin\phi\cos\phi}{r^5}\|u\|-\frac{\alpha_1(1+v_1)\beta_2+\alpha_2(1+v_2)\beta_1\eta_2}{(\beta_1+\beta_2)/2}\frac{\sin\phi}{r^2}\|\theta\|\right\}dS$$
$$-\frac{\mu_1\mu_2(\eta_2-1)}{(\mu_1+k_1\mu_2)}\frac{\partial\|w\|}{\partial y} \tag{5.34b}$$

$$F_z = -\frac{\mu_1\mu_2}{2\pi(\mu_1+k_1\mu_2)}\int\limits_{\Sigma}\frac{1+\eta_2}{r^3}\|w\|dS+\frac{\mu_1\mu_2(\eta_2-1)}{(\mu_1+k_1\mu_2)}\left(\frac{\partial\|u\|}{\partial x}+\frac{\partial\|v\|}{\partial y}\right) \tag{5.34c}$$

$$F_\theta = \frac{\beta_1\beta_2}{2\pi(\beta_1+\beta_2)}\int\limits_{\Sigma}\frac{1}{r^3}\|\theta\|dS \tag{5.34d}$$

where F_x, F_y, F_z and F_θ are finite functions of $(x,y)\in\Sigma$.

In the neighborhood of point P, for a small S_Σ, the displacement and temperature discontinuities are related only to the coordinates x in the normal plane through point P as well as the position of point P along the contour of the crack [33, 36]

$$\begin{cases}\|u\| = A_1(P)\xi^{\alpha_1},\\ \|v\| = A_2(P)\xi^{\alpha_2},\\ \|w\| = A_3(P)\xi^{\alpha_3},\\ \|\theta\| = A_4(P)\xi^{\alpha_4},\end{cases} \tag{5.35}$$

where A_1, A_2, A_3 and A_4 are all complex constants [33], and $0<\mathrm{Re}[\alpha_i]<1$, inserting Eq. (5.35) into Eq. (5.34), and using the following identities [33]

$$\lim_{\varepsilon\to 0}\int\limits_{\Sigma}\frac{\|u\|}{r^3}d\xi d\eta = -2A_1(P)\pi\alpha_1\cot(\pi\alpha_1)\xi^{\alpha_1-1} \tag{5.36a}$$

$$\lim_{\varepsilon\to 0}\int\limits_{\Sigma}\frac{(\xi-x)^2\|u\|}{r^5}d\xi d\eta = -\frac{4}{3}A_1(P)\pi\alpha_1\cot(\pi\alpha_1)\xi^{\alpha_1-1} \tag{5.36b}$$

$$\lim_{\varepsilon\to 0}\int\limits_{\Sigma}\frac{(\xi-x)(\eta-y)\|u\|}{r^5}d\xi d\eta = 0 \tag{5.36c}$$

$$\lim_{\varepsilon\to 0}\int\limits_{\Sigma}\frac{(\eta-y)^2\|u\|}{r^5}d\xi d\eta = -\frac{2}{3}A_1(P)\pi\alpha_1\cot(\pi\alpha_1)\xi^{\alpha_1-1} \tag{5.36d}$$

$$\lim_{\varepsilon \to 0} \int_{\Sigma} \frac{(\eta - y)\|\theta\|}{r^3} d\xi d\eta = 0 \tag{5.36e}$$

$$\lim_{\varepsilon \to 0} \int_{\Sigma} \frac{(\xi - x)\|\theta\|}{r^3} d\xi d\eta = -2A_4(P)\cot(\pi\alpha_4)\xi^{\alpha_4} \tag{5.36f}$$

assuming $\xi \to 0$, yields

$$\begin{cases} \cot(\pi\beta)A_1(P) + \gamma A_3(P) = 0 \\ \cot(\pi\alpha_2)A_2(P) = 0 \\ \gamma A_1(P) - \cot(\pi\beta)A_3(P) = 0 \\ \cot(\pi\alpha_4)A_4(P) = 0 \end{cases} \tag{5.37}$$

where $\beta = \alpha_1 = \alpha_3$, $\gamma = (1 - \eta_2)/(1 + \eta_2)$. As the constants $A_i(P)$ are generally assumed to be non-zero, from Eq. (5.37), the characteristic equations to determine the indices of singular behavior, α_i, are found as

$$\begin{cases} \cot(\pi\alpha_2) = 0 \\ \cot(\pi\alpha_4) = 0 \\ \cot(\pi\beta) = \pm\gamma \mathrm{i} \end{cases} \tag{5.38}$$

where $\mathrm{i} = \sqrt{-1}$, and the acceptable roots are

$$\alpha_2 = \alpha_4 = \frac{1}{2}, \ \beta_{1,2} = \frac{1}{2} \pm \mathrm{i}\varepsilon, \ \varepsilon = \frac{1}{2\pi}\ln \eta_2 \tag{5.39}$$

The roots show that both the displacement discontinuity $\|v\|$ and temperature discontinuity $\|\theta\|$ have the classical singularity index 1/2, while $\|u\|$ or $\|w\|$ has the same singularity index $\frac{1}{2} + \mathrm{i}\varepsilon$ as for pure elastic materials. Considering that the displacement and temperature discontinuities in the neighborhood of the interfacial crack tip, S_Σ, they should have the following forms

$$\begin{cases} \|u\| = \mathrm{Re}[A_1(P)\xi^{\alpha}], \\ \|v\| = A_2(P)\xi^{1/2}, \\ \|w\| = \mathrm{Re}[A_3(P)\xi^{\alpha}], \\ \|\theta\| = A_4(P)\xi^{1/2}, \end{cases} \tag{5.40}$$

where $\alpha = \beta_1 = \frac{1}{2} + \mathrm{i}\varepsilon$, $A_2(P)$ and $A_4(P)$ are arbitrary real constants, while $A_1(P)$ and $A_3(P)$ are arbitrary complex constants which satisfy the relation of $A_1(P) - \mathrm{i}A_3(P)$. So they can also be written as

$$\|w\| + i\|u\| = [A_R(P) + iA_I(P)]\xi^\alpha = A(P)\xi^\alpha \tag{5.41}$$

It should be pointed out that the above results are necessary for studying the singular stress fields near the interface crack front exactly and guaranteeing a unique solution for the hypersingular, integral-differential equations expressed above.

5.2.2.3 Singular Stress and Heat Flux Fields Ahead of the Interfacial Crack Front

The stress and heat flux fields of an arbitrary point in the upper domain induced by the interface crack have been derived already in Eq. (5.34). In order to obtain the exact expressions for the singular stress and heat flux fields of an arbitrary point $\Xi(x, y, z)$ ahead of the periphery of the interface crack, the local nature coordinate system (P, n, m) and the local polar coordinates (ρ, ω) with $P\Xi$ as the negative normal axis are introduced as shown in Fig. 5.3. We then have

$$x = -\rho\cos\omega,\ y = 0,\ z = \rho\sin\omega,\ R = \sqrt{(\xi + \rho\cos\omega)^2 + \eta^2 + (\rho\sin\omega)^2} \tag{5.42}$$

For simplicity, we only study the singular stress and heat flux fileds of Ξ_0 in the interface crack front area ($\omega = 0$). Substituting Eq. (5.42) into Eq. (5.31) and combining with the singular index obtained in Eq. (5.39) as well as the form of the expressions for the displacement and temperature discontinuities in Eq. (5.40), one can obtain the following dominant-part integrals [33]

$$\lim_{\Sigma_\varepsilon \to 0} \int_{S_\varepsilon} \frac{\|u(\xi,\eta)\|}{R^3} dS = \frac{\mathrm{i}\pi}{2\cosh(\pi\varepsilon)} \left[\overline{A(P)}(1 - 2\mathrm{i}\varepsilon)\rho^{-1/2-\mathrm{i}\varepsilon} - A(P)(1 + 2\mathrm{i}\varepsilon)\rho^{-1/2+\mathrm{i}\varepsilon}\right] \tag{5.43a}$$

$$\lim_{\Sigma_\varepsilon \to 0} \int_{S_\varepsilon} \frac{(\xi - x)^2\|u(\xi,\eta)\|}{R^5} dS = \frac{\mathrm{i}\pi}{3\cosh(\pi\varepsilon)} \left[\overline{A(P)}(1 - 2\mathrm{i}\varepsilon)\rho^{-1/2-\mathrm{i}\varepsilon} - A(P)(1 + 2\mathrm{i}\varepsilon)\rho^{-1/2+\mathrm{i}\varepsilon}\right] \tag{5.43b}$$

$$\lim_{\Sigma_\varepsilon \to 0} \int_{S_\varepsilon} \frac{(\xi - x)(\eta - y)\|u(\xi,\eta)\|}{R^5} dS = 0 \tag{5.43c}$$

$$\lim_{\Sigma_\varepsilon \to 0} \int_{S_\varepsilon} \frac{\|v(\xi,\eta)\|}{R^3} dS = \pi A_2(P)\rho^{-1/2} \tag{5.43d}$$

$$\lim_{\Sigma_\varepsilon \to 0} \int_{S_\varepsilon} \frac{(\eta - y)^2 \|v(\xi,\eta)\|}{R^5} dS = \frac{\pi}{3} A_2(P)\rho^{-1/2} \tag{5.43e}$$

$$\lim_{\Sigma_\varepsilon \to 0} \int_{S_\varepsilon} \frac{(\xi - x)(\eta - y)\|v(\xi,\eta)\|}{R^5} dS = 0 \tag{5.43f}$$

$$\lim_{\Sigma_\varepsilon \to 0} \int_{S_\varepsilon} \frac{\|w(\xi,\eta)\|}{R^3} dS = \frac{\pi}{2\cosh(\pi\varepsilon)} \left[\overline{A(P)}(1 - 2\mathrm{i}\varepsilon)\rho^{-1/2-\mathrm{i}\varepsilon} + A(P)(1 + 2\mathrm{i}\varepsilon)\rho^{-1/2+\mathrm{i}\varepsilon}\right] \tag{5.43g}$$

$$\lim_{\Sigma_\varepsilon \to 0} \int_{S_\varepsilon} \frac{\|\theta(\xi,\eta)\|}{R^3} dS = \pi A_4(P)\rho^{-1/2} \tag{5.43h}$$

$$\lim_{\Sigma_\varepsilon \to 0} \int_{S_\varepsilon} \frac{(\xi - x)\|\theta(\xi,\eta)\|}{R^3} dS = -2A_4(P)\rho^{1/2} \tag{5.43i}$$

$$\lim_{\Sigma_\varepsilon \to 0} \int_{S_\varepsilon} \frac{(\eta - y)\|\theta(\xi,\eta)\|}{R^3} dS = 0 \tag{5.43j}$$

where $\overline{A(P)}$ is the conjugate of $A(P)$. Substituting the above integrals into Eq. (5.31), we can get the singular stress and heat flux fields in the neighborhood of point P

$$\begin{aligned} \sigma_{zx}(\rho,0) = {} & \mathrm{i}\frac{\mu_1\mu_2(1+\eta_2)}{4(\mu_1 + k_1\mu_2)\cosh(\pi\varepsilon)} \left[\overline{A(P)}(1 - 2\mathrm{i}\varepsilon)\rho^{-1/2-\mathrm{i}\varepsilon} - A(P)(1 + 2\mathrm{i}\varepsilon)\rho^{-1/2+\mathrm{i}\varepsilon}\right] \\ & + \frac{\mu_1\mu_2}{\pi(\mu_1 + k_1\mu_2)} \frac{\alpha_1(1+\nu_1)\beta_2 + \alpha_2(1+\nu_2)\beta_1\eta_2}{(\beta_1+\beta_2)/2} A_4(P)\rho^{1/2} \end{aligned} \tag{5.44a}$$

$$\sigma_{yz}(\rho,0) = \frac{\mu_1\mu_2}{2(\mu_1+\mu_2)} A_2(P)\rho^{-1/2} \tag{5.44b}$$

$$\begin{aligned} \sigma_z(\rho,0) = {} & \frac{\mu_1\mu_2(1+\eta_2)}{4(\mu_1 + k_1\mu_2)\cosh(\pi\varepsilon)} \\ & \left[\overline{A(P)}(1 - 2\mathrm{i}\varepsilon)\rho^{-1/2-\mathrm{i}\varepsilon} + A(P)(1 + 2\mathrm{i}\varepsilon)\rho^{-1/2+\mathrm{i}\varepsilon}\right] \end{aligned} \tag{5.44c}$$

$$h_z(\rho,0) = -\frac{\beta_1\beta_2}{2(\beta_1+\beta_2)}A_4(P)\rho^{-1/2} \tag{5.44d}$$

Clearly both $\sigma_{yz}(\rho,0)$ and $h_z(\rho,0)$ have the classical square root singularity, $r^{-1/2}$, while the stresses $\sigma_{zx}(\rho,0)$ and $\sigma_z(\rho,0)$ are oscillatory with the singularity of $r^{-1/2+\mathrm{i}\varepsilon}$. It is also noted that the temperature term induces the stress $\sigma_{zx}(\rho,0)$ while the influence will vanish as the point approaching the crack front. In other words, the temperature term does not contribute to the singular stress, $\sigma_{zx}(\rho,0)$.

5.2.3 Stress Intensity Factor and Energy Release Rate

Based on the obtained stress and heat flux fields in the neighborhood of the crack front, and using the definition of stress intensity factors for interfacial cracks by Hutchinson et al. [43], we can get the stress and heat flux intensity factors at the crack front

$$K_{\mathrm{I}} = \lim_{\rho\to 0}\sqrt{2\pi\rho}\rho^{-\mathrm{i}\varepsilon}\sigma_z(\rho,0) \tag{5.45a}$$

$$K_{\mathrm{II}} = \lim_{\rho\to 0}\sqrt{2\pi\rho}\rho^{-\mathrm{i}\varepsilon}\sigma_{zx}(\rho,0) \tag{5.45b}$$

$$K = K_{\mathrm{I}} + \mathrm{i}K_{\mathrm{II}} = \lim_{\rho\to 0}\sqrt{2\pi\rho}\rho^{-\mathrm{i}\varepsilon}[\sigma_z(\rho,0)+\mathrm{i}\sigma_{zx}(\rho,0)] \tag{5.45c}$$

$$K_{\mathrm{III}} = \lim_{\rho\to 0}\sqrt{2\pi\rho}\sigma_{yz}(\rho,0) \tag{5.45d}$$

$$K_h = \lim_{\rho\to 0}\sqrt{2\pi\rho}h_z(\rho,0) \tag{5.45e}$$

They are expressed in terms of the displacement and temperature discontinuities as

$$K = \frac{\sqrt{2\pi}\mu_1\mu_2(1+\eta_2)(1+2\mathrm{i}\varepsilon)}{2(\mu_1+k_1\mu_2)\cosh(\pi\varepsilon)}\lim_{\rho\to 0}\frac{\|w(\rho,0)\|+\mathrm{i}\|u(\rho,0)\|}{\rho^{1/2+\mathrm{i}\varepsilon}} \tag{5.46a}$$

$$K_{\mathrm{III}} = \frac{\sqrt{2\pi}\mu_1\mu_2}{2(\mu_1+\mu_2)}\lim_{\rho\to 0}\frac{\|v(\rho,0)\|}{\sqrt{\rho}} \tag{5.46b}$$

$$K_h = -\frac{\sqrt{2\pi}\beta_1\beta_2}{2(\beta_1+\beta_2)}\lim_{\rho\to 0}\frac{\|\theta(\rho,0)\|}{\sqrt{\rho}} \tag{5.46c}$$

When the bimaterial is reduced to a homogeneous solid, the stress intensity factors are the same as those for homogeneous materials [39].

It is worth noting that the oscillatory singularity appeared in the stress intensity factors is a physically unrealistic phenomenon, thus the well-defined quantity, strain energy release rate having a relevance to conventional fracture mechanics, is necessary to be evaluated. From the above expressions for intensity factors in Eq. (5.46), we can get the displacement jumps across the crack surfaces at the interface crack front edge in the form

$$\Delta U(\rho,0) = \|w(\rho,0)\| + \mathrm{i}\|u(\rho,0)\| = \sqrt{\frac{2}{\pi}} \frac{(\mu_1 + k_1\mu_2)\cosh(\pi\varepsilon)}{\mu_1\mu_2(1+\eta_2)(1+2\mathrm{i}\varepsilon)} K\rho^{1/2+\mathrm{i}\varepsilon} \tag{5.47a}$$

$$\|v(\rho,0)\| = \sqrt{\frac{2}{\pi}}\left(\frac{1}{\mu_1} + \frac{1}{\mu_2}\right) K_{\mathrm{III}}\sqrt{\rho} \tag{5.47b}$$

Based on the virtual crack closure method and following the same procedure proposed in [33, 34], let the interfacial crack front advance locally a very small new crack surface area, ΔS, at point *P*, as shown in Fig. 5.4, the local energy release rate for a unit area of interface to debond can be readily evaluated in the integral form utilizing the obtained tractions and displacement fields ahead of the interface crack front

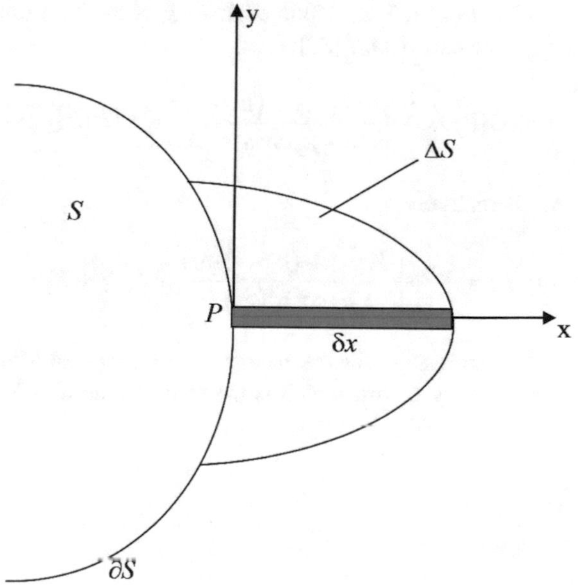

Fig. 5.4 An interface crack with virtual incremental area ΔS of the crack surface [39]

$$G(P) = \lim_{\Delta S \to 0} \frac{1}{2\Delta S} \int_{\Delta S} \{[\sigma_z(x,y) + \mathrm{i}\sigma_{zx}(x,y)] \cdot \Delta U(x,y) + \sigma_{yz}(x,y)\|v(x,y)\|\} dS \tag{5.48}$$

The infinitesimal δx normal to the planar interfacial crack front edge ∂S is used to describe the virtual extension of the crack front. The total changes of the crack surface are expressed as

$$\Delta S = \int_{\partial S(y)} \delta x dy \tag{5.49}$$

Substituting Eq. (5.49) into Eq. (5.48), one can get

$$G(P) = \lim_{\Delta S \to 0} \frac{1}{2\int_{\partial s(y)} \delta x dy} \int_{\partial s(y)} \int_{0}^{\delta x} \left[\frac{K(P)\overline{K(P)}}{4\pi(1+2i\varepsilon)\cosh \pi\varepsilon} \left(\frac{1+k_1}{\mu_1} + \frac{1+k_2}{\mu_2} \right) \left(\frac{\delta x - x}{x} \right)^{1/2+i\varepsilon} + \frac{(\mu_1+\mu_2)K_{\mathrm{III}}^2(P)}{\pi\mu_1\mu_2} \sqrt{\frac{\delta x - x}{x}} \right] dxdy \tag{5.50}$$

where the inner integral of the first term is recognized as the complex beta function $B(1/2 + \mathrm{i}\varepsilon, 3/2 - \mathrm{i}\varepsilon)$ and the inner integral of the second term as the real beta function $B(1/2, 3/2)$. Upon evaluating of the beta function by the gamma function $\Gamma(z)$, one can obtain [39]:

$$G(P) = \frac{(\mu_1 + k_1\mu_2) + (\mu_2 + k_2\mu_1)}{16\mu_1\mu_2 \cosh^2(\pi\varepsilon)} K(P)\overline{K(P)} + \frac{\mu_1 + \mu_2}{4\mu_1\mu_2} K_{\mathrm{III}}^2(P) \tag{5.51a}$$

or, alternatively

$$G(P) = \frac{(\mu_1 + k_1\mu_2) + (\mu_2 + k_2\mu_1)}{16\mu_1\mu_2 \cosh^2(\pi\varepsilon)} \left[K_{\mathrm{I}}^2(P) + K_{\mathrm{II}}^2(P)\right] + \frac{\mu_1 + \mu_2}{4\mu_1\mu_2} K_{\mathrm{III}}^2(P) \tag{5.51b}$$

The expression for the energy release rate shows that it does not include the heat flux intensity factor, which is the same as the elastic one [33].

5.3 Interface Crack Problems in Thermopiezoelectric Materials

Due to their excellent piezoelectric, dielectric and pyroelectric properties, piezothermoelastic materials are being widely used in smart structural systems as sensors, actuators, transducers and intelligent structures, etc. Because of the pyroelectric effect, the response characteristics may change considerably when piezoelectric structures work in environment where temperature varies notably. Since temperature variation in piezoelectric materials will seriously affect the overall sensing and controlling performance of a distributed control system [44], a wide and thorough understanding of the mechanical-electro-thermal coupling behavior in piezoelectric materials is essential to better use of piezothermoelastic-based intelligent structures.

For homogeneous materials, Mindlin [45] gave the equations describing small vibrations of piezoelectric plates to reveal the relationship between the thermal, elastic and electric fields. The analytical theory for piezothermoelastic materials has also been comprehensively studied [46–48], and different methods were applied to obtain the general solutions for linear or nonlinear problems of piezothermoelasticity [49–52]. As it is very important to understand the fracture behaviors of a piezothermoelastic solid with defects under thermal loadings, many efforts have been made in this regard. Yu and Qin [53] analyzed the singularities of the near crack-tip thermoelectroelastic fields. Chen [54] derived a general solution of 3D piezothermoelasticity for both static and dynamic cases. Gao and Wang [55] dealt with 2D N-collinear permeable cracks in a thermopiezoelectric medium. Chen et al. [56] gave the explicit solution for a penny-shaped crack subjected to an arbitrarily point-temperature loading by utilizing the elementary functions. Making use of Stroh's formalism and conformal mapping technique, Qin [57] presented the Green function for a thermopiezoelectric material containing an elliptic hole by using Stroh's formalism and conformal mapping technique. Based on the given Green function, the boundary element method (BEM) was used to analyze various 2D crack problems [58–60]. In addition, many researchers have investigated 2D and 3D crack problems in piezothermoelastic materials under different thermal conditions of stationary or dynamic cases [61–74].

In order to get a stronger electro-mechanical-thermal coupling effect, laminated piezoelectric structures are often used. Therefore, interface cracking problems are frequently encountered in piezoelectric structures. Shen et al. [75] combined the extended version of Eshelby-Stroh's formulation and the method of analytical continuation to study interface cracks and obtained a general, explicit, closed form solution. Later, Shen and Kuang [76] analyzed an electrically impermeable, interface crack in an infinite, piezothermoelastic bi-material under a remote heat flux. Qin and Mai [77] gave the Green functions for thermoelectroelastic bi-materials subjected to a temperature discontinuity in terms of the Stroh formalism in a 2D piezoelectric plate. Qin and Mai [78] employed the Lekhnitskii-Eshelby-Stroh formalism to study an interface crack in a thermopiezoelectric bi-material with the assumption of a contact zone model. Later, Gao and Wang [79] used the same classical interface crack model proposed by Shen and Kuang [76] to investigate an electrically permeable, interface crack. In the frame of this crack model, the solution processed the oscillatory singularities at the crack tips as in elastic materials found in [80]. Herrmann and Loboda [81, 82] adopted the contact-zone model and studied interface cracks under applied thermoelectromechanical loadings in piezothermoelastic bi-materials using electrically permeable and impermeable conditions, respectively. Ueda [83] adopted the Fourier transform technique to study a crack in a piezoelectric laminate under uniform electric and temperature fields. Herrmann and Loboda [84] extended their proposed contact-zone crack model to study moving interface crack problems. Hou and Leung [85] obtained 3D Green's functions for two-phase piezothermoelastic bi-materials expressed in terms of harmonic functions.

Similar as discussed in Sect. 5.1, the displacement discontinuity method are extended to include the electric potential, and temperature discontinuities to provide the fundamental solutions of interface crack problem in piezoelectric biomaterials [86].

5.3.1 Basic Equations

For a stationary process, in the absence of body forces, free electric charges, electric current and body heat source, when the *xoy*-plane is parallel to the plane of isotropy in the Cartesian coordinates (*x*, *y*, *z*), the corresponding constitutive relations of linear, transversely isotropic, piezothermoelastic materials are given by Mindlin [44] in the form

$$\begin{aligned}
\sigma_x &= c_{11}\frac{\partial u}{\partial x} + c_{12}\frac{\partial v}{\partial y} + c_{13}\frac{\partial w}{\partial z} + e_{31}\frac{\partial \varphi}{\partial z} - \lambda_{11}\theta, \\
\sigma_y &= c_{12}\frac{\partial u}{\partial x} + c_{11}\frac{\partial v}{\partial y} + c_{13}\frac{\partial w}{\partial z} + e_{31}\frac{\partial \varphi}{\partial z} - \lambda_{11}\theta, \\
\sigma_z &= c_{13}\frac{\partial u}{\partial x} + c_{13}\frac{\partial v}{\partial y} + c_{33}\frac{\partial w}{\partial z} + e_{33}\frac{\partial \varphi}{\partial z} - \lambda_{33}\theta, \\
\tau_{yz} &= c_{44}\left(\frac{\partial v}{\partial z} + \frac{\partial w}{\partial y}\right) + e_{15}\frac{\partial \varphi}{\partial y}, \\
\tau_{zx} &= c_{44}\left(\frac{\partial u}{\partial z} + \frac{\partial w}{\partial x}\right) + e_{15}\frac{\partial \varphi}{\partial x}, \\
\tau_{xy} &= c_{66}\left(\frac{\partial u}{\partial y} + \frac{\partial v}{\partial x}\right),
\end{aligned} \tag{5.52a}$$

$$
\begin{aligned}
D_x &= e_{15}\left(\frac{\partial u}{\partial z}+\frac{\partial w}{\partial x}\right)-\varepsilon_{11}\frac{\partial \varphi}{\partial x},\\
D_y &= e_{15}\left(\frac{\partial v}{\partial z}+\frac{\partial w}{\partial y}\right)-\varepsilon_{11}\frac{\partial \varphi}{\partial y},\\
D_z &= e_{31}\left(\frac{\partial u}{\partial x}+\frac{\partial v}{\partial y}\right)+e_{33}\frac{\partial w}{\partial z}-\varepsilon_{33}\frac{\partial \varphi}{\partial z}+p_3\theta,
\end{aligned}
\tag{5.52b}
$$

$$
h_x=-\beta_{11}\frac{\partial \theta}{\partial x},\quad h_y=-\beta_{11}\frac{\partial \theta}{\partial y},\quad h_z=-\beta_{33}\frac{\partial \theta}{\partial z},
\tag{5.52c}
$$

where u, v, w, ϕ and θ are elastic displacements, electric potential and temperature change (with respect to the reference temperature), respectively, and can be referred to as extended displacements; c_{ij}, e_{ij}, ε_{ij}, λ_{ii} and p_3 are elastic, piezoelectric, dielectric, thermal modules and pyroelectric constants, respectively, and β_{ij} are the coefficients of heat conduction. The relation $c_{66}=(c_{11}-c_{12})/2$ holds for piezoelectric materials with transverse isotropy. Compared with (5.2), although the anisotropy of piezothermoelectricity makes the problem more complicated, a similar solution method can be employed here.

The equilibrium equations are given by [50]

$$
\begin{aligned}
&\frac{\partial \sigma_x}{\partial x}+\frac{\partial \tau_{xy}}{\partial y}+\frac{\partial \tau_{zx}}{\partial z}=0,\\
&\frac{\partial \tau_{xy}}{\partial x}+\frac{\partial \sigma_y}{\partial y}+\frac{\partial \tau_{yz}}{\partial z}=0,\\
&\frac{\partial \tau_{zx}}{\partial x}+\frac{\partial \tau_{yz}}{\partial y}+\frac{\partial \sigma_z}{\partial z}=0,
\end{aligned}
\tag{5.53a}
$$

$$
\frac{\partial D_x}{\partial x}+\frac{\partial D_y}{\partial y}+\frac{\partial D_z}{\partial z}=0,
\tag{5.53b}
$$

$$
\frac{\partial h_x}{\partial x}+\frac{\partial h_y}{\partial y}+\frac{\partial h_z}{\partial z}=0,
\tag{5.53c}
$$

where σ_{ij}, D_i, and h_i, with $i, j = x, y, z$, are the components of stress, electric displacement and heat flux, respectively, which are all referred to as extended stresses.

5.3.2 Fundamental Solutions for Unit-Point Extended Displacement Discontinuities

Consider a three-dimensional, transversely isotropic, piezothermoelastic bi-material with the two bonded materials occupying the upper (denoted as Material 1) and lower domains (denoted as Material 2), respectively. The poling directions are all perpendicular to the interface. A Cartesian coordinate system is set up with the *xoy* plane lying in the interface. Again, a penny-shaped crack of radius *a* is centered at the origin on the interface, as illustrated in Fig. 5.1. The upper and lower surfaces of the crack are denoted by S^+ and S^-, respectively, and

$$x = r\cos\phi,\ y = r\sin\phi \tag{5.54}$$

The extended displacement discontinuities across the faces of the crack can be expressed as

$$\begin{aligned} &\|u\| = u(x,y,0^+) - u(x,y,0^-), \|v\| = v(x,y,0^+) - v(x,y,0^-),\\ &\|w\| = w(x,y,0^+) - w(x,y,0^-), \|\varphi\| = \varphi(x,y,0^+) - \varphi(x,y,0^-),\\ &\|\theta\| = \theta(x,y,0^+) - \theta(x,y,0^-), (x,y) \in S. \end{aligned} \tag{5.55}$$

According to Zhao et al. [40] (1988) and Zhao and Liu [41], if we let the radius of the crack *a* approach zero, the fundamental solutions corresponding to unit-concentrated extended displacement discontinuities can be obtained, as discussed in the preceding section.

Using the operator theory and the generalized Almansi's theorem, Ding et al. [50] derived the general solutions for piezothermoelasticity satisfying the equilibrium and constitutive equations in Eqs. (5.52) and (5.53). According to the general solutions given by Ding et al. [50], we have

$$\begin{aligned} &U = -\Lambda\left(\mathrm{i}\Psi_0 + \sum_{j=1}^{4}\Psi_j\right), \quad w_m = \sum_{j=1}^{4}\alpha_{mj}\frac{\partial\Psi_j}{\partial z_j}, \quad \theta = \alpha_{34}\frac{\partial^2\Psi_4}{\partial z_4^2},\\ &\sigma_1 = \sum_{j=1}^{4}\gamma_{3j}\frac{\partial^2\Psi_j}{\partial z_j^2}, \quad \sigma_2 = -2c_{66}\Lambda^2\left(\mathrm{i}\Psi_0 + \sum_{j=1}^{4}\Psi_j\right),\\ &\sigma_{zm} = \sum_{j=1}^{4}\gamma_{mj}\frac{\partial^2\Psi_j}{\partial z_j^2}, \quad h_x = -\beta_{33}\alpha_{34}\frac{\partial^3\Psi_4}{\partial x\partial z_4^3},\\ &h_y = -\beta_{33}\alpha_{34}\frac{\partial^3\Psi_4}{\partial y\partial z_4^2}, \quad h_z = -\beta_{33}s_4\alpha_{34}\frac{\partial^3\Psi_4}{\partial z_4^3},\\ &\tau_m = \Lambda\left(\sum_{j=1}^{4}s_j\gamma_{mj}\frac{\partial\Psi_j}{\partial z_j} - \mathrm{i}s_0\rho_m\frac{\partial\Psi_0}{\partial z_0}\right), \quad (m = 1,\ 2), \end{aligned} \tag{5.56}$$

The corresponding notations for the components in the Cartesian coordinates (x, y, z) and cylindrical coordinates (r, ϕ, z) are both introduced as [50]

$$\begin{aligned}
&z_j = s_j z, \quad \Lambda = \frac{\partial}{\partial x} + \mathrm{i}\frac{\partial}{\partial y},\\
&U = u + \mathrm{i}v = \mathrm{e}^{\mathrm{i}\phi}(u_r + \mathrm{i}u_\phi), \quad w_1 = w, \quad w_2 = \varphi,\\
&\sigma_1 = \sigma_x + \sigma_y = \sigma_r + \sigma_\phi,\\
&\sigma_2 = \sigma_x - \sigma_y + 2\mathrm{i}\tau_{xy} = \mathrm{e}^{i\phi}(\sigma_r - \sigma_\phi + 2\mathrm{i}\tau_{r\phi}),\\
&\sigma_{z1} = \sigma_z, \quad \sigma_{z2} = D_z,\\
&\tau_m = \tau_{xm} + \mathrm{i}\tau_{ym} = \mathrm{e}^{\mathrm{i}\phi}(\tau_{rm} + \mathrm{i}\tau_{\phi m})\\
&\tau_{x1} = \tau_{zx}, \quad \tau_{y1} = \tau_{yz}, \quad \tau_{r1} = \tau_{zr}, \quad \tau_{\phi 1} = \tau_{\phi z},\\
&\tau_{x2} = D_x, \quad \tau_{y2} = D_y, \quad \tau_{r2} = D_r, \quad \tau_{\phi 2} = D_\phi,
\end{aligned} \tag{5.57}$$

where $z_j = s_j z$ (j = 0,1,2,3,4), with all material-related constants concerned given in [86].

The harmonic functions $\Psi_j (j = 0, 1, 2, 3, 4)$ satisfy

$$\left(\Delta + \frac{\partial^2}{\partial z_j^2}\right)\Psi_j = 0, \tag{5.58}$$

where

$$\Delta = \frac{\partial^2}{\partial x^2} + \frac{\partial^2}{\partial y^2} \text{ in the Cartesian coordinates}(x, y, z), \tag{5.59a}$$

and

$$\Delta = \frac{\partial^2}{\partial r^2} + \frac{\partial}{r\partial r} + \frac{\partial^2}{r^2\partial\phi^2} \text{ in cylindrical coordinates}(r, \phi, z). \tag{5.59b}$$

5.3.2.1 Fundamental Solution for a Unit-Point Temperature Discontinuity

The boundary condition in this case can be rewritten as:

$$\begin{aligned}
&\|u(\xi,\eta)\| = 0, \quad \|v(\xi,\eta)\| = 0, \quad \|w(\xi,\eta)\| = 0, \quad \|\varphi(\xi,\eta)\| = 0,\\
&\quad \|\theta(\xi,\eta)\| = \delta(\xi,\eta),
\end{aligned} \tag{5.60a}$$

where $\delta(\xi, \eta)$ is the Dirac delta function. It can be seen that this is non-torsional axisymmetric problem, thus all the quantities depend only on (r, z) while independent of angle ϕ. Hence, the general solutions can be simplified as:

$$
\begin{aligned}
&u_r = -\sum_{j=1}^{4} \frac{\partial \Psi_j}{\partial r}, \quad w_m = \sum_{j=1}^{4} \alpha_{mj} \frac{\partial \Psi_j}{\partial z_j}, \\
&h_z = -s_4 \beta_{33} \alpha_{34} \frac{\partial^3 \Psi_4}{\partial z_4^3}, \quad \theta = \alpha_{34} \frac{\partial^2 \Psi_4}{\partial z_4^2}, \\
&\sigma_{zm} = \sum_{j=1}^{4} \gamma_{mj} \frac{\partial^2 \Psi_j}{\partial z_j^2}, \quad \sigma_{zr} = \sum_{j=1}^{4} s_j \gamma_{1j} \frac{\partial^2 \Psi_j}{\partial r \partial z_j},
\end{aligned} \tag{5.61a}
$$

and

$$
\left(\frac{\partial^2}{\partial r^2} + \frac{\partial}{r \partial r} + \frac{\partial^2}{\partial z_j^2} \right) \Psi_j = 0. \tag{5.61b}
$$

Employing the zero-order Hankel transform, the potential functions can be set as:

$$
\Psi_j^+(r, z) = \int_0^\infty \xi A_j^+(\xi) e^{-\xi z_j} J_0(\xi r) d\xi, \text{ for } z > 0 \tag{5.62a}
$$

$$
\Psi_j^-(r, z) = \int_0^\infty \xi A_j^-(\xi) e^{\xi z_j} J_0(\xi r) d\xi, \text{ for } z < 0 \tag{5.62b}
$$

where the superscript "+" and "−" denote the upper and lower domains, respectively.

When the unit temperature discontinuity is applied on the interface crack, the corresponding boundary conditions are as follows:

$$
\begin{cases} u_r^+ - u_r^- = 0, \\ w^+ - w^- = 0, \\ \varphi^+ - \varphi^- = 0, \\ \theta^+ - \theta^- = \delta, \end{cases} \quad \begin{cases} \sigma_{zr}^+ - \sigma_{zr}^- = 0, \\ \sigma_z^+ - \sigma_z^- = 0, \\ D_z^+ - D_z^- = 0, \\ h_z^+ - h_z^- = 0, \end{cases} \tag{5.63}
$$

where $0 \leq r \leq a$ belongs to the crack region. Inserting Eq. (5.61a) into Eq. (5.62), and the resultant expressions into Eq. (5.63), one can get:

$$\begin{cases} \sum_{j=1}^{4} A_j^+(\xi) = \sum_{j=1}^{4} A_j^-(\xi), \\ -\sum_{j=1}^{4} \alpha_{1j}^+ A_j^+(\xi) = \sum_{j=1}^{4} \alpha_{1j}^- A_j^-(\xi), \\ -\sum_{j=1}^{4} \alpha_{2j}^+ A_j^+(\xi) = \sum_{j=1}^{4} \alpha_{2j}^- A_j^-(\xi), \\ \xi^2 \left[\alpha_{34}^+ A_4^+(\xi) - \alpha_{34}^- A_4^-(\xi)\right] = \frac{1}{2\pi}, \end{cases} \tag{5.64a}$$

$$\begin{cases} \sum_{j=1}^{4} s_j^+ \gamma_{1j}^+ A_j^+(\xi) = -\sum_{j=1}^{4} s_j^- \gamma_{1j}^- A_j^-(\xi), \\ \sum_{j=1}^{4} \gamma_{1j}^+ A_j^+(\xi) = \sum_{j=1}^{4} \gamma_{1j}^- A_j^-(\xi), \\ \sum_{j=1}^{4} \gamma_{2j}^+ A_j^+(\xi) = \sum_{j=1}^{4} \gamma_{2j}^- A_j^-(\xi), \\ \beta_{33}^+ s_4^+ \alpha_{34}^+ A_4^+(\xi) = -\beta_{33}^- s_4^- \alpha_{34}^- A_4^-(\xi). \end{cases} \tag{5.64b}$$

Solving Eq. (5.64), the eight coefficients are determined:

$$A_j^+ = A_j^{+*} \frac{1}{\xi^2}, \quad A_j^- = A_j^{-*} \frac{1}{\xi^2}, \quad j = 1, 2, 3, 4, \tag{5.65}$$

where A_j^{+*} and A_j^{-*} are constants, and then the potential functions are obtained:

$$\psi_j^+(r, z) = -A_j^{+*} \ln(R_j + z_j), \tag{5.66a}$$

$$\psi_j^-(r, z) = -A_j^{-*} \ln(R_j - z_j), \tag{5.66b}$$

where $R_j = \sqrt{r^2 + z_j^2}$. With the obtained potential functions, the corresponding extended displacements and stresses in the whole space can all be determined. For instance, the extended stresses of an arbitrary field point in the upper domain are obtained as

$$\sigma_z^+ = \sum_{j=1}^{4} A_j^{+*} \gamma_{1j}^+ \frac{z_j}{R_j^3}, \tag{5.67a}$$

$$D_z^+ = \sum_{j=1}^{4} A_j^{+*} \gamma_{2j}^+ \frac{z_j}{R_j^3}, \tag{5.67b}$$

$$\sigma_{zr}^+ = \sum_{j=1}^{4} A_j^{+*} s_j^+ \gamma_{1j}^+ \frac{r}{R_j^3}, \tag{5.67c}$$

$$\sigma_{zx}^{+} = \sum_{j=1}^{4} A_j^{+*} s_j^{+} \gamma_{1j}^{+} \frac{x}{R_j^3}, \tag{5.67d}$$

$$\sigma_{yz}^{+} = \sum_{j=1}^{4} A_j^{+*} s_j^{+} \gamma_{1j}^{+} \frac{y}{R_j^3}, \tag{5.67e}$$

$$h_z^{+} = A_4^{+*} s_4^{+} \beta_{33}^{+} \alpha_{34}^{+} \left(\frac{3z_4^2}{R_4^5} - \frac{1}{R_4^3} \right). \tag{5.67f}$$

Through the same approach, one can obtain the fundamental solutions satisfying the boundary conditions of other unit point, extended displacement discontinuities, respectively. For simplicity, the detailed derivation is omitted which can be found in [86].

5.3.3 *Boundary Integral-Differential Equations for an Interfacial Crack in Piezothermoelastic Materials*

Consider an arbitrarily shaped, planar crack, S, lying on the interface of two different piezoelectric materials. With the aid of the fundamental solutions obtained in Sect. 5.2.1.2, one can get the fundamental solutions for the interface crack S. By virtue of the obtained fundamental solutions and superposition, the corresponding extended stresses caused by the interface crack, S at an arbitrary, internal field point (x, y, z) in the upper domain can be expressed in terms of the extended displacement discontinuities across the interface crack as

$$\begin{aligned}
\sigma_{zx}^{+} = \int_{S^+} \Bigg\{ & \left(\sum_{j=1}^{3} p_{1j}^{+*} s_j^{+} \gamma_{1j}^{+} \left[3\left(\frac{1}{R_j^3} - \frac{z_j^2}{R_j^5} \right) \frac{(\xi - x)^2}{r^2} - \frac{1}{R_j^3} \right] - s_0^{+} \rho_1^{+} p_2^{+*} \left[3\left(\frac{1}{R_0^3} - \frac{z_0^2}{R_0^5} \right) \frac{(\eta - y)^2}{r^2} - \frac{1}{R_0^3} \right] \right) \\
& \|u\| + 3\frac{(\xi - x)(\eta - y)}{r^2} \left[\sum_{j=1}^{3} p_{1j}^{+*} s_j^{+} \gamma_{1j}^{+} \left(\frac{1}{R_j^3} - \frac{z_j^2}{R_j^5} \right) + s_0^{+} \rho_1^{+} p_2^{+*} \left(\frac{1}{R_0^3} - \frac{z_0^2}{R_0^5} \right) \right] \|v\| \\
& - \sum_{j=1}^{3} B_j^{+*} s_j^{+} \gamma_{1j}^{+} \frac{3(\xi - x) z_j}{R_j^5} \|w\| - \sum_{j=1}^{3} C_j^{+*} s_j^{+} \gamma_{1j}^{+} \frac{3(\xi - x) z_j}{R_j^5} \|\varphi\| \\
& - \sum_{j=1}^{4} A_j^{+*} s_j^{+} \gamma_{1j}^{+} \frac{(\xi - x)}{R_j^3} \|\theta\| \Bigg\} dS(\xi, \eta),
\end{aligned} \tag{5.68a}$$

$$
\begin{aligned}
\sigma_{yz}^{+} = \int\limits_{S^+} & \left\{ 3\frac{(\xi-x)(\eta-y)}{r^2}\left[\sum_{j=1}^{3} p_{1j}^{+*} s_j^{+} \gamma_{1j}^{+}\left(\frac{1}{R_j^3}-\frac{z_j^2}{R_j^5}\right)+s_0^{+}\rho_1^{+}p_2^{+*}\left(\frac{1}{R_0^3}-\frac{z_0^2}{R_0^5}\right)\right]\|u\| \right. \\
& + \left(\sum_{j=1}^{3} p_{1j}^{+*} s_j^{+} \gamma_{1j}^{+}\left[3\left(\frac{1}{R_j^3}-\frac{z_j^2}{R_j^5}\right)\frac{(\eta-y)^2}{r^2}-\frac{1}{R_j^3}\right]\right. \\
& \left. - s_0^{+}\rho_1^{+}p_2^{+*}\left[3\left(\frac{1}{R_0^3}-\frac{z_0^2}{R_0^5}\right)\frac{(\xi-x)^2}{r^2}-\frac{1}{R_0^3}\right]\right)\|v\| \\
& - \sum_{j=1}^{3} B_j^{+*} s_j^{+} \gamma_{1j}^{+}\frac{3(\eta-y)z_j}{R_j^5}\|w\| - \sum_{j=1}^{3} C_j^{+*} s_j^{+} \gamma_{1j}^{+}\frac{3(\eta-y)z_j}{R_j^5}\|\varphi\| \\
& \left. - \sum_{j=1}^{4} A_j^{+*} s_j^{+} \gamma_{1j}^{+}\frac{(\eta-y)}{R_j^3}\|\theta\| \right\} dS(\xi,\eta),
\end{aligned}
\tag{5.68b}
$$

$$
\begin{aligned}
\sigma_{z}^{+} = \int\limits_{S^+} & \left\{ -\sum_{j=1}^{3} p_{1j}^{+*}\gamma_{1j}^{+}\frac{3(\xi-x)z_j}{R_j^5}\|u\| - \sum_{j=1}^{3} p_{1j}^{+*}\gamma_{1j}^{+}\frac{3(\eta-y)z_j}{R_j^5}\|v\| \right. \\
& + \sum_{j=1}^{3} B_j^{+*}\gamma_{1j}^{+}\left(\frac{3z_j^2}{R_j^5}-\frac{1}{R_j^3}\right)\|w\| + \sum_{j=1}^{3} C_j^{+*}\gamma_{1j}^{+}\left(\frac{3z_j^2}{R_j^5}-\frac{1}{R_j^3}\right)\|\varphi\| \\
& \left. + \sum_{j=1}^{4} A_j^{+*}\gamma_{1j}^{+}\frac{z_j}{R_j^3}\|\theta\| \right\} dS(\xi,\eta),
\end{aligned}
\tag{5.68c}
$$

$$
\begin{aligned}
D_{z}^{+} = \int\limits_{S^+} & \left\{ -\sum_{j=1}^{3} p_{1j}^{+*}\gamma_{2j}^{+}\frac{3(\xi-x)z_j}{R_j^5}\|u\| - \sum_{j=1}^{3} p_{1j}^{+*}\gamma_{2j}^{+}\frac{3(\eta-y)z_j}{R_j^5}\|v\| \right. \\
& + \sum_{j=1}^{3} B_j^{+*}\gamma_{2j}^{+}\left(\frac{3z_j^2}{R_j^5}-\frac{1}{R_j^3}\right)\|w\| + \sum_{j=1}^{3} C_j^{+*}\gamma_{2j}^{+}\left(\frac{3z_j^2}{R_j^5}-\frac{1}{R_j^3}\right)\|\varphi\| \\
& \left. + \sum_{j=1}^{4} A_j^{+*}\gamma_{2j}^{+}\frac{z_j}{R_j^3}\|\theta\| \right\} dS(\xi,\eta),
\end{aligned}
\tag{5.68d}
$$

$$
h_z^{+} = -A_4^{+*} s_4^{+} \beta_{33}^{+} \alpha_{34}^{+} \int\limits_{S^+} \left(\frac{1}{R_4^3}-\frac{3z_4^2}{R_4^5}\right)\|\theta\| dS(\xi,\eta),
\tag{5.68e}
$$

where $R_j = \sqrt{(\xi-x)^2+(\eta-y)^2+z_j^2}$, $r = \sqrt{(\xi-x)^2+(\eta-y)^2}$, $z_j = s_j z$.

5.3.4 Hyper-Singular Integral-Differential Equations

As the field point approaches the crack face, the integrals in Eq. (5.68) will become hyper-singular. These integrals must be evaluated using finite-part integrals, and the following integral formulas are applied [33]:

$$I_1 = \lim_{z\to 0} \int_{S^+} \|u(\xi,\eta)\| \frac{(\xi - x)z}{R^5} d\xi d\eta = \frac{2\pi}{3} \frac{\partial \|u(x,y)\|}{\partial x}, \tag{5.69a}$$

$$I_2 = \lim_{z\to 0} \int_{S^+} \|w(\xi,\eta)\| \frac{z^2}{R^5} d\xi d\eta = 0, \tag{5.69b}$$

After some algebraic manipulations, the boundary integral-differential equations for an arbitrarily shaped, interface crack are obtained:

$$\begin{aligned} p_x(x,y) = &- \int_{S^+} \Big\{ \big[L_{11}\big(3\cos^2\phi - 1\big) - L_{12}\big(3\sin^2\phi - 1\big)\big] \frac{1}{r^3} \|u(\xi,\eta)\| \\ &+ 3(L_{11} + L_{12}) \frac{\sin\phi\cos\phi}{r^3} \|v(\xi,\eta)\| - L_{13} \frac{\cos\phi}{r^2} \|\theta(\xi,\eta)\| \Big\} dS(\xi,\eta) \\ &+ 2\pi \left(L_{14} \frac{\partial \|w(x,y)\|}{\partial x} + L_{15} \frac{\partial \|\varphi(x,y)\|}{\partial x} \right), \end{aligned} \tag{5.70a}$$

$$\begin{aligned} p_y(x,y) = &- \int_{S^+} \Big\{ \big[L_{11}\big(3\sin^2\phi - 1\big) - L_{12}\big(3\cos^2\phi - 1\big)\big] \frac{1}{r^3} \|v(\xi,\eta)\| \\ &+ 3(L_{11} + L_{12}) \frac{\sin\phi\cos\phi}{r^3} \|u(\xi,\eta)\| - L_{13} \frac{\sin\phi}{r^2} \|\theta(\xi,\eta)\| \Big\} dS(\xi,\eta) \\ &+ 2\pi \left(L_{14} \frac{\partial \|w(x,y)\|}{\partial y} + L_{15} \frac{\partial \|\varphi(x,y)\|}{\partial y} \right), \end{aligned} \tag{5.70b}$$

$$\begin{aligned} p_z(x,y) = &\int_{S^+} \big(L_{31} \|w(\xi,\eta)\| + L_{32} \|\varphi(\xi,\eta)\|\big) \frac{1}{r^3} dS(\xi,\eta) \\ &+ 2\pi L_1 \left(\frac{\partial \|u(x,y)\|}{\partial x} + \frac{\partial \|v(x,y)\|}{\partial y} \right), \end{aligned} \tag{5.70c}$$

$$\omega = -\int\limits_{S^+} (L_{41}\|w(\xi,\eta)\| + L_{42}\|\varphi(\xi,\eta)\|)\frac{1}{r^3}dS(\xi,\eta) - 2\pi L_2\left(\frac{\partial\|u(x,y)\|}{\partial x} + \frac{\partial\|v(x,y)\|}{\partial y}\right), \tag{5.70d}$$

$$h_z(x,y) = -L_5\int\limits_{S^+}\frac{1}{r^3}\|\theta(\xi,\eta)\|dS(\xi,\eta), \tag{5.70e}$$

where p_x, p_y, and p_z are the mechanical tractions; ω and h_z are the applied electric displacement and heat flux, respectively, and all of them are called the extended tractions, and the related coefficients are:

$$\begin{aligned}
L_{11} &= \sum_{j=1}^{3} p_{1j}^{+*} s_j^+ \gamma_{1j}^+, \; L_{12} = s_0^+ \rho_1^+ p_2^{+*}, \; L_{13} = \sum_{j=1}^{4} A_j^{+*} s_j^+ \gamma_{1j}^+, \\
L_{14} &= \sum_{j=1}^{3} B_j^{+*} s_j^+ \gamma_{1j}^+, \; L_{15} = \sum_{j=1}^{3} C_j^{+*} s_j^+ \gamma_{1j}^+, \; L_5 = A_4^{+*} s_4^+ \beta_{33}^+ \alpha_{34}^+, \\
L_1 &= \sum_{j=1}^{3} p_{1j}^{+*} \gamma_{1j}^+, \; L_{31} = \sum_{j=1}^{3} B_j^{+*} \gamma_{1j}^+, \; L_{32} = \sum_{j=1}^{3} C_j^{+*} \gamma_{1j}^+, \\
L_2 &= \sum_{j=1}^{3} p_{1j}^{+*} \gamma_{2j}^+, \; L_{41} = \sum_{j=1}^{3} B_j^{+*} \gamma_{2j}^+, \; L_{42} = \sum_{j=1}^{3} C_j^{+*} \gamma_{2j}^+,
\end{aligned} \tag{5.71}$$

where $r = \sqrt{(\xi - x)^2 + (\eta - y)^2}$, $\cos\phi = (\xi - x)/r$, $\sin\phi = (\eta - y)/r$, $z_j = s_j z$.

It should be pointed out that Eq. (5.70) are the expressions of the electric and thermal impermeable boundary conditions in the complete form. The electric potential discontinuity or the temperature discontinuity will be zero for electrically or thermally permeable boundary condition. In addition, there are also electrically semi-permeable boundary condition [87] and thermally semi-permeable boundary condition [74, 88]. In these kinds of boundary conditions, the electric displacement in the crack cavity can be determined by the electric potential discontinuity and crack opening displacement as $D_z^c = -\kappa^c\|\varphi\|/\|w\|$, while the heat flux in the crack cavity can be determined by the temperature discontinuity and crack opening displacement as $h_z^c = -\beta^c\|\theta\|/\|w\|$, where ε^c and β^c represent the dielectric and heat conduction coefficients in the crack interior, respectively. The difference in boundary conditions is obvious, for brevity, the detailed discussion about the influence of different electric and thermal boundary conditions was omitted here. Details can be found in [86]. Here we only present the results for electrically and thermally impermeable boundary conditions. When the bi-material becomes homogeneous, the differential terms will disappear. Therefore, the boundary integral-differential equations are reduced to hyper-singular boundary integral equations.

5.3.4.1 Solution Method for the Extended Displacement Discontinuity $\|W\| + G\|\varphi\|$

Combining Eqs. (5.70c) and (5.70d), one obtains

$$\left(\frac{L_1}{L_2}L_{41} - L_{31}\right)\int\limits_{S^+}\{\|w\| + g\|\varphi\|\}\frac{1}{r^3}dS = -p_z + \frac{L_1}{L_2}\omega, \tag{5.72}$$

where

$$g = \frac{L_1 L_{42} - L_2 L_{32}}{L_1 L_{41} - L_2 L_{31}}, \tag{5.73}$$

Equation (5.72) is the hyper-singular, boundary integral equation for the extended displacement discontinuity $\|w\| + g\|\varphi\|$. For the same crack in an isotropic, thermoelastic bi-material, the boundary integral equation of the displacement discontinuity $\|W\|$ in the z-direction takes the same form [39]

$$\frac{E}{8\pi(1-v^2)}\int\limits_{S^+}\frac{\|W\|}{r^3}dS = -t_z(x,y), \tag{5.74}$$

where v and E are, respectively, the Poisson's ratio and Young's modulus, and $t_z(x, y)$ is the prescribed traction along the z-direction on the crack surface. Using

$$\frac{E}{8\pi(1-v^2)} = \frac{L_1}{L_2}L_{41} - L_{31}, \tag{5.75a}$$

$$-t_z(x,y) = -p_z + \frac{L_1}{L_2}\omega \tag{5.75b}$$

We can find that Eqs. (5.72) and (5.74) are in an identical form, and therefore, they have the same solution

$$\|w\| + g\|\varphi\| = \|W\| \tag{5.76}$$

Equation (5.76) indicates that the solution of the extended displacement discontinuity $||w|| + g||\varphi||$ can be directly obtained from the corresponding thermoelastic solution.

5.3.4.2 Extended Stress $\sigma_Z - L_1D_z/L_2$ and Extended Intensity Factor K_{I1}

The stress for isotropic thermoelastic bi-materials in the crack plane is given by [39]:

$$\sigma_z = \frac{E}{8\pi(1-\nu^2)} \int\limits_{S^+} \frac{||W||}{R^3} dS, \tag{5.77}$$

where $R = \sqrt{(\xi - x)^2 + (\eta - y)^2} > 0$.

From Eqs. (5.72) and (5.77), it can be seen that the extended stress $\sigma_z - L_1D_z/L_2$ has the classical singularity $r^{-1/2}$ near the interface crack front in the piezothermoelastic bi-material.

In a purely elastic problem, the Mode I stress intensity factor is defined as:

$$K_{\mathrm{I}}^{M} = \lim_{r\to 0} \sqrt{2\pi r}\sigma_z, \tag{5.78a}$$

and can be expressed in terms of the displacement discontinuity

$$K_{\mathrm{I}}^{M} = \frac{E}{8\pi(1-\nu^2)} \lim_{\rho\to 0} \frac{||W||}{\sqrt{\rho}}. \tag{5.78b}$$

In the sam fashion, the first Mode I extended intensity factor in piezothermoelastic medium is defined as [86]:

$$K_{\mathrm{I}1} = \lim_{r\to 0} \sqrt{2\pi r}(\sigma_z - L_1D_z/L_2), \tag{5.79a}$$

Substituting Eqs. (5.72) into (5.79a), one obtains:

$$K_{\mathrm{I}1} = \left(\frac{L_1}{L_2}L_{41} - L_{31}\right) \lim_{\rho\to 0} \frac{||w|| + g||\varphi||}{\sqrt{\rho}}, \tag{5.79b}$$

Finally, utilizing Eqs. (5.76) and (5.77) and comparing Eqs. (5.78b) and (5.79b) leads to:

$$K_{\mathrm{I}1} = K_{\mathrm{I}}^{M}. \tag{5.80}$$

The Mode I1 intensity factor $K_{\mathrm{I}1}$ is one of the parameters characterizing the extended stresses near the interface crack front.

5.3.5 *Solution Method of the Integral-Differential Equations*

Equations (5.70a, b and f) can be rewritten as:

$$\begin{aligned}
&\int_{S^+} \Big\{ \left[L_{11}(3\cos^2\phi - 1) - L_{12}(3\sin^2\phi - 1) \right] \frac{1}{r^3} \|u\| \\
&\quad + 3(L_{11} + L_{12}) \frac{\sin\phi\cos\phi}{r^3} \|v\| - L_{13} \frac{\cos\phi}{r^2} \|\theta\| \Big\} dS(\xi, \eta) \\
&\quad - 2\pi L_{14} \frac{\partial}{\partial x} \left(\|w\| + \frac{L_{15}}{L_{14}} \|\varphi\| \right) = -p_x,
\end{aligned} \tag{5.81a}$$

$$\begin{aligned}
&\int_{S^+} \Big\{ \left[L_{11}(3\sin^2\phi - 1) - L_{12}(3\cos^2\phi - 1) \right] \frac{1}{r^3} \|v\| \\
&\quad + 3(L_{11} + L_{12}) \frac{\sin\phi\cos\phi}{r^3} \|u\| - L_{13} \frac{\sin\phi}{r^2} \|\theta\| \Big\} dS(\xi, \eta) \\
&\quad - 2\pi L_{14} \frac{\partial}{\partial y} \left(\|w\| + \frac{L_{15}}{L_{14}} \|\varphi\| \right) = -p_y,
\end{aligned} \tag{5.81b}$$

$$-L_5 \int_{S^+} \frac{1}{r^3} \|\theta(\xi, \eta)\| dS = -h_z(x, y). \tag{5.81c}$$

Combining Eqs. (5.75a) and (5.75b) yields

$$\begin{aligned}
&(L_{31} + fL_{41}) \int_{S^+} \left(\|w\| + \frac{L_{32} + fL_{42}}{L_{31} + fL_{41}} \|\varphi\| \right) \frac{1}{r^3} dS \\
&\quad + 2\pi(L_1 + fL_2) \left(\frac{\partial \|u\|}{\partial x} + \frac{\partial \|v\|}{\partial y} \right) = p_z + f\omega,
\end{aligned} \tag{5.82}$$

In Eqs. (5.81) and (5.82), assuming

$$\frac{L_{32} + fL_{42}}{L_{31} + fL_{41}} = \frac{L_{15}}{L_{14}}, \tag{5.83}$$

the constant f is then solved as

$$f = \frac{L_{15}L_{31} - L_{32}L_{14}}{L_{42}L_{14} - L_{41}L_{15}}. \tag{5.84}$$

Therefore, Eqs. (5.81) and (5.82) can be rewritten as:

$$\begin{aligned}\frac{\varsigma p_x}{L_{14}} = \int_{S^+} \Bigg\{ & \left[\frac{\varsigma L_{11}}{L_{14}}\left(1 - 3\cos^2\phi\right) - \frac{\varsigma L_{12}}{L_{14}}\left(1 - 3\sin^2\phi\right)\right]\frac{1}{r^3}\|u\| \\ & - 3\frac{\varsigma(L_{11} + L_{12})}{L_{14}}\frac{\sin\phi\cos\phi}{r^3}\|v\| + \frac{\varsigma L_{13}}{L_{14}}\frac{\cos\phi}{r^2}\|\theta\|\Bigg\} dS(\xi,\eta) \\ & + 2\pi\frac{\partial\|w^*\|}{\partial x},\end{aligned} \tag{5.85a}$$

$$\begin{aligned}\frac{\varsigma p_y}{L_{14}} = \int_{S^+} \Bigg\{ & \left[\frac{\varsigma L_{11}}{L_{14}}\left(1 - 3\sin^2\phi\right) - \frac{\varsigma L_{12}}{L_{14}}\left(1 - 3\cos^2\phi\right)\right]\frac{1}{r^3}\|v\| \\ & - 3\frac{\varsigma(L_{11} + L_{12})}{L_{14}}\frac{\sin\phi\cos\phi}{r^3}\|u\| + \frac{\varsigma L_{13}}{L_{14}}\frac{\sin\phi}{r^2}\|\theta\|\Bigg\} dS(\xi,\eta) \\ & + 2\pi\frac{\partial\|w^*\|}{\partial y},\end{aligned} \tag{5.85b}$$

$$\frac{(p_z + f\omega)}{(L_1 + fL_2)} = \frac{L_{31} + fL_{41}}{\varsigma(L_1 + fL_2)} \int_{S^+} \frac{\|w^*\|}{r^3} dS + 2\pi\left(\frac{\partial\|u\|}{\partial x} + \frac{\partial\|v\|}{\partial y}\right), \tag{5.85c}$$

$$h_z(x,y) = -L_5 \int_{S^+} \frac{1}{r^3}\|\theta(\xi,\eta)\| dS, \tag{5.85d}$$

where ς is a constant to be determined, and $\|w^*\|$ is the extended displacement discontinuity defined as

$$\|w^*\| = \varsigma\left(\|w\| + \frac{L_{15}}{L_{14}}\|\varphi\|\right). \tag{5.86}$$

On the other hand, the boundary integral-differential equations of the same crack in a 3D, two-phase, isotropic, thermoelastic bi-material are given by Zhao et al. [39], as

$$
\begin{aligned}
-\frac{2\pi(\mu_1+k_1\mu_2)}{\mu_1\mu_2(\eta_2-1)}p_1 = \int\limits_{S^+} \Bigg[& \left(\frac{2\eta_1-\eta_2-1}{\eta_2-1}\frac{1}{r^3}+\frac{3(1-\eta_1+\eta_2)}{\eta_2-1}\frac{\cos^2\phi}{r^3}\right)\|\tilde{u}\| \\
& +\frac{3(1-\eta_1+\eta_2)}{\eta_2-1}\frac{\sin\phi\cos\phi}{r^3}\|\tilde{v}\| \\
& -\frac{\alpha_1(1+v_1)\beta_2+\alpha_2(1+v_2)\beta_1\eta_2}{(\beta_1+\beta_2)/2}\frac{\cos\phi}{r^2}\|\theta(\xi,\eta)\|\Bigg\}\Bigg]dS \\
& +2\pi\frac{\partial\|\tilde{w}\|}{\partial x},
\end{aligned}
\tag{5.87a}
$$

$$
\begin{aligned}
-\frac{2\pi(\mu_1+k_1\mu_2)}{\mu_1\mu_2(\eta_2-1)}p_2 = \int\limits_{S^+} \Bigg[& \left(\frac{2\eta_1-\eta_2-1}{\eta_2-1}\frac{1}{r^3}+\frac{3(1-\eta_1+\eta_2)}{\eta_2-1}\frac{\sin^2\phi}{r^3}\right)\|\tilde{v}\| \\
& +\frac{3(1-\eta_1+\eta_2)}{\eta_2-1}\frac{\sin\phi\cos\phi}{r^3}\|\tilde{u}\| \\
& -\frac{\alpha_1(1+v_1)\beta_2+\alpha_2(1+v_2)\beta_1\eta_2}{(\beta_1+\beta_2)/2}\frac{\sin\phi}{r^2}\|\theta(\xi,\eta)\|\Bigg\}\Bigg]dS \\
& +2\pi\frac{\partial\|\tilde{w}\|}{\partial y},
\end{aligned}
\tag{5.87b}
$$

$$
-\frac{2\pi(\mu_1+k_1\mu_2)}{\mu_1\mu_2(1-\eta_2)}p_3 = \int\limits_{S^+}\frac{1+\eta_2}{1-\eta_2}\frac{\|\tilde{w}\|}{r^3}dS+2\pi\left(\frac{\partial\|\tilde{u}\|}{\partial x}+\frac{\partial\|\tilde{v}\|}{\partial y}\right), \tag{5.87c}
$$

$$
h_z(x,y) = -\frac{\beta_1\beta_2}{2\pi(\beta_1+\beta_2)}\int\limits_{S^+}\frac{1}{r^3}\|\tilde{\theta}\|dS, \tag{5.87d}
$$

where

$$
\eta_1=\frac{\mu_1+k_1\mu_2}{\mu_1+\mu_2},\ \eta_2=\frac{\mu_1+k_1\mu_2}{\mu_2+k_2\mu_1},\ k_1=3-4v_1,\ k_2=3-4v_2, \tag{5.88}
$$

and p_1, p_2 and p_3 are the tractions along the *x*-, *y*- and *z*-directions on the crack faces in the thermal-elastic medium, h_z is the heat flux along the *z*-direction. μ_1, v_1, μ_2 and v_2 are, respectively, the shear modulus and Poisson's ratio of Material 1 in the upper domain and Material 2 in the lower domain; α_1, β_1, α_2 and β_2 are, respectively, coefficient of linear thermal expansion, and coefficient of thermal conductivity of the two bonded half spaces. In addition, $\|\tilde{u}\|$, $\|\tilde{v}\|$, $\|\tilde{w}\|$ and $\|\tilde{\theta}\|$ are the displacement discontinuities in the *x*-, *y*- and *z*-directions and temperature discontinuity, respectively.

Comparing the coefficients in Eqs. (5.85) and (5.87), and letting:

$$\frac{\varsigma(L_{11}+2L_{12})}{L_{14}}=\frac{2\eta_1-\eta_2-1}{\eta_2-1}, \tag{5.89a}$$

$$-\frac{\varsigma(L_{11}+L_{12})}{L_{14}}=\frac{1-\eta_1+\eta_2}{\eta_2-1}, \tag{5.89b}$$

$$\frac{(L_{31}+fL_{41})}{\varsigma(L_1+fL_2)}=\frac{1+\eta_2}{1-\eta_2}, \tag{5.89c}$$

$$\frac{\varsigma L_{13}}{L_{14}}=-\frac{\alpha_1(1+\nu_1)\beta_2+\alpha_2(1+\nu_2)\beta_1\eta_2}{(\beta_1+\beta_2)/2}, \tag{5.89d}$$

$$L_5=\frac{\beta_1\beta_2}{2\pi(\beta_1+\beta_2)}, \tag{5.89e}$$

and

$$p_x=\wp p_1,\ p_y=\wp p_2,\ \ell(p_z+f\omega)=\wp p_3, \tag{5.89f}$$

the corresponding material constants are obtained:

$$\varsigma=\sqrt{\frac{L_{14}}{L_{11}}\frac{L_{31}+fL_{41}}{L_1+fL_2}}, \tag{5.90a}$$

$$\eta_1=-\frac{2\varsigma L_{12}}{\varsigma L_{11}+L_{14}}, \tag{5.90b}$$

$$\eta_2=\frac{\varsigma L_{11}-L_{14}}{\varsigma L_{11}+L_{14}}, \tag{5.90c}$$

$$-\frac{\alpha_1(1+\nu_1)\beta_2+\alpha_2(1+\nu_2)\beta_1\eta_2}{(\beta_1+\beta_2)/2}=\frac{\varsigma L_{13}}{L_{14}}, \tag{5.90d}$$

$$\frac{\beta_1\beta_2}{2\pi(\beta_1+\beta_2)}=L_5. \tag{5.90e}$$

From Eqs. (5.87) and (5.89f), one can obtain the coefficients:

$$\wp=-\frac{2\pi L_{14}(\mu_1+k_1\mu_2)}{\mu_1\mu_2(\eta_2-1)\varsigma}, \tag{5.91a}$$

$$\ell=-\frac{L_{14}}{\varsigma(L_1+fL_2)}. \tag{5.91b}$$

Comparing Eqs. (5.85) and (5.87), it can be seen that the boundary integral-differential equations for interface cracks between the piezothermoelastic media and thermoelastic media have the identical forms. Therefore, the corresponding solutions are the same, namely

$$\|u\| = \|\tilde{u}\|,\ \|v\| = \|\tilde{v}\|,\ \|w^*\| = \|\tilde{w}\|,\ \|\theta\| = \|\tilde{\theta}\|. \tag{5.92}$$

Equation (5.92) shows that the solution for the extended displacement discontinuity of the interface cracks in piezothermoelastic, bi-materials can be obtained directly from the corresponding solution of the isotropic, thermoelastic bi-materials.

5.3.5.1 Singular Behavior Near the Interface Crack Front

The singular behavior of interface crack in a 3D, two-phase, isotropic, thermoelastic bi-material was analyzed by Zhao et al. [39]. Following the same procedure, the singularity of the interface crack in a 3D transversely isotropic, piezothermoelastic bi-material is presented here. An arbitrary point P is chosen on the crack edge Γ of crack S. The edge Γ is smooth at point P. Without loss of generality, the Cartesian coordinate system $oxyz$ is oriented so that the x-direction and y-direction are normal and tangent to Γ, respectively. A small circular area Σ in S with a radius ε is centered at point P, as shown in Fig. 5.3.

The hyper-singular parts of Eq. (5.85) shall be finite in Σ for finite, prescribed, extended loadings, namely:

$$\begin{aligned} F_x = & \int_{\Sigma} \left\{ \left[\frac{L_{11} + 2L_{12}}{r^3} + 3\frac{(L_{12} - L_{11})\cos^2\phi}{r^3} \right] \|u\| \right. \\ & \left. - \frac{3(L_{11} + L_{12})\sin\phi\cos\phi}{r^3}\|v\| + \frac{L_{13}\cos\phi}{r^2}\|\theta\| \right\} dS(\xi, \eta) \\ & + 2\pi \frac{L_{14}}{\varsigma} \frac{\partial \|w^*\|}{\partial x}, \end{aligned} \tag{5.93a}$$

$$\begin{aligned} F_y = & \int_{\Sigma} \left\{ \left[\frac{L_{11} + 2L_{12}}{r^3} + 3\frac{(L_{12} - L_{11})\sin^2\phi}{r^3} \right] \|v\| \right. \\ & \left. - \frac{3(L_{11} + L_{12})\sin\phi\cos\phi}{r^3}\|u\| + \frac{L_{13}\sin\phi}{r^2}\|\theta\| \right\} dS(\xi, \eta) \\ & + 2\pi \frac{L_{14}}{\varsigma} \frac{\partial \|w^*\|}{\partial y}, \end{aligned} \tag{5.93b}$$

$$F_{MD} = \frac{L_{31} + fL_{41}}{\varsigma} \int\limits_{\Sigma} \frac{\|w^*\|}{r^3} dS + 2\pi(L_1 + fL_2)\left(\frac{\partial\|u\|}{\partial x} + \frac{\partial\|v\|}{\partial y}\right), \tag{5.93c}$$

$$F_\theta = L_5 \int\limits_{\Sigma} \frac{1}{r^3}\|\theta\| dS, \tag{5.93d}$$

where F_x, F_y, F_{MD} and F_θ are functions of finite values of $(x, y) \in \Sigma$.

In the neighborhood of point P, for a small S_Σ, the extended displacement discontinuities are related only to the coordinates x in the normal plane through point P as well as the position of point P along the contour of the crack [33, 40]:

$$\begin{cases} \|u\| = A_1(P)\xi^{\alpha_1}, \\ \|v\| = A_2(P)\xi^{\alpha_2}, \\ \|w^*\| = A_3(P)\xi^{\alpha_3}, \\ \|\theta\| = A_4(P)\xi^{\alpha_4}, \end{cases} \tag{5.94}$$

where A_1, A_2, A_3 and A_4 are all complex constants [33], and $0 < \mathrm{Re}[\alpha_i] < 1$. Inserting Eqs. (5.94) into (5.93), and making use of the following integral formulas:

$$\lim_{\varepsilon \to 0} \int\limits_{\Sigma} \frac{\|u\|}{r^3} d\xi d\eta = -2A_1(P)\pi\alpha_1 \cot(\pi\alpha_1)\xi^{\alpha_1 - 1}, \tag{5.95a}$$

$$\lim_{\varepsilon \to 0} \int\limits_{\Sigma} \frac{(\xi - x)^2\|u\|}{r^5} d\xi d\eta = -\frac{4}{3}A_1(P)\pi\alpha_1 \cot(\pi\alpha_1)\xi^{\alpha_1 - 1}, \tag{5.95b}$$

$$\lim_{\varepsilon \to 0} \int\limits_{\Sigma} \frac{(\xi - x)(\eta - y)\|u\|}{r^5} d\xi d\eta = 0, \tag{5.95c}$$

$$\lim_{\varepsilon \to 0} \int\limits_{\Sigma} \frac{(\eta - y)^2\|u\|}{r^5} d\xi d\eta = -\frac{2}{3}A_1(P)\pi\alpha_1 \cot(\pi\alpha_1)\xi^{\alpha_1 - 1}, \tag{5.95d}$$

$$\lim_{\varepsilon \to 0} \int\limits_{\Sigma} \frac{(\eta - y)\|\theta\|}{r^3} d\xi d\eta = 0, \tag{5.95e}$$

$$\lim_{\varepsilon \to 0} \int\limits_{\Sigma} \frac{(\xi - x)\|\theta\|}{r^3} d\xi d\eta = -2A_4(P)\cot(\pi\alpha_4)\xi^{\alpha_4}. \tag{5.95f}$$

Letting $\xi \to 0$, yields

$$\begin{cases} \cot(\pi\beta)A_1(P) + \gamma A_3(P) = 0, \\ \cot(\pi\alpha_2)A_2(P) = 0, \\ \gamma A_1(P) - \cot(\pi\beta)A_3(P) = 0, \\ \cot(\pi\alpha_4)A_4(P) = 0, \end{cases} \tag{5.96}$$

where $\beta = \alpha_1 = \alpha_3$, $\gamma = (1-\eta_2)/(1+\eta_2)$.

As the constants $A_i(P)$ are generally assumed to be non-zero, the characteristics to determine singularity behavior indices α_i are found from Eqs. (5.96):

$$\begin{cases} \cot(\pi\alpha_2) = 0, \\ \cot(\pi\alpha_4) = 0, \\ \cot(\pi\beta) = \pm\gamma\mathrm{i}, \end{cases} \tag{5.97}$$

where $\mathrm{i} = \sqrt{-1}$, and the acceptable roots are:

$$\alpha_2 = \alpha_4 = \frac{1}{2},\ \beta_{1,2} = \frac{1}{2} \pm \mathrm{i}\kappa,\ \kappa = \frac{1}{2\pi}\ln\eta_2. \tag{5.98}$$

The roots show that the displacement discontinuity $\|v\|$ and temperature discontinuity $\|\theta\|$ both have the classical singularity index 1/2, while $\|u\|$ or $\|w^*\|$ has the singularity index $1/2 + \mathrm{i}\kappa$ as the case in purely elastic materials [33]. Taking into consideration that the extended displacement discontinuities are in the neighborhood of the interface crack tip S_Σ, they should have the following forms

$$\begin{cases} \|u\| = \mathrm{Re}[A_1(P)\xi^\alpha], \\ \|v\| = A_2(P)\xi^{1/2}, \\ \|w^*\| = \mathrm{Re}[A_3(P)\xi^\alpha], \\ \|\theta\| = A_4(P)\xi^{1/2}, \end{cases} \tag{5.99}$$

where $\alpha = \beta_1 = 1/2 + \mathrm{i}\varepsilon$, $A_2(P)$ and $A_4(P)$ are arbitrary real constants, whilst $A_1(P)$ and $A_3(P)$ are arbitrary complex constants which satisfy the relation of $A_1(P) = \mathrm{i}A_3(P)$. So they can also be written as

$$\|w^*\| + i\|u\| = [A_R(P) + iA_I(P)]\xi^\alpha = A(P)\xi^\alpha. \tag{5.100}$$

It should be pointed out that the above results are necessary for studying the singular extended stress fields near the interface crack front exactly and for a unique solution for the hyper-singular, integral-differential Eq. (5.70).

5.3.5.2 Singular Fields Around Interfacial Cracks in Piezoethermoelastic Materials

Based on the expressions of the extended stress fields at an arbitrary field point in the upper domain induced by the interface crack given in Eq. (5.68), the same treatment proposed in [39] is adopted to study the singular extended stress fields ahead of the crack front. In order to get the exact expressions for the singular extended stress fields at an arbitrary point $\Xi(x, y, z)$ near point P of the periphery Γ of the interface crack, the local nature coordinate system *Pnm,* and the local polar coordinates, $P\rho\omega$, with $P\Xi$ as the negative normal axis are introduced as shown in Fig. 5.2. The transformational relation between the two coordinate systems are

$$x = -\rho \cos \omega,\ y = 0,\ z = \rho \sin \omega,\ R = \sqrt{(\xi + \rho \cos \omega)^2 + \eta^2 + (\rho \sin \omega)^2}. \tag{5.101}$$

For brevity, we only study the singular extended stresses at Ξ_0 in the interface crack front area ($\omega = 0$). Substituting Eq. (5.101) into Eq. (5.68) and combining with the singular index in Eq. (5.98) as well as the form of the expressions for the extended displacement discontinuities in Eq. (5.99), one can have the following dominant-part integrals [33, 39]:

$$\begin{aligned} &\lim_{\Sigma_\varepsilon \to 0} \int_{S_\varepsilon} \frac{\|u(\xi, \eta)\|}{R^3} dS \\ &= \frac{i\pi}{2\cosh(\pi\kappa)} \left[\overline{A(P)}(1 - 2i\kappa)\rho^{-1/2 - i\kappa} - A(P)(1 + 2i\kappa)\rho^{-1/2 + i\kappa} \right], \end{aligned} \tag{5.102a}$$

$$\begin{aligned} &\lim_{\Sigma_\varepsilon \to 0} \int_{S_\varepsilon} \frac{(\xi - x)^2 \|u(\xi, \eta)\|}{R^5} dS \\ &= \frac{i\pi}{3\cosh(\pi\kappa)} \left[\overline{A(P)}(1 - 2i\kappa)\rho^{-1/2 - i\kappa} - A(P)(1 + 2i\kappa)\rho^{-1/2 + i\kappa} \right], \end{aligned} \tag{5.102b}$$

$$\lim_{\Sigma_\varepsilon \to 0} \int_{S_\varepsilon} \frac{(\xi - x)(\eta - y) \|u(\xi, \eta)\|}{R^5} dS = 0, \tag{5.102c}$$

$$\lim_{\Sigma_\varepsilon \to 0} \int_{S_\varepsilon} \frac{\|v(\xi, \eta)\|}{R^3} dS = \pi A_2(P)\rho^{-1/2}, \tag{5.102d}$$

$$\lim_{\Sigma_\varepsilon \to 0} \int_{S_\varepsilon} \frac{(\eta - y)^2 \|v(\xi, \eta)\|}{R^5} dS - \frac{\pi}{3} A_2(P)\rho^{-1/2}, \tag{5.102e}$$

$$\lim_{\Sigma_\varepsilon \to 0} \int_{S_\varepsilon} \frac{(\xi - x)(\eta - y)\|v(\xi, \eta)\|}{R^5} dS = 0, \tag{5.102f}$$

$$\begin{aligned} &\lim_{\Sigma_\varepsilon \to 0} \int_{S_\varepsilon} \frac{\|w^*(\xi, \eta)\|}{R^3} dS \\ &= \frac{\pi}{2\cosh(\pi\kappa)} \left[\overline{A(P)}(1 - 2\mathrm{i}\kappa)\rho^{-1/2-\mathrm{i}\kappa} + A(P)(1 + 2\mathrm{i}\kappa)\rho^{-1/2+\mathrm{i}\kappa} \right], \end{aligned} \tag{5.102g}$$

$$\lim_{\Sigma_\varepsilon \to 0} \int_{S_\varepsilon} \frac{\|\theta(\xi, \eta)\|}{R^3} dS = \pi A_4(P)\rho^{-1/2}, \tag{5.102h}$$

$$\lim_{\Sigma_\varepsilon \to 0} \int_{S_\varepsilon} \frac{(\xi - x)\|\theta(\xi, \eta)\|}{R^3} dS = -2A_4(P)\rho^{1/2}, \tag{5.102i}$$

$$\lim_{\Sigma_\varepsilon \to 0} \int_{S_\varepsilon} \frac{(\eta - y)\|\theta(\xi, \eta)\|}{R^3} dS = 0, \tag{5.102j}$$

where $\overline{A(P)}$ is the conjugate of $A(P)$, Σ_ε denotes the fan-shaped integral area with P as the origin and ε as the radius. Substituting Eqs. (5.102) into (5.68), the extended stress fields of point $P(\rho, 0)$ ahead of the crack front are obtained as

$$\begin{aligned} \sigma_{zx}(\rho, 0) = &\frac{\mathrm{i}L_{14}e^{\pi\kappa}}{2(1 - \eta_2)\varsigma} \left[\overline{A(P)}(1 - 2\mathrm{i}\kappa)\rho^{-1/2-\mathrm{i}\kappa} - A(P)(1 + 2\mathrm{i}\kappa)\rho^{-1/2+\mathrm{i}\kappa} \right] \\ &+ \frac{2L_{13}}{(1 - \eta_2)} A_4(P)\rho^{1/2}, \end{aligned} \tag{5.103a}$$

$$\sigma_{yz}(\rho, 0) = \frac{\pi L_{14}\eta_1}{(1 - \eta_2)\varsigma} A_2(P)\rho^{-1/2}, \tag{5.103b}$$

$$\begin{aligned} &\ell[\sigma_z(\rho, 0) + f D_z(\rho, 0)] \\ &= \frac{2\pi L_{14}e^{\pi\kappa}}{(1 - \eta_1)\varsigma} \left[\overline{A(P)}(1 - 2\mathrm{i}\kappa)\rho^{-1/2-\mathrm{i}\kappa} + A(P)(1 + 2\mathrm{i}\kappa)\rho^{-1/2+\mathrm{i}\kappa} \right], \end{aligned} \tag{5.103c}$$

$$h_z(\rho, 0) = -\pi L_5 A_4(P)\rho^{-1/2}, \tag{5.103d}$$

which are analogous to the expressions in the isotropic, thermoelastic bi-materials [39].

5.3.6 Extended Stress Intensity Factors

In isotropic thermoelastic bi-materials, the stress and heat flux intensity factors are defined as [39]:

$$K_{\rm I} = \lim_{\rho \to 0} \sqrt{2\pi\rho}\, \rho^{-{\rm i}\varepsilon} \sigma_z(\rho, 0), \tag{5.104a}$$

$$K_{\rm II} = \lim_{\rho \to 0} \sqrt{2\pi\rho}\, \rho^{-{\rm i}\varepsilon} \sigma_{zx}(\rho, 0), \tag{5.104b}$$

$$K_{\rm III} = \lim_{\rho \to 0} \sqrt{2\pi\rho}\, \sigma_{yz}(\rho, 0), \tag{5.104c}$$

$$K_h = \lim_{\rho \to 0} \sqrt{2\pi\rho}\, h_z(\rho, 0), \tag{5.104d}$$

and can be expressed in terms of the displacement and temperature discontinuities:

$$K_{\rm I} + {\rm i}K_{\rm II} = \sqrt{2\pi} \frac{\mu_1 \mu_2 (1 + 2{\rm i}\varepsilon) {\rm e}^{\pi\varepsilon}}{(\mu_1 + k_1 \mu_2)} \lim_{\rho \to 0} \frac{\|\tilde{w}(\rho, 0)\| + {\rm i}\|\tilde{u}(\rho, 0)\|}{\rho^{1/2 + {\rm i}\varepsilon}}, \tag{5.105a}$$

$$K_{\rm III} = \frac{\sqrt{2\pi}\mu_1 \mu_2}{2(\mu_1 + \mu_2)} \lim_{\rho \to 0} \frac{\|\tilde{v}(\rho, 0)\|}{\sqrt{\rho}}, \tag{5.105b}$$

$$K_h = -\frac{\sqrt{2\pi}\beta_1 \beta_2}{2(\beta_1 + \beta_2)} \lim_{\rho \to 0} \frac{\|\tilde{\theta}(\rho, 0)\|}{\sqrt{\rho}}. \tag{5.105c}$$

Similarly, the extended stress intensity factors in piezothermoelastic bi-materials can be defined as [86]:

$$K_{\rm I2} = \lim_{\rho \to 0} \sqrt{2\pi\rho}\, \rho^{-{\rm i}\varepsilon} \ell[\sigma_z(\rho, 0) + f\omega(\rho, 0)], \tag{5.106a}$$

$$K_{\rm II} = \lim_{\rho \to 0} \sqrt{2\pi\rho}\, \rho^{-{\rm i}\varepsilon} \sigma_{zx}(\rho, 0), \tag{5.106b}$$

$$K_{\rm III} = \lim_{\rho \to 0} \sqrt{2\pi\rho}\, \sigma_{yz}(\rho, 0), \tag{5.106c}$$

$$K_h = \lim_{\rho \to 0} \sqrt{2\pi\rho}\, h_z(\rho, 0). \tag{5.106d}$$

After inserting Eqs. (5.103) into (5.106), one can obtain the intensity factors expressed in terms of the extended displacement discontinuities:

$$K_{I2} + \mathrm{i}K_{II} = 2\sqrt{2\pi}\pi \frac{L_{14}(1+2\mathrm{i}\kappa)\mathrm{e}^{\pi\kappa}}{(1-\eta_2)\varsigma} \lim_{\rho\to 0} \frac{\|w\| + L_{15}\|\varphi\|/L_{14} + \mathrm{i}\|u\|}{\rho^{1/2+\mathrm{i}\kappa}}, \tag{5.107a}$$

$$K_{III} = \sqrt{2\pi}\pi \frac{L_{14}\eta_1}{(1-\eta_2)\varsigma} \lim_{\rho\to 0} \frac{\|v(\rho,0)\|}{\sqrt{\rho}}, \tag{5.107b}$$

$$K_h = -\sqrt{2\pi}\pi L_5 \lim_{\rho\to 0} \frac{\|\theta(\rho,0)\|}{\sqrt{\rho}}, \tag{5.107c}$$

The Mode I2 intensity factor K_{I2} is another new fracture parameter near the crack tip in piezothermoelastic bi-materials. And the new fracture parameters can be used in the fracture criterion for piezothermoelastic interface crack problem. It can be observed that the extended stress intensity factors are in the equivalent form with the piezoelectric ones [37].

5.4 Fundamental Solutions for Magnetoelectrothermoelastic Bi-Materials

Magnetoelectrothermoelastic materials has an extra magnetic field coupled with electric and thermoelastic field in comparison with piezothermoelastic materials. Electromagnetic coupling can be found in most electric conductors when magnetic field is applied. Integrity of magnetoelectrothermoelastic materials under coupled multifield environment is essential to the application of these materials in advanced control and actuation of modern microelectromechanical systems (MEMSs). For homogeneous materials, a lot of researches have been conducted [89–98]. However, laminated structures are often used to get stronger, multiphysical coupling effect. Therefore, interface problems occur frequently, and the interface crack problems of dissimilar mangetoelectroelastic bi-materials were investigated by many researchers [99–106]. Here, employing the similar approaches as discussed in the preceding sections, we present the fundamental solution framework for manetoelectrothermoelastic materials.

When the *xy*-plane is parallel to the plane of isotropy in the Cartesian coordinates (x,y,z), the corresponding constitutive relations of transversely isotropic, thermo-magneto-electroelastic materials are [107, 108]

$$\begin{aligned}
\sigma_x &= c_{11}\frac{\partial u}{\partial x}+c_{12}\frac{\partial v}{\partial y}+c_{13}\frac{\partial w}{\partial z}+e_{31}\frac{\partial \varphi}{\partial z}+f_{31}\frac{\partial \psi}{\partial z}-\lambda_{11}\theta\\
\sigma_y &= c_{12}\frac{\partial u}{\partial x}+c_{11}\frac{\partial v}{\partial y}+c_{13}\frac{\partial w}{\partial z}+e_{31}\frac{\partial \varphi}{\partial z}+f_{31}\frac{\partial \psi}{\partial z}-\lambda_{11}\theta\\
\sigma_z &= c_{13}\frac{\partial u}{\partial x}+c_{13}\frac{\partial v}{\partial y}+c_{33}\frac{\partial w}{\partial z}+e_{33}\frac{\partial \varphi}{\partial z}+f_{33}\frac{\partial \psi}{\partial z}-\lambda_{33}\theta\\
\tau_{yz} &= c_{44}\left(\frac{\partial v}{\partial z}+\frac{\partial w}{\partial y}\right)+e_{15}\frac{\partial \varphi}{\partial y}+f_{15}\frac{\partial \psi}{\partial y}\\
\tau_{zx} &= c_{44}\left(\frac{\partial u}{\partial z}+\frac{\partial w}{\partial x}\right)+e_{15}\frac{\partial \varphi}{\partial x}+f_{15}\frac{\partial \psi}{\partial x}\\
\tau_{xy} &= c_{66}\left(\frac{\partial u}{\partial y}+\frac{\partial v}{\partial x}\right)
\end{aligned} \tag{5.108a}$$

$$\begin{aligned}
D_x &= e_{15}\left(\frac{\partial u}{\partial z}+\frac{\partial w}{\partial x}\right)-\varepsilon_{11}\frac{\partial \varphi}{\partial x}-g_{11}\frac{\partial \psi}{\partial x}\\
D_y &= e_{15}\left(\frac{\partial v}{\partial z}+\frac{\partial w}{\partial y}\right)-\varepsilon_{11}\frac{\partial \varphi}{\partial y}-g_{11}\frac{\partial \psi}{\partial y}\\
D_z &= e_{31}\left(\frac{\partial u}{\partial y}+\frac{\partial v}{\partial x}\right)+e_{33}\frac{\partial w}{\partial z}-\varepsilon_{33}\frac{\partial \varphi}{\partial z}-g_{33}\frac{\partial \psi}{\partial z}+p_3\theta
\end{aligned} \tag{5.108b}$$

$$\begin{aligned}
B_x &= f_{15}\left(\frac{\partial u}{\partial z}+\frac{\partial w}{\partial x}\right)-g_{11}\frac{\partial \varphi}{\partial x}-\mu_{11}\frac{\partial \psi}{\partial x}\\
B_y &= f_{15}\left(\frac{\partial v}{\partial z}+\frac{\partial w}{\partial y}\right)-g_{11}\frac{\partial \varphi}{\partial y}-\mu_{11}\frac{\partial \psi}{\partial y}\\
B_z &= f_{31}\left(\frac{\partial u}{\partial y}+\frac{\partial v}{\partial x}\right)+f_{33}\frac{\partial w}{\partial z}-g_{33}\frac{\partial \varphi}{\partial z}-\mu_{33}\frac{\partial \psi}{\partial z}+m_3\theta
\end{aligned} \tag{5.108c}$$

$$h_x=-\beta_{11}\frac{\partial \theta}{\partial x},\quad h_y=-\beta_{11}\frac{\partial \theta}{\partial y},\quad h_z=-\beta_{33}\frac{\partial \theta}{\partial z} \tag{5.108d}$$

where u, v, w, φ, ψ and θ are the mechanical elastic displacements, electric potential, magnetic potential and temperature change (with respect to the reference temperature), respectively, and referred to as extended displacements; c_{ij}, e_{ij}, f_{ij}, ε_{ij}, g_{ij}, μ_{ij}, λ_{ii}, p_3 and m_3 are elastic, piezoelectric, piezomagnetic, dielectric, electro-magnetic, magnetic, thermal modulus, pyroelectric and pyromagnetic constants, respectively, and β_{ij} are the coefficients of heat conduction. The relation $c_{66} = (c_{11} - c_{12})/2$ holds for materials with transverse isotropy.

In the absence of body force, electric charge, electric current and body heat source, the governing equations of TMEE materials in the Cartesian coordinate system o-xyz are given by Shechtman et al. [110]

$$\begin{aligned}&\frac{\partial \sigma_x}{\partial x}+\frac{\partial \tau_{xy}}{\partial y}+\frac{\partial \tau_{zx}}{\partial z}=0\\&\frac{\partial \tau_{xy}}{\partial x}+\frac{\partial \sigma_y}{\partial y}+\frac{\partial \tau_{yz}}{\partial z}=0\\&\frac{\partial \tau_{zx}}{\partial x}+\frac{\partial \tau_{yz}}{\partial y}+\frac{\partial \sigma_z}{\partial z}=0\end{aligned} \tag{5.109a}$$

$$\frac{\partial D_x}{\partial x}+\frac{\partial D_y}{\partial y}+\frac{\partial D_z}{\partial z}=0 \tag{5.109b}$$

$$\frac{\partial B_x}{\partial x}+\frac{\partial B_y}{\partial y}+\frac{\partial B_z}{\partial z}=0 \tag{5.109c}$$

$$\frac{\partial h_x}{\partial x}+\frac{\partial h_y}{\partial y}+\frac{\partial h_z}{\partial z}=0 \tag{5.109d}$$

where σ_{ij}, D_i, B_i, and h_i, with $i, j = x, y, z$, are the components of stress, electric displacement, magnetic induction and heat flux, respectively, which are referred to as extended stresses here.

Assuming a penny-shaped crack with radius a centered at the origin of the coordinate system is oriented at the interface of two bonded dissimilar magneto-electro-thermo-elastic materials perpendicular to the poling direction, as is shown in Fig. 5.5. The crack lies in the plane xoy, and two bonded solids are assumed to be perfectly combined in the interface and occupy the upper and lower half-space respectively. The upper and lower surfaces of the crack are denoted by S^+ and S^-, respectively, and

$$x = r\cos\phi,\ y = r\sin\phi \tag{5.110}$$

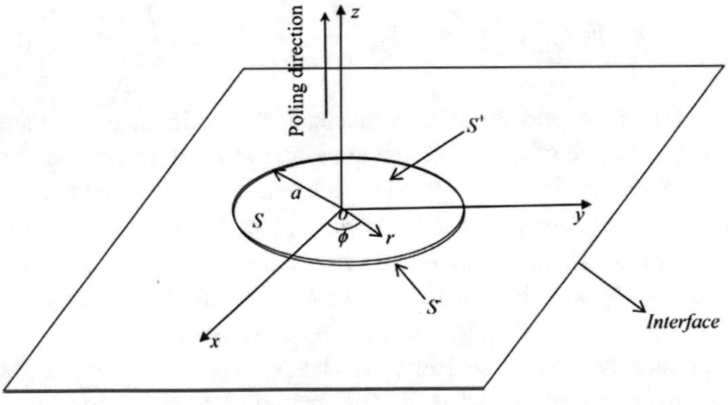

Fig. 5.5 A penny-shaped interface crack of radius a perpendicular to the poling direction [86]

The EDDs across the surfaces of the crack can be expressed as:

$$\begin{aligned}&\|u\| = u(x,y,0^+) - u(x,y,0^-),\ \|v\| = v(x,y,0^+) - v(x,y,0^-)\\&\|w\| = w(x,y,0^+) - w(x,y,0^-),\ \|\varphi\| = \varphi(x,y,0^+) - \varphi(x,y,0^-)\\&\|\psi\| = \psi(x,y,0^+) - \psi(x,y,0^-),\ \|\theta\| = \theta(x,y,0^+) - \theta(x,y,0^-),\ (x,y) \in S\end{aligned} \tag{5.111}$$

In analogy with Zhao et al. [40] and Zhao and Liu [41], let the radius of the crack a approach zero, the fundamental solutions corresponding to a unit concentrated EDD can all be obtained. It is required that the fundamental solution should satisfy the governing equations of thermo-magneto-electro-elasticity [37] and the following conditions [109], respectively:

$$\lim_{a\to 0}\int_S \{\|u\|, \|v\|, \|w\|, \|\varphi\|, \|\psi\|, \|\theta\|\}\mathrm{d}S = \{1,0,0,0,0,0\} \tag{5.112a}$$

$$\lim_{a\to 0}\int_S \{\|u\|, \|v\|, \|w\|, \|\varphi\|, \|\psi\|, \|\theta\|\}\mathrm{d}S = \{0,1,0,0,0,0\} \tag{5.112b}$$

$$\lim_{a\to 0}\int_S \{\|u\|, \|v\|, \|w\|, \|\varphi\|, \|\psi\|, \|\theta\|\}\mathrm{d}S = \{0,0,1,0,0,0\} \tag{5.112c}$$

$$\lim_{a\to 0}\int_S \{\|u\|, \|v\|, \|w\|, \|\varphi\|, \|\psi\|, \|\theta\|\}\mathrm{d}S = \{0,0,0,1,0,0\} \tag{5.112d}$$

$$\lim_{a\to 0}\int_S \{\|u\|, \|v\|, \|w\|, \|\varphi\|, \|\psi\|, \|\theta\|\}\mathrm{d}S = \{0,0,0,0,1,0\} \tag{5.112e}$$

$$\lim_{a\to 0}\int_S \{\|u\|, \|v\|, \|w\|, \|\varphi\|, \|\psi\|, \|\theta\|\}\mathrm{d}S = \{0,0,0,0,0,1\} \tag{5.112f}$$

Using the operator theory and the generalized Almansi's theorem, Chen et al. [95] derived the general solution satisfying Eqs. (5.108) and (5.112). According to this general solution, we have

$$
\begin{aligned}
&U = \Lambda\left(\mathrm{i}\Psi_0 + \sum_{j=1}^{5}\Psi_j\right), \quad w_m = \sum_{j=1}^{5} s_j k_{mj}\frac{\partial \Psi_j}{\partial z_j}, \quad \theta = k_{45}\frac{\partial^2 \Psi_5}{\partial z_5^2} \\
&\sigma_1 = 2\sum_{j=1}^{5}\left(c_{66} - \omega_{1j}s_j^2\right)\frac{\partial^2 \Psi_j}{\partial z_j^2}, \quad \sigma_2 = 2c_{66}\Lambda^2\left(\mathrm{i}\Psi_0 + \sum_{j=1}^{5}\Psi_j\right) \\
&\sigma_{zm} = \sum_{j=1}^{5}\omega_{mj}\frac{\partial^2 \Psi_j}{\partial z_j^2}, \quad h_x = -\beta_{33}k_{45}\frac{\partial^3 \Psi_5}{\partial x \partial z_5^3} \\
&h_y = -\beta_{33}k_{45}\frac{\partial^3 \Psi_5}{\partial y \partial z_5^2}, \quad h_z = -\beta_{33}s_5 k_{45}\frac{\partial^3 \Psi_5}{\partial z_5^3} \\
&\tau_m = \Lambda\left(\mathrm{i}s_0\rho_m\frac{\partial \Psi_0}{\partial z_0} + \sum_{j=1}^{5}s_j\omega_{mj}\frac{\partial \Psi_j}{\partial z_j}\right), \qquad m = 1,\ 2,\ 3
\end{aligned}
\tag{5.113}
$$

The corresponding notations for all components in the Cartesian coordinates (x, y, z) and cylindrical coordinates (r, ϕ, z) are defined as:

$$
\begin{aligned}
&z_j = s_j z, \quad \Lambda = \frac{\partial}{\partial x} + \mathrm{i}\frac{\partial}{\partial y} \\
&U = u + \mathrm{i}v = \mathrm{e}^{\mathrm{i}\phi}\left(u_r + \mathrm{i}u_\varphi\right) \\
&w_1 = w, \quad w_2 = \varphi, \quad w_3 = \psi \\
&\sigma_1 = \sigma_x + \sigma_y = \sigma_r + \sigma_\phi \\
&\sigma_2 = \sigma_x - \sigma_y + 2\mathrm{i}\tau_{xy} = \mathrm{e}^{i\phi}\left(\sigma_r - \sigma_\phi + 2\mathrm{i}\tau_{r\phi}\right) \\
&\sigma_{z1} = \sigma_z, \quad \sigma_{z2} = D_z, \quad \sigma_{z3} = B_z \\
&\tau_m = \tau_{xm} + \mathrm{i}\tau_{ym} = \mathrm{e}^{\mathrm{i}\phi}\left(\tau_{rm} + \mathrm{i}\tau_{\phi m}\right) \\
&\tau_{x1} = \tau_{zx}, \quad \tau_{y1} = \tau_{yz}, \quad \tau_{r1} = \tau_{zr}, \quad \tau_{\phi 1} = \tau_{\phi z} \\
&\tau_{x2} = D_x, \quad \tau_{y2} = D_y, \quad \tau_{r2} = D_r, \quad \tau_{\phi 2} = D_\phi \\
&\tau_{x3} = B_x, \quad \tau_{y3} = B_y, \quad \tau_{r3} = B_r, \quad \tau_{\phi 3} = B_\phi
\end{aligned}
\tag{5.114}
$$

where all the concerning material related constants can be found in [109].

The harmonic functions Ψ_j (j = 0, 1, 2, 3, 4, 5) satisfy

$$
\left(\Delta + \frac{\partial^2}{\partial z_i^2}\right)\Psi_j = 0 \tag{5.115}
$$

where

$$
\Delta = \frac{\partial^2}{\partial x^2} + \frac{\partial^2}{\partial y^2}\ \text{in the Cartesian coordinates}(x, y, z) \tag{5.116a}
$$

and

$$\Delta = \frac{\partial^2}{\partial r^2} + \frac{\partial}{r\partial r} + \frac{\partial^2}{r^2\partial\phi^2} \text{in cylindrical coordinates}(r, \phi, z). \tag{5.116b}$$

The fundamental solution for the interface crack problem in thermoelectro-magnetic bi-materials can be obtained using similar procedures in the preceding section, and details can be found in [109].

5.5 Fundamental Solutions for Interface Crack Problems in Quasi-Crystalline Materials

The first quasicrystal (QC) was found by Shechtman et al. [110]. Since then, the atomic structure and physical properties of the QC material have been attracting increasing attentions. As a promising solid structure, QCs possess a series of ideal properties such as low adhesion, low coefficient of friction, low porosity, low thermal conductivity, high abrasion resistance, and high resistivity [111, 112], and have been adopted progressively in high-tech industries, such as the automotive, aerospace and energy industries. Due to its quasi-periodic symmetry, concepts of high-dimensional space have been introduced instead of the classical crystallographic theory to describe the physical properties of QC materials. The phonon field represents the lattice vibrations while the phason field depicts the quasi-periodic rearrangement of atoms, and both fields are used to describe the elasticity of QCs. As a typical QC material, 1D, hexagonal QCs exhibit just one quasi-periodic axis, while the perpendicular plane of the axis exhibits the classical crystalline properties. The properties of QCs are very sensitive to defects, which are inevitable in QC materials. Extensive research has been performed on quasi-crystals with various forms of defects [113–135]. Green's functions for modelling the interface defects in quasi-crystal structures has recently been proposed as laminated structures is increasingly introduced to elevate the coupling effect of multiphysical fields [136, 137]. Employing the similar approaches in the preceding sections, we present the fundamental solutions for interface crack problems in quasi-crystals in this section.

In the absence of body forces, the constitutive relations for 1D hexagonal QCs with thermo-electro effect, referred to the Cartesian coordinate (x, y, z) with xoy coincident with the periodic plane and the z-axis identical to the quasi-periodic direction, can be expressed as [138, 139]

$$
\begin{aligned}
\sigma_{xx} &= c_{11}\frac{\partial u_x}{\partial x} + c_{12}\frac{\partial u_y}{\partial y} + c_{13}\frac{\partial u_z}{\partial z} + R_1\frac{\partial w_z}{\partial z} + e_{31}\frac{\partial \varphi}{\partial z} - \beta_1\theta,\\
\sigma_{yy} &= c_{12}\frac{\partial u_x}{\partial x} + c_{11}\frac{\partial u_y}{\partial y} + c_{13}\frac{\partial u_z}{\partial z} + R_1\frac{\partial w_z}{\partial z} + e_{31}\frac{\partial \varphi}{\partial z} - \beta_1\theta,\\
\sigma_{zz} &= c_{13}\frac{\partial u_x}{\partial x} + c_{13}\frac{\partial u_y}{\partial y} + c_{33}\frac{\partial u_z}{\partial z} + R_2\frac{\partial w_z}{\partial z} + e_{33}\frac{\partial \varphi}{\partial z} - \beta_3\theta,\\
\sigma_{yz} &= c_{44}\left(\frac{\partial u_z}{\partial y} + \frac{\partial u_y}{\partial z}\right) + R_3\frac{\partial w_z}{\partial y} + e_{15}\frac{\partial \varphi}{\partial y},\\
\sigma_{zx} &= c_{44}\left(\frac{\partial u_z}{\partial x} + \frac{\partial u_x}{\partial z}\right) + R_3\frac{\partial w_z}{\partial x} + e_{15}\frac{\partial \varphi}{\partial x},\\
\sigma_{xy} &= c_{66}\left(\frac{\partial u_x}{\partial y} + \frac{\partial u_y}{\partial x}\right),
\end{aligned}
\tag{5.117a}
$$

$$
\begin{aligned}
H_{zy} &= R_3\left(\frac{\partial u_z}{\partial y} + \frac{\partial u_y}{\partial z}\right) + K_2\frac{\partial w_z}{\partial y} + e'_{15}\frac{\partial \varphi}{\partial y},\\
H_{zx} &= R_3\left(\frac{\partial u_z}{\partial x} + \frac{\partial u_x}{\partial z}\right) + K_2\frac{\partial w_z}{\partial x} + e'_{15}\frac{\partial \varphi}{\partial x},\\
H_{zz} &= R_1\left(\frac{\partial u_x}{\partial x} + \frac{\partial u_y}{\partial y}\right) + R_2\frac{\partial u_z}{\partial z} + K_1\frac{\partial w_z}{\partial z} + e'_{33}\frac{\partial \varphi}{\partial z},
\end{aligned}
\tag{5.117b}
$$

$$
\begin{aligned}
D_x &= e_{15}\left(\frac{\partial u_z}{\partial x} + \frac{\partial u_x}{\partial z}\right) + e'_{15}\frac{\partial w_z}{\partial x} - \zeta_{11}\frac{\partial \varphi}{\partial x},\\
D_y &= e_{15}\left(\frac{\partial u_z}{\partial y} + \frac{\partial u_y}{\partial z}\right) + e'_{15}\frac{\partial w_z}{\partial y} - \zeta_{11}\frac{\partial \varphi}{\partial y},
\end{aligned}
\tag{5.117c}
$$

$$
\begin{aligned}
D_z &= e_{31}\left(\frac{\partial u_x}{\partial x} + \frac{\partial u_y}{\partial y}\right) + e_{33}\frac{\partial u_z}{\partial z} + e'_{33}\frac{\partial w_z}{\partial z} - \zeta_{33}\frac{\partial \varphi}{\partial z} + p_3\theta,\\
q_x &= -K_{11}\frac{\partial \theta}{\partial x}, q_y = -K_{11}\frac{\partial \theta}{\partial y}, q_z = -K_{33}\frac{\partial \theta}{\partial z},
\end{aligned}
\tag{5.117d}
$$

where u_i, w_i, and φ are, respectively, phonon displacements, phason displacements, and electric potential; σ_{ij}, H_{ij}, and c_{ij} (K_i) are, respectively, phonon stresses, phason stresses, and elastic stiffness constants; R_i are phonon-phason relevant elastic constants; β_i are thermal constants; p_3 denotes the pyroelectric constant; θ represents the temperature variation and $\theta = 0$ corresponds to a reference state; D_i, E_i, and ζ_{ii} are electric displacements, electric fields, and dielectric coefficients, respectively; e_{ij} and e'_{ij} are piezoelectric coefficients; and K_{11} and K_{33} are coefficients of thermal conductivity.

Without body forces and free charges, the equilibrium equations for 1D hexagonal piezoelectric QCs can be expressed as

$$\frac{\partial \sigma_{xx}}{\partial x} + \frac{\partial \sigma_{xy}}{\partial y} + \frac{\partial \sigma_{xz}}{\partial z} = 0, \quad \frac{\partial \sigma_{yx}}{\partial x} + \frac{\partial \sigma_{yy}}{\partial y} + \frac{\partial \sigma_{yz}}{\partial z} = 0, \tag{5.118a}$$

$$\frac{\partial \sigma_{zx}}{\partial x} + \frac{\partial \sigma_{zy}}{\partial y} + \frac{\partial \sigma_{zz}}{\partial z} = 0, \quad \frac{\partial H_{zx}}{\partial x} + \frac{\partial H_{zy}}{\partial y} + \frac{\partial H_{zz}}{\partial z} = 0, \tag{5.118b}$$

$$\frac{\partial D_x}{\partial x} + \frac{\partial D_y}{\partial y} + \frac{\partial D_z}{\partial z} = 0, \quad \frac{\partial q_x}{\partial x} + \frac{\partial q_y}{\partial y} + \frac{\partial q_z}{\partial z} = 0. \tag{5.118c}$$

Comparing the basic equations for 1D QC materials to those for 3D transversely isotropic magnetoelectrothermoelastic (METE) materials [109], we can find the special equivalent relations as listed in Table 5.1. That is to say, if the variables and the coefficients in the governing equations for METEs are replaced respectively by those for 1D QCs based on the equivalent relations in Table 5.1, we can obtain the governing equations for 1D QCs.

In column 3 of Table 5.1, u, v, w, φ, ψ and θ are the mechanical elastic displacements, electric potential, magnetic potential and temperature change, respectively, for METE material. c_{ij}, e_{ij}, f_{ij}, ε_{ij}, g_{ij}, μ_{ij}, λ_{ii}, p_3 and m_3 are elastic, piezoelectric, piezomagnetic, dielectric, electromagnetic, magnetic, thermal modulus, pyroelectric and pyromagnetic constants, respectively, and β_{ij} are the coefficients of heat conduction. With the above analogy relation, one can get the solutions for 1D QC materials directly from those for METEs [109].

Table 5.1 The analogy relation between 1D quasicrystal and magnetoelectrothermoelastic materials [145]

Material	1D QC	3D METE
Extended displacements	u_x, u_y, u_z	u, v, w
	w_z, φ, θ	ψ, φ, θ
Extended stresses	$\sigma_{ij}, i,j = x, y, z$	$\sigma_{ij}, i,j = x, y, z$
	D_x, D_y, D_z	D_x, D_y, D_z
	H_{zx}, H_{zy}, H_{zz}	B_x, B_y, B_z
	q_x, q_y, q_z	h_x, h_y, h_z
Coefficients	c_{ij}, R_1, R_2, R_3	$c_{ij}, f_{31}, f_{33}, f_{15}$
	$e_{15}, e_{31}, e_{33}, e'_{15}, e'_{33}$	$e_{15}, e_{31}, e_{33}, -g_{11}, -g_{33}$
	$K_{11}, K_{33}, 0$	$\beta_{11}, \beta_{33}, m_3$
	β_1, β_3, p_3	$\lambda_{11}, \lambda_{33}, p_3$
	$K_1, K_2, \zeta_{11}, \zeta_{33}$	$-\mu_{33}, -\mu_{11}, \varepsilon_{11}, \varepsilon_{33}$

5.5.1 Fundamental Solutions for Unit-Point EDDs

Consider a penny-shaped crack with radius a centered at the origin of the coordinate system lies at the interface of two bonded dissimilar QC materials perpendicular to the quasi-periodic direction, as illustrated in Fig. 5.4. The crack is located in the periodic plane xoy. The two solids are assumed to be perfectly bonded along the interface except the cracked segment and occupy the upper and lower half-space, respectively. The upper and lower surfaces of the crack are denoted by S^{+} and S^{-}, respectively, and the relations between the Cartesian coordinates and cylindrical coordinates are

$$\begin{cases} x = r\cos\phi, \\ y = r\sin\phi, \\ R^2 = r^2 + z^2 = x^2 + y^2 + z^2. \end{cases} \tag{5.119}$$

The EDDs across the interface crack faces can be expressed as:

$$\begin{aligned} &\|u_x\| = u_x(x,y,0^+) - u_x(x,y,0^-),\ \|u_y\| = u_y(x,y,0^+) - u_y(x,y,0^-), \\ &\|u_z\| = u_z(x,y,0^+) - u_z(x,y,0^-),\ \|w_z\| = w_z(x,y,0^+) - w_z(x,y,0^-), \\ &\|\varphi\| = \varphi(x,y,0^+) - \varphi(x,y,0^-),\ \|\theta\| = \theta(x,y,0^+) - \theta(x,y,0^-),\ (x,y) \in S. \end{aligned} \tag{5.120}$$

For unit-point phonon displacement discontinuity in z-direction, the fundamental solutions are given by Zhao et al. [140]

$$\sigma_z^+ = \sum_{j=1}^{4} A_j^{+*}\gamma_{1j}^+ \left(\frac{3z_j^2}{R_j^5} - \frac{1}{R_j^3}\right), \tag{5.121a}$$

$$H_{zz}^+ = \sum_{j=1}^{4} A_j^{+*}\gamma_{2j}^+ \left(\frac{3z_j^2}{R_j^5} - \frac{1}{R_j^3}\right), \tag{5.121b}$$

$$D_z^+ = \sum_{j=1}^{4} A_j^{+*}\gamma_{3j}^+ \left(\frac{3z_j^2}{R_j^5} - \frac{1}{R_j^3}\right), \tag{5.121c}$$

$$\sigma_{zr}^+ = 3\sum_{j=1}^{4} A_j^{+*}\beta_{1j}^+ \frac{rz_j}{R_j^5}, \tag{5.121d}$$

$$\sigma_{zx}^+ = 3\sum_{j=1}^{4} A_j^{+*}\beta_{1j}^+ \frac{xz_j}{R_j^5}, \tag{5.121e}$$

$$\sigma_{yz}^{+} = 3\sum_{j=1}^{4} A_j^{+*}\beta_{1j}^{+}\frac{yz_j}{R_j^5}, \tag{5.121f}$$

$$q_z^{+} = 0, \tag{5.121g}$$

where $R_j = \sqrt{r^2 + z_j^2}$, and the coefficients A_j^{+*} can be obtained by solving the following equations [140]:

$$\begin{cases} \sum_{j=1}^{5} A_j^{+*} = \sum_{j=1}^{5} A_j^{-*}, \\ -\sum_{j=1}^{5}\left[\alpha_{1j}^{+}A_j^{+*} + \alpha_{1j}^{-}A_j^{-*}\right] = \frac{1}{2\xi}, \\ -\sum_{j=1}^{5}\alpha_{2j}^{+}A_j^{+*} = \sum_{j=1}^{5}\alpha_{2j}^{-}A_j^{-*}, \\ -\sum_{j=1}^{5}\alpha_{3j}^{+}A_j^{+*} = \sum_{j=1}^{5}\alpha_{3j}^{-}A_j^{-*}, \\ \alpha_{45}^{+}A_5^{+} = \alpha_{45}^{-}A_5^{-}, \end{cases} \quad \begin{cases} \sum_{j=1}^{5}\beta_{1j}^{+}A_j^{+*} = -\sum_{j=1}^{5}\beta_{1j}^{-}A_j^{-*}, \\ \sum_{j=1}^{5}\gamma_{1j}^{+}A_j^{+*} = \sum_{j=1}^{5}\gamma_{1j}^{-}A_j^{-*}, \\ \sum_{j=1}^{5}\gamma_{2j}^{+}A_j^{+*} = \sum_{j=1}^{5}\gamma_{2j}^{-}A_j^{-*}, \\ \sum_{j=1}^{5}\gamma_{3j}^{+}A_j^{+*} = \sum_{j=1}^{5}\gamma_{3j}^{-}A_j^{-*}, \\ K_{33}^{+}s_5^{+}A_5^{+*}\alpha_{45}^{+} = -K_{33}^{-}s_5^{-}A_5^{-*}\alpha_{45}^{-}. \end{cases} \tag{5.122}$$

For unit-point phason displacement discontinuity in z-direction, the fundamental solutions are given by Zhao et al. [140]:

$$\sigma_z^{+} = \sum_{j=1}^{4} B_j^{+*}\gamma_{1j}^{+}\left(\frac{3z_j^2}{R_j^5} - \frac{1}{R_j^3}\right), \tag{5.123a}$$

$$H_{zz}^{+} = \sum_{j=1}^{4} B_j^{+*}\gamma_{2j}^{+}\left(\frac{3z_j^2}{R_j^5} - \frac{1}{R_j^3}\right), \tag{5.123b}$$

$$D_z^{+} = \sum_{j=1}^{4} B_j^{+*}\gamma_{3j}^{+}\left(\frac{3z_j^2}{R_j^5} - \frac{1}{R_j^3}\right), \tag{5.123c}$$

$$\sigma_{zr}^{+} = 3\sum_{j=1}^{4} B_j^{+*}\beta_{1j}^{+}\frac{rz_j}{R_j^5}, \tag{5.123d}$$

$$\sigma_{zx}^{+} = 3\sum_{j=1}^{4} B_j^{+*}\beta_{1j}^{+}\frac{xz_j}{R_j^5}, \tag{5.123e}$$

$$\sigma_{yz}^{+} = 3\sum_{j=1}^{4} B_j^{+*}\beta_{1j}^{+}\frac{yz_j}{R_j^5}, \tag{5.123f}$$

$$q_z^{+} = 0, \tag{5.123g}$$

where the coefficients B_j^{+*} can be obtained by solving the following equations:

$$\begin{cases} \sum_{j=1}^{5} B_j^{+*} = \sum_{j=1}^{5} B_j^{-*}, \\ -\sum_{j=1}^{5} \alpha_{1j}^{+} B_j^{+*} = \sum_{j=1}^{5} \alpha_{1j}^{-} B_j^{-*}, \\ -\sum_{j=1}^{5} \left[\alpha_{2j}^{+} B_j^{+*} + \alpha_{2j}^{-} B_j^{-*}\right] = \frac{1}{2\pi}, \\ -\sum_{j=1}^{5} \alpha_{3j}^{+} B_j^{+*} = \sum_{j=1}^{5} \alpha_{3j}^{-} B_j^{-*}, \\ \alpha_{45}^{+} B_j^{+*} = \alpha_{45}^{-} B_j^{-*}, \end{cases} \quad \begin{cases} \sum_{j=1}^{5} \beta_{1j}^{+} B_j^{+*} = -\sum_{j=1}^{5} \beta_{1j}^{-} B_j^{-*}, \\ \sum_{j=1}^{5} \gamma_{1j}^{+} B_j^{+*} = \sum_{j=1}^{5} \gamma_{1j}^{-} B_j^{-*}, \\ \sum_{j=1}^{5} \gamma_{2j}^{+} B_j^{+*} = \sum_{j=1}^{5} \gamma_{2j}^{-} B_j^{-*}, \\ \sum_{j=1}^{5} \gamma_{3j}^{+} B_j^{+*} = \sum_{j=1}^{5} \gamma_{3j}^{-} B_j^{-*}, \\ K_{33}^{+} s_5^{+} B_5^{+*} \alpha_{45}^{+} = -K_{33}^{-} s_5^{-} B_5^{-*} \alpha_{45}^{-}. \end{cases} \tag{5.124}$$

For unit-point electric potential discontinuity, the fundamental solutions are given by [140]:

$$\sigma_z^{+} = \sum_{j=1}^{4} C_j^{+*}\gamma_{1j}^{+}\left(\frac{3z_j^2}{R_j^5} - \frac{1}{R_j^3}\right), \tag{5.125a}$$

$$H_{zz}^{+} = \sum_{j=1}^{4} C_j^{+*}\gamma_{2j}^{+}\left(\frac{3z_j^2}{R_j^5} - \frac{1}{R_j^3}\right), \tag{5.125b}$$

$$D_z^{+} = \sum_{j=1}^{4} C_j^{+*}\gamma_{3j}^{+}\left(\frac{3z_j^2}{R_j^5} - \frac{1}{R_j^3}\right), \tag{5.125c}$$

$$\sigma_{zr}^{+} = 3\sum_{j=1}^{4} C_j^{+*}\beta_{1j}^{+}\frac{rz_j}{R_j^5}, \tag{5.125d}$$

$$\sigma_{zx}^{+} = 3\sum_{j=1}^{4} C_j^{+*}\beta_{1j}^{+}\frac{xz_j}{R_j^5}, \tag{5.125e}$$

$$\sigma_{yz}^{+} = 3\sum_{j=1}^{4} C_j^{+*}\beta_{1j}^{+}\frac{yz_j}{R_j^5}, \tag{5.125f}$$

$$q_z^{+} = 0, \tag{5.125g}$$

where coefficients C_j^{+*} can be obtained by solving the following equations:

$$\begin{cases} \sum_{j=1}^{5} C_j^{+*} = \sum_{j=1}^{5} C_j^{-*}, \\ -\sum_{j=1}^{5} \alpha_{1j}^{+} C_j^{+*} = \sum_{j=1}^{5} \alpha_{1j}^{-} C_j^{-*}, \\ -\sum_{j=1}^{5} \alpha_{2j}^{+} C_j^{+*} = \sum_{j=1}^{5} \alpha_{2j}^{-} C_j^{-*}, \\ -\sum_{j=1}^{5} \left[\alpha_{3j}^{+} C_j^{+*} + \alpha_{3j}^{-} C_j^{-*}\right] = \frac{1}{2\pi}, \\ \alpha_{45}^{+} C_5^{+*} = \alpha_{45}^{-} C_5^{-*}, \end{cases} \qquad \begin{cases} \sum_{j=1}^{5} \beta_{1j}^{+} C_j^{+*} = -\sum_{j=1}^{5} \beta_{1j}^{-} C_j^{-*}, \\ \sum_{j=1}^{5} \gamma_{1j}^{+} C_j^{+*} = \sum_{j=1}^{5} \gamma_{1j}^{-} C_j^{-*}, \\ \sum_{j=1}^{5} \gamma_{2j}^{+} C_j^{+*} = \sum_{j=1}^{5} \gamma_{2j}^{-} C_j^{-*}, \\ \sum_{j=1}^{5} \gamma_{3j}^{+} C_j^{+*} = \sum_{j=1}^{5} \gamma_{3j}^{-} C_j^{-*}, \\ K_{33}^{+} s_5^{+} C_5^{+*} \alpha_{45}^{+} = -K_{33}^{-} s_5^{-} C_5^{-*} \alpha_{45}^{-}. \end{cases} \tag{5.126}$$

For unit-point temperature discontinuity, the fundamental solutions are given by Zhao et al. [140]:

$$\sigma_z^{+} = \sum_{j=1}^{5} D_j^{+*}\gamma_{1j}^{+}\frac{z_j}{R_j^3}, \tag{5.127a}$$

$$H_{zz}^{+} = \sum_{j=1}^{5} D_j^{+*}\gamma_{2j}^{+}\frac{z_j}{R_j^3}, \tag{5.127b}$$

$$D_z^{+} = \sum_{j=1}^{5} D_j^{+*}\gamma_{3j}^{+}\frac{z_j}{R_j^3}, \tag{5.127c}$$

$$\sigma_{zr}^{+} = \sum_{j=1}^{5} D_j^{+*}\beta_{1j}^{+}\frac{r}{R_j^3}, \tag{5.127d}$$

$$\sigma_{zx}^{+} = \sum_{j=1}^{5} D_j^{+*}\beta_{1j}^{+}\frac{x}{R_j^3}, \tag{5.127e}$$

$$\sigma_{yz}^{+} = \sum_{j=1}^{5} D_j^{+*} \beta_{1j}^{+} \frac{y}{R_j^3}, \tag{5.127f}$$

$$q_z^{+} = K_{33}^{+} s_5^{+} \alpha_{45}^{+} D_5^{+*} \left(\frac{3z_5^2}{R_5^5} - \frac{1}{R_5^3} \right), \tag{5.127g}$$

where the coefficients D_j^{+*} can be obtained by solving the following equations:

$$\begin{cases} \sum_{j=1}^{5} D_j^{+*} = \sum_{j=1}^{5} D_j^{-*}, \\ -\sum_{j=1}^{5} \alpha_{1j}^{+} D_j^{+*} = \sum_{j=1}^{5} \alpha_{1j}^{-} D_j^{-*}, \\ -\sum_{j=1}^{5} \alpha_{2j}^{+} D_j^{+*} = \sum_{j=1}^{5} \alpha_{2j}^{-} D_j^{-*}, \\ -\sum_{j=1}^{5} \alpha_{3j}^{+} D_j^{+*} = \sum_{j=1}^{5} \alpha_{3j}^{-} D_j^{-*}, \\ \alpha_{45}^{+} D_5^{+*} - \alpha_{45}^{-} D_5^{-*} = \frac{1}{2\pi}, \end{cases} \quad \begin{cases} \sum_{j=1}^{5} \beta_{1j}^{+} D_j^{+*} = -\sum_{j=1}^{5} \beta_{1j}^{-} D_j^{-*}, \\ \sum_{j=1}^{5} \gamma_{1j}^{+} D_j^{+*} = \sum_{j=1}^{5} \gamma_{1j}^{-} D_j^{-*}, \\ \sum_{j=1}^{5} \gamma_{2j}^{+} D_j^{+*} = \sum_{j=1}^{5} \gamma_{2j}^{-} D_j^{-*}, \\ \sum_{j=1}^{5} \gamma_{3j}^{+} D_j^{+*} = \sum_{j=1}^{5} \gamma_{3j}^{-} D_j^{-*}, \\ K_{33}^{+} s_5^{+} D_5^{+*} \alpha_{45}^{+} = -K_{33}^{-} s_5^{-} D_5^{-*} \alpha_{45}^{-}. \end{cases} \tag{5.128}$$

Meanwhile, the extended stresses for unit-point, in-plane displacement discontinuities are given as [140]:

$$\sigma_z^{+} = \sin\phi \cdot \sum_{j=1}^{4} p_{1j}^{+*} \gamma_{1j}^{+} \frac{3rz_j}{R_j^5}, \tag{5.129a}$$

$$H_{zz}^{+} = \sin\phi \cdot \sum_{j=1}^{4} p_{1j}^{+*} \gamma_{2j}^{+} \frac{3rz_j}{R_j^5}, \tag{5.129b}$$

$$D_z^{+} = \sin\phi \cdot \sum_{j=1}^{4} p_{1j}^{+*} \gamma_{3j}^{+} \frac{3rz_j}{R_j^5}, \tag{5.129c}$$

$$\sigma_{zr}^{+} = \sin\phi \cdot \left(\sum_{j=1}^{4} \beta_{1j}^{+} p_{1j}^{+*} \frac{2r^2 - z_j^2}{R_j^5} + s_0^{+} \rho_1^{+} p_2^{+*} \frac{1}{R_0^3} \right), \tag{5.129d}$$

$$\sigma_{z\varphi}^{+} = -\cos\phi \cdot \left(s_0^{+} \rho_1^{+} p_2^{+*} \frac{2r^2 - z_0^2}{R_0^5} + \sum_{j=1}^{4} \beta_{1j}^{+} p_{1j}^{+*} \frac{1}{R_j^3} \right), \tag{5.129e}$$

$$\sigma_{zx}^{+} = 3\sin\phi\cos\phi\cdot\left[\sum_{j=1}^{4} p_{1j}^{+*}\beta_{1j}^{+}\left(\frac{1}{R_j^3}-\frac{z_j^2}{R_j^5}\right)+s_0^{+}\rho_1^{+}p_2^{+*}\left(\frac{1}{R_0^3}-\frac{z_0^2}{R_0^5}\right)\right], \tag{5.129f}$$

$$\begin{aligned}\sigma_{yz}^{+} = {} & \sum_{j=1}^{4} p_{1j}^{+*}\beta_{1j}^{+}\cdot\left[3\left(\frac{1}{R_j^3}-\frac{z_j^2}{R_j^5}\right)\sin^2\phi-\frac{1}{R_j^3}\right] \\ & - s_0^{+}\rho_1^{+}p_2^{+*}\left[3\left(\frac{1}{R_0^3}-\frac{z_0^2}{R_0^5}\right)\cos^2\phi-\frac{1}{R_0^3}\right],\end{aligned} \tag{5.129g}$$

$$q_z^{+} = 0, \tag{5.129h}$$

for the unit-point phonon displacement discontinuity in y-direction, and

$$\sigma_z^{+} = \cos\phi\cdot\sum_{j=1}^{4} p_{1j}^{+*}\gamma_{1j}^{+}\frac{3rz_j}{R_j^5}, \tag{5.130a}$$

$$H_{zz}^{+} = \cos\phi\cdot\sum_{j=1}^{4} p_{1j}^{+*}\gamma_{2j}^{+}\frac{3rz_j}{R_j^5}, \tag{5.130b}$$

$$D_z^{+} = \cos\phi\cdot\sum_{j=1}^{4} p_{1j}^{+*}\gamma_{3j}^{+}\frac{3rz_j}{R_j^5}, \tag{5.130c}$$

$$\sigma_{zr}^{+} = \cos\phi\cdot\left(\sum_{j=1}^{4}\beta_{1j}^{+}p_{1j}^{+*}\frac{2r^2-z_j^2}{R_j^5}+s_0^{+}\rho_1^{+}p_2^{+*}\frac{1}{R_0^3}\right), \tag{5.130d}$$

$$\sigma_{z\varphi}^{+} = \sin\phi\cdot\left(s_0^{+}\rho_1^{+}p_2^{+*}\frac{2r^2-z_0^2}{R_0^5}+\sum_{j=1}^{4}\beta_{1j}^{+}p_{1j}^{+*}\frac{1}{R_j^3}\right), \tag{5.130e}$$

$$\sigma_{yz}^{+} = 3\sin\phi\cos\phi\cdot\left[\sum_{j=1}^{4} p_{1j}^{+*}\beta_{1j}^{+}\left(\frac{1}{R_j^3}-\frac{z_j^2}{R_j^5}\right)+s_0^{+}\rho_1^{+}p_2^{+*}\left(\frac{1}{R_0^3}-\frac{z_0^2}{R_0^5}\right)\right], \tag{5.130f}$$

$$\begin{aligned}\sigma_{zx}^{+} = {} & \sum_{j=1}^{4} p_{1j}^{+*}\beta_{1j}^{+}\cdot\left[3\left(\frac{1}{R_j^3}-\frac{z_j^2}{R_j^5}\right)\cos^2\phi-\frac{1}{R_j^3}\right] \\ & - s_0^{+}\rho_1^{+}p_2^{+*}\left[3\left(\frac{1}{R_0^3}-\frac{z_0^2}{R_0^5}\right)\sin^2\phi-\frac{1}{R_0^3}\right]\end{aligned} \tag{5.130g}$$

$$q_z^{+} = 0 \tag{5.130h}$$

for the unit-point phonon displacement discontinuity in x-direction. The coefficients can be obtained by solving the following equations [140]:

$$\left\{\begin{array}{l} p_2^{-*} - p_2^{+*} + \sum_{j=1}^{5}\left[p_{1j}^{-*} - p_{1j}^{+*}\right] = \frac{1}{\pi}, \\ p_2^{-*} - p_2^{+*} - \sum_{j=1}^{5}\left[p_{1j}^{-*} - p_{1j}^{+*}\right] = 0, \\ -\sum_{j=1}^{5}\alpha_{1j}^{+} p_{1j}^{+*} = \sum_{j=1}^{5}\alpha_{1j}^{-} p_{1j}^{-*}, \\ -\sum_{j=1}^{5}\alpha_{2j}^{+} p_{1j}^{+*} = \sum_{j=1}^{5}\alpha_{2j}^{-} p_{1j}^{-*}, \\ -\sum_{j=1}^{5}\alpha_{3j}^{+} p_{1j}^{+*} = \sum_{j=1}^{5}\alpha_{3j}^{-} p_{1j}^{-*}, \\ \alpha_{45}^{+} p_{15}^{+*} = \alpha_{45}^{-} p_{15}^{-*}, \end{array}\right. \quad \left\{\begin{array}{l} s_0^{+}\rho_1^{+} p_2^{+*} + s_0^{-}\rho_1^{-} p_2^{-*} = 0, \\ \sum_{j=1}^{5}\left[\beta_{1j}^{+} p_{1j}^{+*} + \beta_{1j}^{-} p_{1j}^{-*}\right] = 0, \\ \sum_{j=1}^{5}\gamma_{1j}^{+} p_{1j}^{+*} = \sum_{j=1}^{5}\gamma_{1j}^{-} p_{1j}^{-*}, \\ \sum_{j=1}^{5}\gamma_{2j}^{+} p_{1j}^{+*} = \sum_{j=1}^{5}\gamma_{2j}^{-} p_{1j}^{-*}, \\ \sum_{j=1}^{5}\gamma_{3j}^{+} p_{1j}^{+*} = \sum_{j=1}^{5}\gamma_{3j}^{-} p_{1j}^{-*}, \\ -K_{33}^{+} s_5^{+}\alpha_{45}^{+} p_{15}^{+*} = K_{33}^{-} s_5^{-}\alpha_{45}^{-} p_{15}^{-*}. \end{array}\right. \tag{5.131}$$

5.6 Application of General Solution in the Problem of an Interface Crack of Arbitrary Shape

The fundamental, general solutions obtained in the previous sections can be readily used to obtain the analytical solutions for a regularly-shaped, interface crack in a smart medium under general, multiphysical loading, as long as the integral of the point force solution over the crack face has a closed form result. In other words, the obtained, point loading results works as the Green functions, and the exact solution can be obtained directly through integration over the crack area Σ provided. For irregularly shaped interface crack, numerical method can be developed based on boundary integral equations of a general triangular element in the crack plane [141]. Detailed derivations, discussions and related results for irregularly shaped, interface crack problems in thermoelastic bimaterials can be found in [141].

Similar approaches can be applied to interface crack problems in piezothermoelastic materials and magnetoelectrothermoelastic materials, as illustrated in [142, 143]. Figures 5.6 and 5.7 show some of the results from the numerical calculations for an elliptical, interface crack under fixed multiphysical loading in piezoelectric materials and magnetoelectrothermoelastic materials. The results were obtained using the triangular elements and the numerical algorithms developed in [142, 143]. Clearly, the energy release rate at the minor axis tip of the elliptical crack shows a much higher value than that at the major axis tip, indicating that the crack will prefer to grow in the direction of the minor axis and tends to become a penny-shaped one in the end. This finding coincides with what repeatedly seen in classical texts of linear elastic fracture mechanics [144]. It is worth noting that when

Fig. 5.6 Energy release rate at the tips of major and minor axes under fixed loadings versus the ellipticity ratio *a*/*b* under electric and thermal semi-permeable boundary conditions for an elliptical interface crack in a thermopiezoelectric bi-material [142]

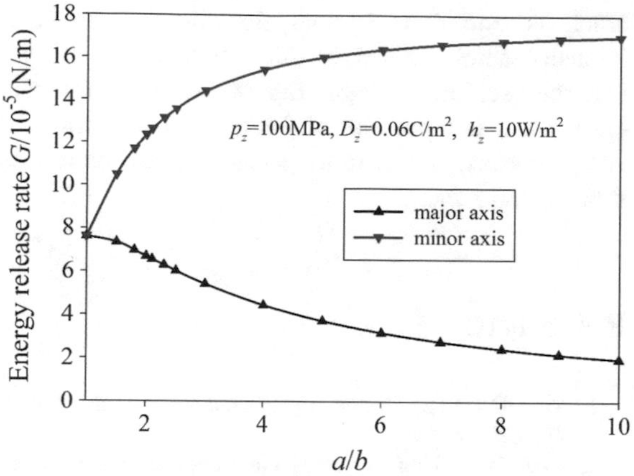

Fig. 5.7 Energy release rate at the tips of major and minor axes of an interface, elliptical crack under combined loadings versus the ellipticity ratio *a*/*b* for magnetoelectrothermoelastic materials. When a/b = 1, the results represent the solution for a penny-shaped crack [143]

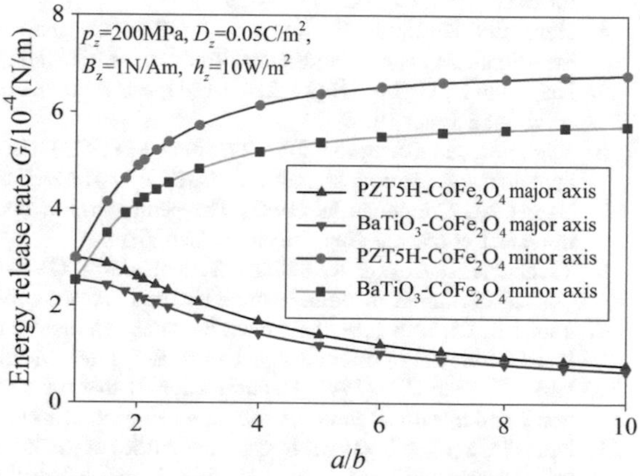

the ratio of major to minor axis, *a*/*b*, equals 1, the results will become the solution for penny-shaped crack, which can be directly obtained from the analytical, fundamental solutions through direct integration [141, 142]. Similar results for quasi-crystals can be found in [145].

5.7 Summary

The displacement and temperature discontinuity boundary integral-differential equation method is developed to analyze an interface crack in an isotropic thermoelastic bi-material, and the fundamental solutions for a unit point displacement and temperature discontinuities on the interface are proposed and the corresponding hyper-singular integral-differential equations for an arbitrarily shaped interfacial crack are obtained. By analyzing the singular behavior, the stress and heat flux intensity factors as well as the energy release rate are obtained. It can be observed that the oscillatory singularity of the stress intensity factors are the same as the elastic one, independent of the thermal properties. It is also worth noting that the energy release rate is in the same form as the elastic one, independent of the heat flux intensity factor.

References

1. Sih GC (1962) On the singular character of thermal stress near a crack tip. J Appl Mech 29:587–589
2. Chen WH, Ting K (1985) Finite element analysis of mixed-mode thermoelastic fracture problems. Nucl Eng Des 90:55–65
3. Brown EJ, Erdogan F (1968) Thermal stresses in bonded materials containing cuts on the interface. Int J Eng Sci 6:517–529
4. Herrmann K, Braun H, Kemeny P (1979) Comparison of experimental and numerical investigations concerning thermal cracking of dissimilar materials. Int J Fract 15:187–190
5. Herrmann K, Grebner H (1982) Curved thermal crack growth in a bounded brittle two-phase solid. Int J Fract 19:69–74
6. Martin-Moran CJ, Barber JR, Comninou M (1983) The penny-shaped interface crack with heat flow Part 1: perfect contact. ASME J Appl Mech 50:29–36
7. Barber JR, Comninou M (1983) The penny-shaped interface crack with heat flow part 2: imperfect contact. J Appl Mech 50:770–776
8. Takakuda K, Aoyagi K, Shibuya T, Koizumi T (1984) Steady state thermal stresses in a bonded dissimilar medium with an external crack on the interface. JSME 27:2643–2650
9. Yuuki R, Cho SB (1989) Efficient boundary element analysis of stress intensity factors for interface cracks in dissimilar materials. Eng Fract Mech 34:179–188
10. Lee KY, Shul CW (1991) Determination of thermal stress intensity factors for an interface crack under vertical uniform heat flow. Eng Fract Mech 40(6):1067–1074
11. Munz D, Yang YY (1992) Stress singularities at the interface in bonded dissimilar materials under mechanical and thermal loading. J Appl Mech 59:857–861
12. Chao CK, Chang RC (1994) Analytical solutions and numerical examples for thermoelastic interface crack problems in dissimilar anisotropic media. J Thermal Stresses 17:285–299
13. Lee KY, Park SJ (1995) Thermal-stress intensity factors for partially insulated interface crack under uniform heat-flow. Eng Fract Mech 50:475–482
14. Wilson RI, Meguid SA (1995) On the determination of mixed mode stress intensity factors of an angled cracks in a disc using FEM. Finite Ele Anal Des 18:433–438
15. Martynyak RM (1999) Thermal opening of an initially closed interface crack under conditions of imperfect thermal contact between its lips. Mater Sci 35:612–622
16. Zhang SC (2000) Thermal stress intensities at an interface crack between two elastic layers. Int J Fract 106:277–290
17. Herrmann KP, Loboda VV (2001) Contact zone models for an interface crack in a thermomechanically loaded anisotropic bimaterial. J Thermal Stresses 24:479–506
18. Kharun IV, Loboda VV (2004) A thermoelastic problem for interface cracks with contact zones. Int J Solids Struct 41:159–175
19. Kharun IV, Loboda VV (2004) A problem of thermoelasticity for a set of interface cracks with contact zones between dissimilar anisotropic materials. Mech Mater 36:585–600
20. Herrmann KP, Loboda VV, Kharun IV (2004) Interface crack with a contact zone in an isotropic bimaterial under thermomechanical loading. Theor Appl Fract Mech 42:335–348

21. Ratnesh K, Chandra JM (2008) Thermal weight functions for bi-material interface crack system using energy principles. Int J Solids Struct 45:6157–6176
22. Pant M, Singh IV, Mishra BK (2010) Numerical simulation of thermo-elastic fracture problems using element free Galerkin method. Int J Mech Sci 52:1745–1755
23. Khandelwal R, Chandra JM (2011) Thermal weight functions and stress intensity factors for bonded dissimilar media using body analogy. J Appl Mech 78:1–9
24. Ma LF, He R, Zhang JS, Shaw B (2013) A simple model for the study of the tolerance of interfacial crack under thermal load. Acta Mech 224:1571–1577
25. Bregman AM, Kassir MK (1974) Thermal fracture of bonded dissimilar media containing a penny-shaped crack. Int J Fract 10:87–98
26. Muskhelishvili NI (1958) Singular integral equations. Springer, Dordrecht
27. Andrzej K, Stanislaw JM (2003) On the three-dimensional problem of an interface crack under uniform heat flow in a bimaterial periodically-layered space. Int J Fract 123:127–138
28. Johnson J, Qu J (2007) An interaction integral method for computing mixed mode stress intensity factors for curved bimaterial interface cracks in non-uniform temperature fields. Eng Fract Mech 74:2282–2291
29. Nomura Y, Ikeda T, Miyazaki N (2010) Stress intensity factor analysis of a three-dimensional interfacial corner between anisotropic bimaterials under thermal stress. Int J Solids Struc 47:1775–1784
30. Guo JH, Yu J, Xing YM (2013) Anti-plane analysis on a finite crack in a one-dimensional hexagonal quasicrystal strip. Mech Res Commun 52:40–45
31. Li L, Wu HP, Bao YM, Lu YL, Chai GZ (2013) Analysis of weight function method for three dimensional interface crack under mechanical and thermal loading. Acta Mech Sinica 34(4):401–409 (In Chinese)
32. Crouch SL (1976) Solution of plane elasticity problems by the displacement discontinuity method. I. Infinite body solution. Int J Numer Methods Eng 10:301–343
33. Tang RJ, Chen MC, Yue JC (1998) Theoretical analysis of three-dimensional interface crack. Sci Chin 41(4):443–448
34. Chen MC, Noda NA, Tang RJ (1999) Application of finite-part integrals to planar interfacial fracture problems in three-dimensional bimaterials. J Appl Mech 66:885–890
35. Chen MC, Gao C, Tang RJ (1999) A study on stress intensity factors and singular stress fields of 3D interface crack. Acta Mech Sinica 20(1):8–15 (In Chinese)
36. Zhao MH, Fang PZ, Shen YP (2004) Boundary integral-differential equations and boundary element method for interfacial cracks in three-dimensional piezoelectric media. Eng Anal Bound Elem 28:753–762
37. Zhao MH, Li N, Fan CY, Xu GT (2008) Analysis method of planar interface cracks of arbitrary shape in three-dimensional transversely isotropic magnetoelectroelastic bimaterials. Int J Solids Struct 45:1804–1824
38. Zhao YF, Zhao MH, Pan EN, Fan CY (2015) Green's functions and extended displacement discontinuity method for interfacial cracks in three-dimensional transversely isotropic magneto-electro-elastic bi-materials. Int J Solids Struct 52:56–71
39. Zhao M, Dang H, Fan C, Chen Z (2016) Analysis of an arbitrarily shaped interface cracks in a three-dimensional isotropic thermoelastic bi-material. Part 1: theoretical solution. Int J Solids Struct 97–98:168–181
40. Zhao MH, Shen YP, Liu YJ, Liu GN (1988) The method of analysis of cracks in three-dimensional transversely isotropic media: boundary integral equation approach. Eng Anal Bound Elem 21:169–178
41. Zhao MH, Liu YJ (1994) Boundary integral equations and the boundary element method for three-dimensional fracture mechanics. Eng Anal Bound Elem 13:333–338
42. Hou PF, Jiang HY, Li QH (2013) Three-dimensional steady-state general solution for isotropic thermoelastic materials with applications 1: general solutions. J Therm Stresses 36:727–747
43. Hutchinson JW, Mear ME, Rice JR (1987) Crack paralleling an interface between dissimilar materials. ASME J Appl Mech 54:828–832

44. Rao SS, Sunar M (1993) Analysis of distributed thermopiezoelectric sensors and actuators inadvanced intelligent structures. AIAA J 31:1280–1086
45. Mindlin RD (1974) Equations of high frequency vibrations of thermopiezoelectric crystal plates. Int J Solids Struct 10:625–637
46. Nowacki W (1978) Some general theorems of thermopiezoelectricity. J Therm Stresses 1:171–182
47. Chandrasekharaiah DS (1988) A generalized linear thermoelasticity theroy for piezoelectric media. Acta Mech 71:39–49
48. Altay GA, Dokmeci MC (2003) Some comments on the higher order theories of piezoelectric, piezothermoelastic and thermopiezoelectric rods and shells. Int J Solids Struct 40:4699–4706
49. Ashida F, Tauchert TR, Noda N (1994) A general solution technique for piezothermoelasticity of hexagonal solids of class 6 mm in cartesian coordinates. Math Mech 74:87–95
50. Ding HJ, Guo FL, Hou PF (2000) A general solution for piezothermoelasticity of transversely isotropic piezoelectric materials and its applications. Int J Eng Sci 38:1415–1440
51. Tarn JQ (2002) A state space formalism for piezothermoelasticity. Int J Solids Struct 39:5173–5184
52. Hou PF, Leung AYT, Chen CP (2009) Three-dimensional fundamental solution for transversely isotropic piezothermoelastic material. Int J Num Methods Eng 78:84–100
53. Yu SW, Qin QH (1996) Damage analysis of thermopiezoelectric properties: part I-crack tip singularities. Theor Appl Fract Mech 25:263–277
54. Chen WQ (2000) On the general solution for piezothermoelasticity for transverse isotropy with application. ASME J Appl Mech 67:705–711
55. Gao CF, Wang MZ (2001) Collinear permeable cracks in thermopiezoelectric materials. Mech Mater 33:1–9
56. Chen WQ, Lim CW, Ding HJ (2005) Point temperature solution for a penny-shaped crack in an infinite transversely isotropic thermo-piezo-elastic medium. Eng Anal Bound Elem 29:524–532
57. Qin QH (1998) Thermoelectroelastic Green’s function for a piezoelectric plate containing an elliptic hole. Mech Mater 30:21–29
58. Qin QH (1999) Thermoelectroelastic analysis of cracks in piezoelectric half-plane by BEM. Com Mech 23:353–360
59. Qin QH, Lu M (2000) BEM for crack-inclusion problems in thermopiezoelectric materials. Eng Fract Mech 69:577–588
60. Qin QH, Mai YW (2002) BEM for crack-hole problems in thermopiezoelectric materials. Eng Fract Mech 69:577–588
61. Shang FL, Wang ZK, Li ZH (1996) Thermal stresses anaysis of a three-dimensional crack in a thermopiezoelectric solid. Eng Fract Mech 55:737–750
62. Shang FL, Kuna M, Scherzer M (2002) Analytical solution for two penny-shaped crack problems in thermo-piezoelectric materials and their finite element comparisons. Int J Fract 117:113–128
63. Shang FL, Kuna M, Scherzer M (2003) Development of finite element techniques for three-dimensional analyses of thermo-piezoelectric materials. ASME J Eng Mater Tech 125:18–21
64. Shang FL, Kuna M (2003) Thermal stress around a penny-shaped crack in a thermopiezoelectric solid. Com Mater Sci 26:197–201
65. Niraula OP, Noda N (2002) Thermal stress analysis in thermopiezoelectric strip with an edge crack. J Therm Stresses 25:389–405
66. Gao CF, Zhao YT, Wang MZ (2002) An exact and explicit treatment of an elliptic hole problem in thermopiezoelectric media. Int J Solids Struct 39:2665–2685
67. Tsamasphyros G, Song ZF (2005) Analysis of a crack in a finite thermopiezoelectric plate under heat flux. Int J Solids Struct 136:143–166

68. Ueda S (2006) Thermal stress intensity factors for a normal crack in a piezoeelctric strip. J Therm Stresses 29:1107–1125
69. Ueda S, Hatano H (2012) T-shaped crack in a piezoelectric material under thermo-electro-mechanical loadings. J Therm Stresses 35:12–29
70. Ueda S, Tani Y (2008) Thermal stress intensity factors for two coplanar cracks in a piezoelectric strip. J Therm Stresses 31:403–415
71. Ueda S, Ikeda Y, Ishii A (2012) Transient thermoelectromecahnical response of a piezoelectric strip with two parallel cracks of different lengths. J Therm Stresses 35:534–549
72. Wang BL, Noda N (2010) Exact thermoelectroelasticity solution for a penny-shaped crack in piezoelectric materials. J Therm Stresses 27:241–251
73. Wang BL, Sun YG, Zhu Y (2011) Fracture of a finite piezoelectric layer with a penny-shaped crack. Int J Fract 172:19–39
74. Zhong XC, Zhang KS (2013) An opening crack model for thermopiezoelectric solids. Eur J Mech A-Solids 41:101–110
75. Shen S, Wang X, Kuang ZB (1995) Fracture mechanics for piezothermoelastic materials. Acta Mech Sinica 16(4):283–293 (In Chinese)
76. Shen S, Kuang ZB (1998) Interface crack in bi-piezothermoelastic media and the interaction with a point heat source. Int J Solids Struct 35:3899–3915
77. Qin QH, Mai YW (1998) Thermoelectroelastic Green's function and its application for bimaterial of piezoelectric materials. Arch Appl Mech 68:433–444
78. Qin QH, Mai YW (1999) A closed crack tip model for interface cracks in thermopiezoelectric materials. Int J Solids Struct 36:2463–2479
79. Gao CF, Wang MZ (2001) A permeable interface crack between dissimilar thermopiezoelectric media. Acta Mech 149:85–95
80. Williams ML (1959) The stresses around a fault or cracks in dissimilar media. Bull Seismol Soc Am 49:199–204
81. Herrmann KP, Loboda VV (2003) Fracture mechanical assessment of interface cracks with contact zones in piezoelectric bimaterials under thermoelectromechanical loadings I. Electrically permeable interface cracks. Int J Solids Struct 40:4191–4217
82. Herrmann KP, Loboda VV (2003) Fracture mechanical assessment of interface cracks with contact zones in piezoelectric bimaterials under thermoelectromechanical loadings II. Electrically impermeable interface cracks. Int J Solids Struct 40:4291–4237
83. Ueda S (2003) Thermally induced fracture of a piezoelectric laminate with a crack normal to interfaces. J Therm Stresses 26:311–331
84. Herrmann KP, Loboda VV (2006) Contact zone approach for a moving interface crack in a piezoelectric bimaterial under thermoelectromechanical loading. Arch Appl Mech 75:665–677
85. Hou PF, Leung AYT (2009) Three-dimensional Green's functions for two-phase transversely isotropic piezothermoelastic media. J Intell Mat Syst Struct 20:11–21
86. Zhao MH, Dang HY, Fan CY, Chen ZT (2017) Extended displacement discontinuity method for an interface crack in a three-dimensional transversely isotropic piezothermoelastic bi-material. Part 1: Theoretical solution. Int J Solids Struct 117:14–25
87. Fan CY, Dang HY, Zhao MH (2014) Nonlinear solution of the PS model for a semi-permeable crack in a 3D piezoelectric medium. Eng Anal Bound Elements 46:23–29
88. Li XF, Lee KY (2015) Effect of heat conduction of penny-shaped crack interior on thermal stress intensity factors. Int J Heat Mass Transf 91:127–134
89. Wang BL, Niraula OP (2007) Transient thermal fracture analysis of transversely isotropic magneto-electro-elastic materials. J Therm Stresses 30:297–317
90. Feng WJ, Pan E, Wang X (2008) Stress analysis of a penny-shaped crack in a magneto-electro-thermo-elastic layer under uniform heat flow and shear loads. J Therm Stresses 31:497–514
91. Aboudi J (2001) Micromechanical analysis of fully coupled electro-magneto-thermo-elastic multiphase composites. Smart Mater Struct 10:867–877

92. Gao CF, Kessler H, Balke H (2003) Fracture analysis of electromagnetic thermoelastic solids. Eur J Mech A-Solid 22:433–442
93. Niraula OP, Wang BL (2006) Thermal stress analysis in magneto-electro-thermo-elasticity with a penny-shaped crack under uniform heat flow. J Therm Stresses 29:423–437
94. Zhong XC, Huang QA (2014) Thermal stress intensity factor for an opening crack in thermomagnetoelectroelastic solids. J Therm Stresses 37:928–946
95. Chen WQ, Lee KY, Ding HJ (2004) General solution for transversely isotropic magneto-electro-thermo-elasticity and the potential theory method. Int J Eng Sci 42:1361–1379
96. Hou PF, Yi T, Wang L (2008) 2D general solution and fundamental solution for orthotropic electro-magneto-thermo-elastic materials. J Therm Stresses 31:807–822
97. Hou PF, Chen HR, He S (2009) Three-dimensional fundamental solution for transversely isotropic electro-magneto-thermo-elastic materials. J Therm Stresses 32:887–904
98. Li Y, Dang HY, Xu GT, Fan CY, Zhao MH (2016) Extended displacement discontinuity boundary integral equation and boundary element method for cracks in thermo-magneto-electro-elastic media. Smart Mater Struct 25:085048
99. Gao CF, Tong P, Zhang TY (2003) Interfacial crack problems in magneto-electric solids. Int J Eng Sci 41:2105–2121
100. Chue CH, Liu TJ (2005) Magneto-electro-elastic antiplane analysis of a bimaterial $BaTiO_3$-$CoFe_2O_4$ composite wedge with an interface crack. Theor Appl Fract Mech 44:275–296
101. Li R, Kardomateas GA (2006) The mode III interface crack in piezo-electro-magneto-elastic dissimilar bimaterials. J Appl Mech Trans ASME 73:220–227
102. Zhong XC, Li XF (2006) A finite length crack propagating along the interface of two dissimilar magnetoelectroelastic materials. Int J Eng Sci 44:1394–1407
103. Ma P, Feng WJ, Su RKL (2011) Fracture assessment of an interface crack between two dissimilar magnetoelectroelastic materials under heat flow and magnetoelectromechanical loadings. Acta Mechanica Solids Sinica 24:0894–9166
104. Herrmann KP, Loboda VV, Khodanen TV (2010) An interface crack with contact zones in a piezoelectric/piezomagnetic bimaterial. Arhive Appl Mech 80:651–670
105. Feng WJ, Ma P, Su RKL (2012) An electrically impermeable and magnetically permeable interface crack with a contact zone in magneto-electroelastic bimaterials under a thermal flux and magnetoelectromechanical loads. Int J Solids Struct 49:3472–3483
106. Ma P, Jing J (2013) Further analysis for fracture behaviors of an interfacial crack between piezoelectric and piezomagnetic layers. Engs Mech 30:327–333
107. Li JY, Dunn ML (1998) Anisotropic coupled-field inclusion and inhomogeneity problems. J Intell Mater Syst Struct 7:404
108. Huang JH, Kuo WS (1997) The analysis of piezoelectric/piezomagnetic composite materials containing ellipsoidal inclusions. J Appl Phys 81:1378
109. Zhao MH, Dang HY, Fan CY, Chen ZT (2017) Analysis of an interface crack of arbitrary shape in a three-dimensional transversely isotropic magnetoelectrothermoelastic bi-material. Part 1: theoretical solution. J Therm Stresses 40(8):929–952
110. Shechtman D, Blech I, Gratias D, Cahn JW (1984) Metallic phase with long-range orientational order and no translational symmetry. Phys Rev Lett 53:1951–1953
111. Dubois JM, Kang SS, Stebut JV (1991) Quasicrystalline low-friction coatings. J Mater Sci Lett 10:537–541
112. Wu JS, Brien V, Brunet P, Dong C, Dubois JM (2000) Scratch-induced surface microstructures on the deformed surface of Al-Cu-Fe icosahedral quasicrystals. Mater Sci Eng 4:294–296
113. Fan TY, Tang ZY, Chen WQ (2012) Theory of linear, nonlinear and dynamic fracture of quasicrystals. Eng Fract Mech 82:185–194
114. Li PD, Li XY, Kang GZ (2015) Crack tip plasticity of half-infinite Dugdale crack embedded in an infinite space of one-dimensional hexagonal quasicrystal. Mech Res Commun 70:72–78

115. Wang YW, Wu TH, Li XY, Kang GZ (2015) Fundamental elastic field in an infinite medium of two-dimensional hexagonal quasicrystal with a planar crack: 3D exact analysis. Int J Solids Struct 66:171–183
116. Tupholme GE (2015) Row of shear cracks moving in one-dimensional hexagonal quasicrystalline materials. Eng Fract Mech 134:451–458
117. Sladek J, Sladek V, Atluri SN (2015) Path-independent integral in fracture mechanics of quasicrystals. Eng Fract Mech 140:61–71
118. Fan CY, Li Y, Xu GT, Zhao MH (2016) Fundamental solutions and analysis of three-dimensional cracks in one-dimensional hexagonal piezoelectric quasicrystals. Mech Res Commun 74:39–44
119. Wang X (2004) Eshelby's problem of an inclusion of arbitrary shape in a decagonal quasicrystalline plane or half-plane. Int J Eng Sci 42:1911–1930
120. Gao Y, Ricoeur A (2012) Three-dimensional analysis of a spheroidal inclusion in a two-dimensional quasicrystal body. Phil Mag 92:4334–4353
121. Guo JH, Zhang ZY, Xing YM (2016) Antiplane analysis for an elliptical inclusion in 1D hexagonal piezoelectric quasicrystal composites. Phil Mag 96:349–369
122. Guo JH, Pan E (2016) Three-phase cylinder model of one-dimensional hexagonal piezoelectric quasicrystal composites. J Appl Mech 83:0810071
123. Fan TY, Trebin HR, Messerschmidt U, Mai YW (2004) Plastic flow coupled with a crack in some one- and two-dimensional quasicrystals. J Phys: Condens Matter 16:5229–5240
124. Li HL, Liu GT (2012) Stroh formulism for icosahdral quasicrystal and its applications. Phys Lett A 376:987–990
125. Lazar M, Agiasofitou E (2014) Fundamentals in generalized elasticity and dislocation theory of quasicrystals: green tensor, dislocation key-formulas and dislocation loops. Phil Mag 94:4080–4101
126. Chen WQ, Ma YL, Ding HJ (2004) On three-dimensional elastic problems of one-dimensional hexagonal quasicrystals bodies. Mech Res Commun 31:633–641
127. Wang X (2006) The general solution of one-dimensional hexagonal quasicrystal. Mech Res Comm 33:576–580
128. Li XY, Fan TY (2008) Exact solutions of two semi-infinite collinear cracks in a strip of one dimensional hexagonal quasicrystal. Appl Math Comput 196:1–5
129. Guo JH, Lu ZX (2011) Exact solution of four cracks originating from an elliptical hole in one-dimensional hexagonal quasicrystals. Appl Math Comput 217:9397–9403
130. Li XY, Li PD (2012) Three-dimensional thermo-elastic general solutions of one-dimensional hexagonal quasi-crystal and fundamental solutions. Phys Lett A 376:2004–2009
131. Li XY, Deng H (2013) On 2D Green's functions for 1D hexagonal quasi-crystals. Phys B 430:45–51
132. Li XY, Wang T, Zheng RF, Kang GZ (2015) Fundamental thermo-electro-elastic solutions for 1D hexagonal QC. Z Angew Math Mech 95:457–468
133. Yang J, Li X (2014) The anti-plane shear problem of two symmetric cracks originating from an elliptical hole in 1D hexagonal piezoelectric QCs. Adv Mater Res 936:127–135
134. Yang J, Li X (2015) Analytic solutions of problem about a circular hole with a straight crack in one-dimensional hexagonal quascrystals with piezoelectric effects. Theor Appl Fract Mech 82:17–24
135. Yu J, Guo JH, Pan E, Xing YM (2015) General solutions of plane problem in one-dimensional quasicrystal piezoelectric materials and its application on fracture mechanics. Appl Math Mech 36:793–814
136. Li PD, Li XY, Zheng RF (2013) Thermo-elastic Green's functions for an infinite bi-material of one-dimensional hexagonal quasi-crystals. Phys Lett A 377:637–642
137. Zhang LL, Wu D, Xu WS, Yang LZ, Ricoeur A, Wang ZB, Gao Y (2016) Green's functions of one-dimensional quasicrystal bi-material with piezoelectric effect. Phys Lett A 380:3222–3228
138. Ding HJ, Yang WG, Hu CZ, Wang RH (1993) Generalized elasticity theory of quasicrystals. Phys Rev B 48:7003–7010

139. Wang X, Pan E (2008) Analytical solutions for some defect problems in 1D hexagonal and 2D octagonal quasicrystals. Pramana J Phys 70:911–933
140. Zhao MH, Dang HY, Fan CY, Chen ZT (2017) Analysis of a three-dimensional arbitrarily shaped interface crack in a one-dimensional hexagonal thermo-electro-elastic quasicrystal bi-material. Part 1: theoretical solution. Eng Fract Mech 179:59–78
141. Dang HY, Zhao MH, Fan CY, Chen ZT (2016) Analysis of an arbitrarily shaped interface cracks in a three-dimensional isotropic thermoelastic bi-material. Part 2: numerical method. Int J Solids Struct 99:48–56
142. Dang HY, Zhao MH, Fan CY, Chen ZT (2017) Extended displacement discontinuity method for an interface crack in a three-dimensional transversely isotropic piezothermoelastic bi-material. Part 2: numerical method. Int J Solids Struct 109:199–209
143. Dang HY, Zhao MH, Fan CY, Chen ZT (2017) Analysis of an interface crack of arbitrary shape in a three-dimensional transversely isotropic magnetoelectrothermoelastic bi-material. Part 2: numerical method. J Therm Stresses 40(8):953–972
144. Anderson TL (1991) Fracture mechanics, fundamentals and applications. CRC Press
145. Dang HY, Zhao MH, Fan CY, Chen ZT (2017) Analysis of a three-dimensional arbitrarily shaped interface crack in a one-dimensional hexagonal thermo-electro-elastic quasicrystal bi-material. Part 2: numerical method. Eng Fract Mech 180:268–281

Chapter 6
Advanced Thermal Fracture Analysis Based on Non-Fourier Heat Conduction Models

6.1 Introduction

In this chapter, the non-Fourier heat conduction models such as the hyperbolic heat conduction, dual phase lag heat conduction, and the memory-dependent fractional heat conduction models are used to deal with crack problems in advanced composite materials. A few typical examples, such as cracks in a half-plane with a thin film coating, partially-insulated crack with thermal insulation interior, circumferential crack in a hollow cylinder and viscoelastic materials are be presented to illustrate the use of the models and the unique features of the heat conduction models revealed in these prolems.

6.2 Hyperbolic Heat Conduction in a Cracked Half-Plane with a Coating

High-rate heat transfer has become a major concern in modern industries especially in material processing, such as the application of pulsed laser heating in additive manufacturing. Recently, very strong substrate/coating interfaces have been obtained via pulsed laser coating of bioceramic/metal nanomaterials on metal substrates [1]. Investigation of the temperature field is essential to calculating thermal stresses within materials fabricated by advanced manufacturing, which is necessary to understand the problem of thermal damage, and accurate heat conduction analysis is of great importance for the structural integrity.

As discussed intensively in the literature, the Fourier, parabolic heat conduction model, although provides sufficient accuracy for many engineering applications, implies infinite thermal wave propagation speed and is ineffective at the very small length and time scales associated with small-scale systems [2, 3]. For many technological applications that involving high thermal energy with extremely short

Z. T. Chen and A. H. Akbarzadeh, *Advanced Thermal Stress Analysis of Smart Materials and Structures*, Structural Integrity 10,
https://doi.org/10.1007/978-3-030-25201-4_6

time, the results predicted using the parabolic heat conduction model differ significantly from the experimental results [4, 5]. Examples include the transient temperature field caused by pulsed laser heating of thin structures, and the measured surface temperature of a slab immediately after an intense thermal shock, which was 300 °C higher than that obtained from the parabolic heat conduction model [6, 7].

As discussed before, when relaxation, or the time lag of heat flux is considered in a thermal process, one has a hyperbolic heat conduction equation, which implies a finite speed for heat transport [8]. Consideration of the hyperbolic heat conduction model becomes important if irreversible physical processes, such as crack or void initiation in a solid, are involved in the process of heat transport. In applications involving high rate heating where extremely small time scales are concerned, it is appropriate to use the hyperbolic heat conduction model [9, 10].

Inherent defects in materials such as dislocations and cracks may disturb the temperature distribution when thermal loading is applied to the material, and singular stress and thermal fields may be developed in the neighborhood of discontinuities. Some studies have been devoted to studying the singular behavior of temperature gradient around crack tip based on the classical Fourier heat conduction model [11–13]. Many researchers have paid attention to the effect of cracks, holes and other defects under thermal loading in advanced materials using the Fourier heat conduction model [14–23].

A few investigations on crack problems in thermo-elastic materials have been made using the hyperbolic heat conduction model. Among them are Manson and Rosakis [24], who derived a solution of the hyperbolic heat conduction equation for a travelling point heat source around a propagating crack tip, and measured the temperature distribution at the tip of a dynamically propagating crack. Tzou [25] investigated the near-tip, thermal field around a moving crack, and evaluated the effect of crack velocity on the thermal shock waves. The transient thermal stresses around a crack in a strip and a half-plane were recently investigated in [26, 27] under thermal impact loading.

During laser manufacturing processes, high-energy pulsed laser beams are rapidly moving along the workpiece to generate various surface topologies as designed, leading to extremely high, local temperature gradients. Occasionally, structures are suffered from damage due to the applied thermal loading. The local thermal stresses near the crack can be elevated by the intensified temperature gradient, which may initiate crack propagation or breakdown of the structure even under normal thermal conditions. An accurate analysis of the intensification of temperature gradient near the crack is essential to predicting the failure behavior under thermal loading [28].

In this section, we present the theoretical framework to investigate the transient temperature field around a crack in a substrate bonded by a coating under a thermal impact using the hyperbolic heat conduction model. Considering that the substrate is much thicker than coating, the crack problem of a half-plane bonded to a coating strip is investigated, approximately equivalent to the actual situation. Fourier and Laplace transforms are employed to establish the singular integral equation about the temperature field. The singular integral equation is solved numerically and the asymptotic fields around the crack tips are obtained. Laplace inversion is then applied to get the temperature field in the time domain. The effect of the parameters of the hyperbolic heat conduction model and the geometric size of the composite on the temperature disturbance is demonstrated. The results of the current problem lead to the existing solution based on parabolic heat conduction model when the relaxation time vanishes. The results based on hyperbolic heat conduction model show much higher dynamic temperature disturbance in the very early stages of impact comparing to the parabolic model. The theoretical framework can be easily extended to investigate the thermomechanical behavior of cracked structures of different geometries or thermal loading conditions.

6.2.1 Basic Equations

Consider a thermoelastic half-plane containing a crack of length $2c$ parallel to the interface between the half-plane and the coating, as shown in Fig. 6.1. The half-plane is initially at the uniform temperature zero, and the free surface of the coating at $y = -(a + b)$ is suddenly heated to a temperature T_0. The crack surfaces are assumed to be thermally insulated, which indicates no heat flux can go through the crack surfaces. This kind of thermal conditions can be observed in saturated, porous materials where the interior of cracks is filled by thermally insulated fluids. In this study, the effects of inertia and thermal-elastic coupling are neglected which leads to an uncoupled, quasi-static problem.

In the heat-transfer process involving high temperature gradients, large heat fluxes or short, transient durations, the heat propagation speed is finite. Fourier's law can be modified with a time lag of heat flux in response to a temperature disturbance in the following form [29]

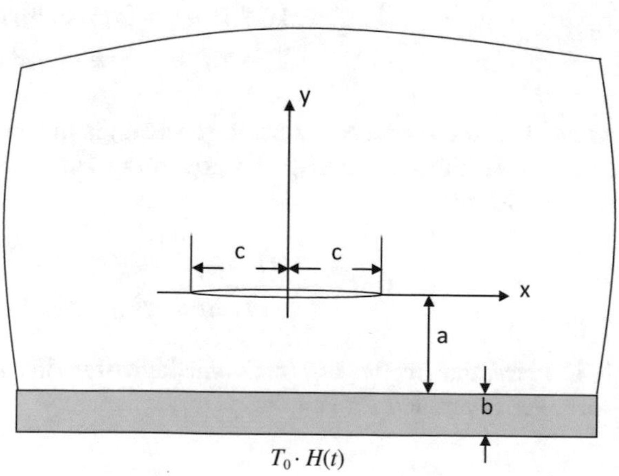

Fig. 6.1 A thin layer (coating) on top of a cracked half-plane substrate under a sudden thermal shock [34]

$$q + \tau \frac{\partial q}{\partial t} = -k \cdot \nabla T \tag{6.1}$$

where q is the heat flux (thermal heat current density), T is the temperature, k is the thermal conductivity of the material, ∇ is the spatial gradient operator, and τ is the so-called relaxation time (a non-negative constant), or build-up period for the commencement of heat flow after a temperature gradient has been imposed on the medium.

The local energy balance equation with vanishing heat source can be expressed as [30]

$$-\nabla q = \rho C \cdot \frac{\partial T}{\partial t} \tag{6.2}$$

where ρ and C are the mass density and the specific heat capacity, respectively. Incorporating Eq. (6.1) with Eq. (6.2) leads to the hyperbolic heat conduction equation for the substrate and the coating,

$$d_i \cdot \Delta T^{(i)} = \frac{\partial T^{(i)}}{\partial t} + \tau_i \frac{\partial^2 T^{(i)}}{\partial t^2}, \quad (i = 1, 2) \tag{6.3}$$

where Δ is Laplace's differential operator, τ_i are the relaxation times for the substrate and the coating, respectively; k_i are the thermal conductivity of the materials $(i = 1, 2)$, and $d_i = \frac{k_i}{\rho_i C_i}$ $(i = 1, 2)$ are the thermal diffusivities for the substrate and the coating, respectively.

It should be noted that the relaxation time for most engineering materials are of the order of 10^{-14} to 10^{-6} s, but experiments have shown that some nonhomogeneous materials have relaxation time up to 10 s which are very important materials often used as thermal insulators [31–33].

Introducing the following dimensionless variables

$$\begin{aligned} (\bar{x}, \bar{y}, \bar{a}, \bar{b}) &= (x, y, a, b)/c \\ \overline{T} = T/T_0, &\quad \bar{t} = t d_0 / c^2 \end{aligned} \tag{6.4}$$

where T_0 is the reference temperature and d_0 is the reference thermal diffusivity (we can choose either d_1 or d_2), the governing Eq. (6.3) have the following dimensionless forms:

$$\nabla^2 T^{(i)} = \frac{\partial T^{(i)}}{\partial t} \frac{d_0}{d_i} + \frac{\tau_i d_0^2}{c^2 d_i} \frac{\partial^2 T^{(i)}}{\partial t^2}, \quad (i = 1, 2) \tag{6.5}$$

It is noted that in the Eq. (6.5) and hereafter, the hat "–" of the dimensionless variables is omitted for simplicity.

The hyperbolic heat Eq. (6.5) is subjected to the following boundary and initial conditions in dimensionless forms

$$T(x,-(a+b)) = T_0, \quad (|x| < \infty, t > 0) \tag{6.6}$$

$$T(x,-a^+) = T(x,-a^-), \quad (|x| < \infty, t > 0) \tag{6.7}$$

$$k_1 \frac{\partial T(x,-a^+)}{\partial y} = k_2 \frac{\partial T(x,-a^-)}{\partial y}, \quad (|x| < \infty, t > 0) \tag{6.8}$$

$$\frac{\partial T(x,0)}{\partial y} = 0, \quad (|x| < 1) \tag{6.9}$$

$$T(x,0^+) = T(x,0^-), \quad (|x| \geq 1) \tag{6.10}$$

$$\frac{\partial T(x,0^+)}{\partial y} = \frac{\partial T(x,0^-)}{\partial y}, \quad (|x| \geq 1) \tag{6.11}$$

$$T = 0, \quad (t = 0) \tag{6.12}$$

$$\frac{\partial T}{\partial t} = 0, \quad (t = 0) \tag{6.13}$$

6.2.2 Temperature Field

Apply Laplace transform to Eq. (6.5):

$$\begin{aligned} T^{(i)*}(x,y,p) &= L\left(T^{(i)}(x,y,t)\right) = \int_0^\infty T^{(i)}(x,y,t)\exp(-pt)dt \\ T^{(i)}(x,y,t) &= L^{-1}\left(T^{(i)*}(x,y,p)\right) = \frac{1}{2\pi i}\int_{Br} T^{(i)*}(x,y,p)\exp(pt)dp \end{aligned} \tag{6.14}$$

where Br stands for the Bromwich path of integration. Considering the initial conditions (6.12) and (6.13), we have:

$$\nabla^2 T^{(i)*} = A_i p T^{(i)*} + B_i p^2 T^{(i)*}, \quad (i = 1,2) \tag{6.15}$$

with A_i and B_i defined as

$$A_i = \frac{d_0}{d_i}, \quad B_i = \frac{\tau_i d_0^2}{c^2 d_i} \quad (i = 1, 2) \tag{6.16}$$

The boundary conditions in the Laplace transform plane (*p*-plane) are:

$$T^*(x, -(a+b)) = T_0/p, \quad (|x| < \infty) \tag{6.17}$$

$$T^*(x, -a^+) = T^*(x, -a^-), \quad (|x| < \infty) \tag{6.18}$$

$$k_1 \frac{\partial T^*(x, -a^+)}{\partial y} = k_2 \frac{\partial T^*(x, -a^-)}{\partial y}, \quad (|x| < \infty) \tag{6.19}$$

$$\frac{\partial T^*(x, 0)}{\partial y} = 0, \quad (|x| < 1) \tag{6.20}$$

$$T^*(x, 0^+) = T^*(x, 0^-), \quad (|x| \geq 1) \tag{6.21}$$

$$\frac{\partial T(x, 0^+)}{\partial y} = \frac{\partial T^*(x, 0^-)}{\partial y}, \quad (|x| \geq 1) \tag{6.22}$$

The appropriate temperature field in the Laplace domain satisfying the boundary condition and regularity condition can be expressed as

$$\begin{aligned} T^{(1)*}(x, y, p) = & \int_{-\infty}^{\infty} E_1(\xi) \exp(-ry) \exp(-ix\xi) d\xi \\ & + W(y, p), \quad \text{for}\, y \geq 0 \end{aligned} \tag{6.23}$$

$$\begin{aligned} T^{(1)*}(x, y, p) = & \int_{-\infty}^{\infty} [E_2(\xi) \exp(ry) + E_3(\xi) \exp(-ry)] \exp(-ix\xi) d\xi \\ & + W(y, p), \quad \text{or} - a \leq y \leq 0 \end{aligned} \tag{6.24}$$

and

$$\begin{aligned} \dot{T}^{(2)*}(x, y, p) = & \int_{-\infty}^{\infty} [D_1(\xi) \exp(ny) + D_2(\xi) \exp(-ny)] \exp(-ix\xi) d\xi \\ & + V(y, p), \quad \text{for} - (a+b) \leq y \leq -a \end{aligned} \tag{6.25}$$

where $E_i(\xi)$ $(i = 1, 2, 3)$ and $D_j(\xi)$ $(j = 1, 2)$ are unknowns to be determined; functions $W(y, p)$, $V(y, p)$, r and n can be found in [34]. Application of Eqs. (6.15)–(6.16) leads to the expressions of $D_2(\xi)$ and $E_i(\xi)$ $(i = 1, 2)$ as the functions of $D_1(\xi)$ [34].

We introduce the temperature density function as:

$$\phi(x) = \frac{\partial T^{(1)*}(x, 0^+)}{\partial x} - \frac{\partial T^{(1)*}(x, 0^-)}{\partial x}, \tag{6.26}$$

It is clear from the boundary conditions (6.21) and (6.22) that

$$\int_{-1}^{1} \phi(t) dt = 0 \tag{6.27}$$

and

$$\phi(x) = 0, \quad (|x| \geq 1) \tag{6.28}$$

Substituting Eqs. (6.23) and (6.24) into Eq. (6.26) considering Eqs. (6A3a–d) and using Fourier inverse transform, we have:

$$D_1(\xi) = \frac{1}{i4\pi\lambda_1\xi} \int_{-1}^{1} \phi(s) \exp(is\xi) ds \tag{6.39}$$

Substituting Eq. (6.29) into Eq. (6.20) and applying the relation (6A3a–d), we get the singular integral equation for $\phi(x)$ as follows

$$\int_{-1}^{1} \phi(t) \left[\frac{1}{t - x} + H(x, t) \right] dt = 2\pi q f, \quad (|x| < 1) \tag{6.30}$$

where the kernel function $H(x, t)$ is given as

$$H(x, t) = \int_{0}^{\infty} \left\{ 1 - \frac{r(\lambda_1 - \lambda_2)}{\lambda_1 \xi} \right\} \sin[\xi(x - t)] d\xi \tag{6.31}$$

and λ_1, λ_2 are defined in [34].

The integral Eq. (6.30) under singled-value condition (6.27) has the following form of solution [34]:

$$\phi(x) = \frac{\Phi(x)}{\sqrt{1 - x^2}}, \quad |x| < 1 \tag{6.32}$$

where $\Phi(x)$ is bounded and continuous on the interval [−1, 1]. From the properties of symmetry or from the condition (6.27), it is seen that $\Phi(x)$ is an odd function of x, i.e.,

$$\Phi(-x) = -\Phi(x) \tag{6.33}$$

Following the numerical techniques of Erdogan [34], Eqs. (6.30) and (6.27) can be solved at discrete points as

$$\sum_{k=1}^{N} \frac{1}{N} \Phi(t_k) \left[\frac{1}{t_k - x_r} + H(x_r, t_k) \right] = 2fq, \quad r = 1, 2, \ldots, N - 1 \tag{6.34}$$

$$\sum_{k=1}^{N} \frac{\pi}{N} \Phi(t_k) = 0 \tag{6.35}$$

$$t_k = \cos[(2k - 1)\pi/2N], \quad k = 1, 2, \ldots, N \tag{6.36}$$

$$x_r = \cos(r\pi/N), \quad r = 1, 2, \ldots, N - 1 \tag{6.37}$$

Once function $\Phi(t)$ is obtained, function $D_1(\xi)$ can be calculated by using the Chebyshev quadrature for integration as

$$D_1(\xi) \cong \frac{1}{4\pi\xi\lambda_1} \sum_{i=1}^{N} w_i \Phi(x_i) \sin(\xi x_i) \tag{6.38}$$

$$x_i = \cos\left(\frac{2i - 1}{2N} \pi \right), \quad i = 1, 2, \ldots, N \tag{6.39}$$

$$w_i = \pi/N \tag{6.40}$$

Substituting Eq. (6.38) into Eqs. (6.23)–(6.25), we can get the temperature field in the p-plane. The temperature in the time domain can be obtained by applying the Laplace inverse transform.

6.2.3 Temperature Gradients

From Eqs. (6.38) and (6.29), we can get the expressions for $D_2(\xi)$ and $E_i(\xi)$. By considering the asymptotic nature of the integrands in Eqs. (6.29) for large values of the integration variable ξ and using the asymptotic formula [35]:

$$\int_{-1}^{1} \frac{F_j(t)}{\sqrt{1-t^2}} e^{i\xi t} dt = \sqrt{\frac{\pi}{2|\xi|}} \left\{ F_j(1) \exp\left[i\left(\xi - \frac{\pi\xi}{4|\xi|} \right) \right] + F_j(-1) \exp\left[-i\left(\xi - \frac{\pi\xi}{4|\xi|} \right) \right] + o\left(\frac{1}{|\xi|} \right) \right\}, \tag{6.41}$$

$$\int_{0}^{\infty} x^{\mu-1} \exp(-sx) \left\{ \begin{matrix} \sin \\ \cos \end{matrix} \right\} (\beta x) dx = \frac{\Gamma(\mu)}{\left(s^2+\beta^2\right)^{\mu/2}} \left\{ \begin{matrix} \sin \\ \cos \end{matrix} \right\} \left(\mu \tan^{-1}\left(\frac{\beta}{s} \right) \right), \quad s > 0, \mu > 0, \tag{6.42}$$

the singular temperature gradients near the crack tip in Laplace domain can be obtained as

$$T^*_{,y}(r,\theta,p) = -\frac{\Phi(1)}{2\sqrt{2r}} \cos\left(\frac{\theta}{2} \right), \tag{6.43}$$

$$T^*_{,x}(r,\theta,p) = \frac{\Phi(1)}{2\sqrt{2r}} \sin\left(\frac{\theta}{2} \right), \tag{6.44}$$

$$T^*_{,r}(r,\theta,p) = -\frac{\Phi(1)}{2\sqrt{2r}} \sin\left(\frac{\theta}{2} \right) \tag{6.45}$$

where (r, θ) are the polar coordinates measured from the crack tip defined by

$$r^2 = (x-1)^2 + y^2, \quad \tan(\theta) = y/(x-1) \tag{6.46}$$

Right at the crack tip, r $\to$ 0, and $\theta = -\pi$, the temperature gradient reaches a maximum value and the intensity factor of the temperature gradient (IFTG) at the crack tip can be defined as [12]

$$K^*_T(p) = \lim_{r\to 0} 2\sqrt{r} T^*_{,r}(r,\theta,p) \Big|_{\theta=-\pi} = \frac{\Phi(1)}{\sqrt{2}} \tag{6.47}$$

By applying the inverse Laplace transform to Eqs. (6.43)–(6.45), the crack-tip temperature gradients in the time domain can be obtained as

$$T_{,y}(r,\theta,t) = -\frac{K_T(t)}{2\sqrt{r}}\cos\left(\frac{\theta}{2}\right), \tag{6.48}$$

$$T_{,x}(r,\theta,t) = \frac{K_T(t)}{2\sqrt{r}}\sin\left(\frac{\theta}{2}\right), \tag{6.49}$$

$$T_{,r}(r,\theta,t) = -\frac{K_T(t)}{2\sqrt{r}}\sin\left(\frac{\theta}{2}\right), \tag{6.50}$$

where the IFTGs in the time domain, $K_T(t)$, is given by

$$K_T(t) = L^{-1}\left(K_T^*(p)\right) \tag{6.51}$$

It can be seen from Eq. (6.34) that the dynamic temperature gradients present an $r^{-1/2}$ singularity at the crack tip, which is in agreement with the corresponding static thermal crack problem [11, 12, 28]. The dynamic effect is merely introduced by the IFTGs, which are time-dependent as shown in Eq. (6.51).

6.2.4 Numerical Results

The temperature field in the time domain can be obtained from Eqs. (6.23)–(6.25) by using the numerical inversion of Laplace transform, as detailed in Miller and Guy [36], with the following parameters: $N = 8 \sim 10, \beta = 0, 0.2 \le \delta \le 0.3$. By choosing the geometric size of the composite to be $a/c = 2b/c = 1$, the material parameters to be $d_2 = 2d_1 = 2.0, k_1 = 1.0, k_2 = 0.5, \tau_1 = 1.0, \tau_2 = 0.4$ and the boundary condition $T_0 = 1.0$, the temperature field can be obtained by solving the singular integral Eq. (6.27) and substituting Eq. 6.28 into Eq. 6.19.

The steady temperature distribution in the cracked half-plane bonded to a coating is shown in Fig. 6.2 as $t \to \infty$. The disturbance of the crack on the temperature field can be observed from the iso-temperature lines, and there is a temperature jump across the crack faces.

The dynamic IFTGs can be obtained from Eqs. (6.47) and (6.51) once the algebraic Eq. (6.34) are solved and the numerical inverse of Laplace transform is performed.

The variation of dynamic IFTGs versus time is shown in Fig. 6.3 for hyperbolic heat conduction model and parabolic model. For hyperbolic heat conduction model,

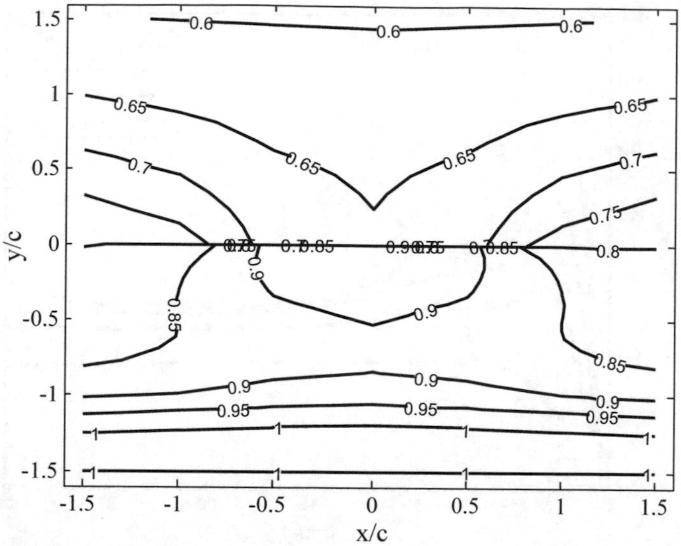

Fig. 6.2 Steady-state temperature distribution in the cracked half-plane with a coating under thermal shock [34]

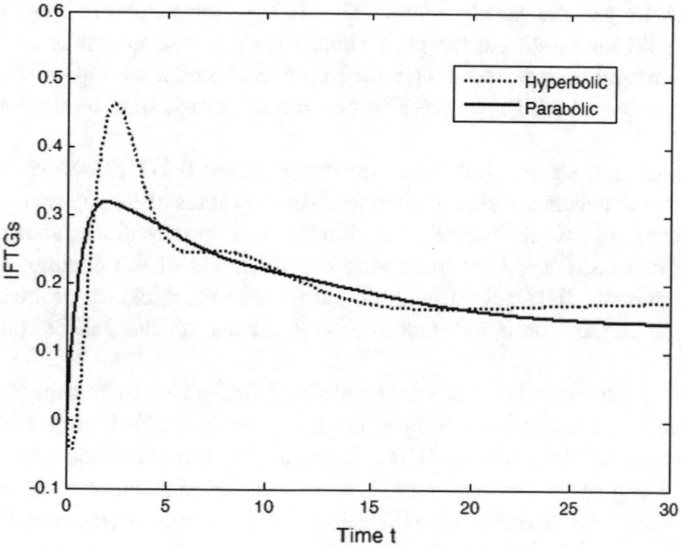

Fig. 6.3 Intensity factors of temperature gradients vary with time [34]

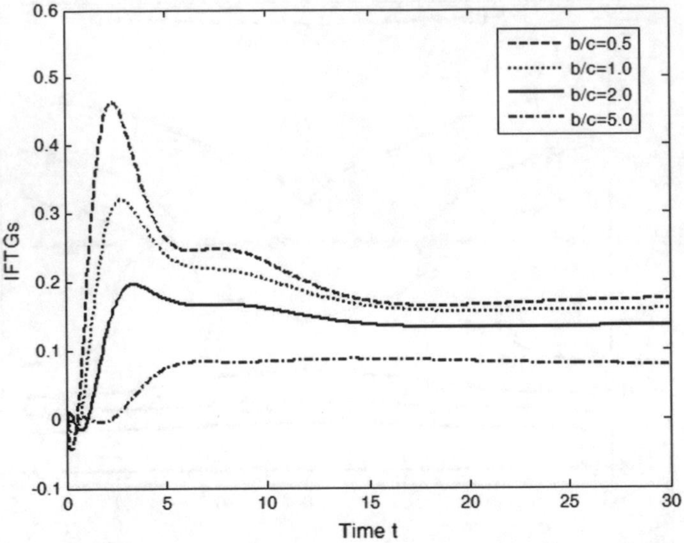

Fig. 6.4 Dynamic IFTGs for different b values when $a/c = 1.0$ [34]

the IFTGs fluctuate and increase with time and reach peak values then oscillate for some time to get the steady value. The IFTGs for parabolic model increase smoothly with time until get the peak value then decrease gradually to the steady value. The magnitudes of the IFTGs for hyperbolic model are bigger than that for parabolic model, which shows the effect of the relaxation time on the temperature field.

The effect of the geometric size b on the dynamic IFTGs is shown in Fig. 6.4 when other parameters are kept unchanged. As the values of b increase the dynamic IFTGs decrease, which means the temperature perturbation in the cracked half-plane can be reduced by increasing the thickness of the coating. Figure 6.7 shows the dynamic IFTGs for different a values with the thickness of the coating be $b/c = 1$. It can be observed that the magnitudes of the IFTGs decrease as a increases.

By setting the relaxation time of the substrate (haf-plane) to be naught, the effect of the relaxation time of the coating τ_1 on the dynamic IFTGs is shown in Fig. 6.5. The magnitude of the dynamic IFTGs decreases as the relaxation time decreases and the limiting case of $\tau_1 = 0$ correspond to the parabolic heat conduction model. It is clearly seen that there is much difference in the very early stages of the thermal loading impact. In other words, the big difference is obvious in the very small time scales, but as the time increases, the values of IFTGs for different relaxation times converge.

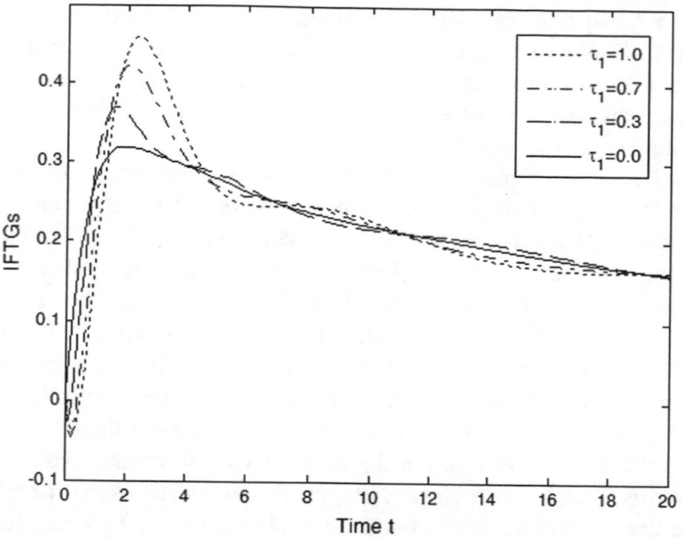

Fig. 6.5 Dynamic IFTGs for different τ_1 values when $\tau_2 = 0.0$ [34]

6.3 Thermoelastic Analysis of a Partially Insulated Crack in a Strip

In the previous section, we investigate the thermal response of a cracked, semi-infinite half-plane with a coating under thermal impact using the hyperbolic heat conduction model. The results exhibit a strong overshooting phenomenon in comparison to the results based on the classical, Fourier's heat conduction model. When the temperature field is obtained, thermoelastic analysis can be readily performed based on the classical thermoelasticity. As the final failure of cracked structure is usually governed by the mechanical stress or strain, thermoelastic analysis of cracked structure based on non-Fourier heat conduction model will provide a more conservative prediction to the reliability of materials and structures under transient thermal disturbances. Here we present a piece of work on the thermoelastic response of a cracked strip of a finite width based on the hyperbolic heat conduction.

Many structural components are often subjected to severe thermal loading, leading to intense thermal stresses in the components, especially around cracks and other defects. Materials become brittle when thermal stresses appear quickly as the result of a high temperature gradient in an unsteady temperature field. Thermal stresses combining with mechanical loadings can give rise to cracking and catastrophic failure of materials and structures [37].

The distribution of thermal stress in the vicinity of a crack in an elastic body has been extensively studied since 1950s using the classical Fourier heat conduction [38–40]. If the effects of both the inertial term and the thermo-elastic coupling term are neglected, superposition can be applied to solve the thermoelastic fracture problem [41, 42]. More general, coupled, thermoelastic theories have been adopted to study the fracture problems in thermoelasticity [43, 44]. In most of the usual engineering applications it is appropriate to use the uncoupled, thermoelastic theory without significant error [45–47].

Enriched research studies have been accumulated on crack problems under thermal loading in advanced materials based on the classical, Fourier heat conduction [14–17, 21]. The Fourier heat conduction model, although with sufficient accuracy for many engineering applications, implies infinite, thermal wave speed and is ineffective at the very small spatial and time scales associated with small-scale systems [2]. For many technological applications that involving high thermal energy with extremely short time, the results obtained from the Fourier heat conduction model differ significantly from the experimental results [4, 5]. The hyperbolic heat conduction model becomes more applicable than the Fourier heat conduction when irreversible physical processes, such as crack or void initiation in a solid, are involved in heat transport [9, 10]. A thermoelastic analysis of a cracked half-plane under a thermal shock impact was recently given by Chen and Hu [27] based on the hyperbolic heat conduction theory.

In this section, we present the transient temperature and thermal stresses around a crack in a thermoelastic strip under a temperature impact loading using the hyperbolic heat conduction. The theoretical framework established in Sect. 6.1 is used to solve the temperature field first; and the resultant thermal field is then applied to solve the transient, thermoelastic crack problem in the strip.

6.3.1 Definition of the Problem

Consider a thermo-elastic strip containing a crack of length $2c$ parallel to the free surface, as shown in Fig. 6.6. The strip is initially at the uniform temperature zero, and its free surface, $y = -h_a$ and $y = h_b$ are suddenly heated to temperature T_a and

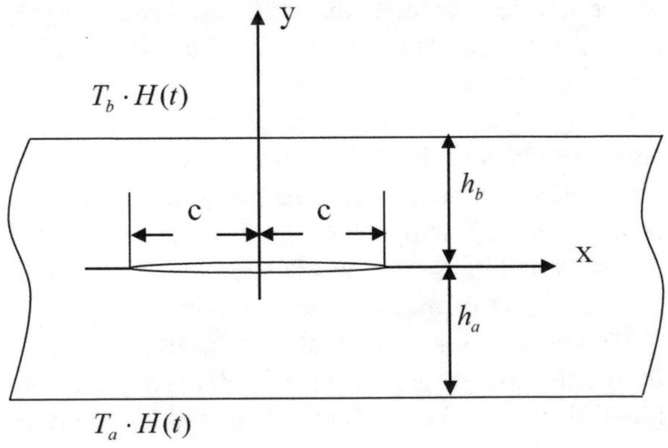

Fig. 6.6 Geometry of the cracked strip and coordinates [26]

T_b, respectively. The crack surfaces are assumed to be partially insulate, which indicates a temperature drop across the crack surfaces is contributed by the thermal resistance R_c of the crack region [21]. The effects of inertia and thermal-elastic coupling are neglected which leads to an uncoupled and quasi-static problem. The hyperbolic heat conduction equations are adopted in the thermal stress analysis.

6.3.1.1 Thermal-Elastic Field Equations

The basic equations of plane thermal stress problems for thermal elastic body are the equilibrium equations:

$$\frac{\partial \sigma_x}{\partial x} + \frac{\partial \sigma_{xy}}{\partial y} = 0, \quad \frac{\partial \sigma_{xy}}{\partial x} + \frac{\partial \sigma_y}{\partial y} = 0 \tag{6.52}$$

the strain-displacement relations:

$$\varepsilon_x = \frac{\partial u}{\partial x}, \quad \varepsilon_y = \frac{\partial v}{\partial y}, \quad \varepsilon_{xy} = \frac{1}{2}\left(\frac{\partial u}{\partial y} + \frac{\partial v}{\partial x}\right) \tag{6.53}$$

the compatibility equation:

$$\frac{\partial^2 \varepsilon_x}{\partial y^2} + \frac{\partial^2 \varepsilon_y}{\partial x^2} = 2\frac{\partial^2 \varepsilon_{xy}}{\partial x \partial y}, \tag{6.54}$$

and the constitutive law:

$$\begin{aligned} \varepsilon_x &= \frac{1}{E}\left(\sigma_x - v\sigma_y\right) + \alpha T, \\ \varepsilon_y &= \frac{1}{E}\left(\sigma_y - v\sigma_x\right) + \alpha T, \\ \varepsilon_{xy} &= \frac{1+v}{E}\sigma_{xy}, \end{aligned} \tag{6.55}$$

where E, v and α are the Young's modulus, the Poisson's ratio and the coefficient of linear thermal expansion, respectively.

Let $U(x, y)$ be the Airy stress function, then the stresses can be expresses in terms of U as

$$\sigma_x = \frac{\partial^2 U}{\partial y^2}, \quad \sigma_y = \frac{\partial^2 U}{\partial x^2}, \quad \sigma_{xy} = \frac{\partial^2 U}{\partial x \partial y}, \tag{6.56}$$

Substituting Eq. (6.56) into the compatibility condition (6.65) and considering the constitutive law (6.56) leads to

$$\nabla^2\nabla^2 U + E \cdot \nabla^2(\alpha T) = 0, \tag{6.57}$$

By introducing the dimensionless variables

$$\begin{aligned} &\overline{U} = U/\left(E\alpha c^2 T_0\right), \quad \bar{\sigma}_{ij} = \sigma_{ij}/(E\alpha T_0), \quad \bar{\varepsilon}_{ij} = \varepsilon_{ij}/(\alpha T_0) \\ &(\bar{u}, \bar{v}) = (u, v)/(\alpha c T_0), \quad (\bar{x}, \bar{y}, \bar{h}) = (x, y, h)/c \\ &\overline{T} = T/T_0, \quad \bar{t} = a/tc^2 \end{aligned} \tag{6.58}$$

where T_0 is the reference temperature.

The governing Eq. (6.57) has the following dimensionless form:

$$\nabla^2\nabla^2 U + \nabla^2(T) = 0, \tag{6.59}$$

where T can be solved by employing the same approach as detailed in Sect. 6.1 via the thermal boundary conditions. It is noted that in Eq. (6.59) and hereafter, the hat "_" of the dimensionless variables is omitted for simplicity.

Considering the following boundary and initial conditions for the thermal field in the dimensionless form

$$T(x, -h_a) = T_a, \quad (|x| < \infty, t > 0) \tag{6.60}$$

$$T(x, h_b) = T_b, \quad (|x| < \infty, t > 0) \tag{6.61}$$

$$\frac{\partial T(x, 0^+)}{\partial y} = \frac{\partial T(x, 0^-)}{\partial y} = H[T(x, 0^+) - T(x, 0^-)], \quad (|x| < 1) \tag{6.62}$$

$$T(x, 0^+) = T(x, 0^-), \quad (|x| \geq 1) \tag{6.63a}$$

$$\frac{\partial T(x, 0^+)}{\partial y} = \frac{\partial T(x, 0^-)}{\partial y}, \quad (|x| \geq 1) \tag{6.63b}$$

$$T = 0, \quad (t = 0) \tag{6.64a}$$

$$\frac{\partial T}{\partial t} = 0, \quad (t = 0) \tag{6.64b}$$

where the quantity, H, is the dimensionless thermal conductivity of the crack surface defined as $H = c/(R_c k)$ [30, 48, 49]. The limiting value of $H = 0$ corresponds to the completely insulated crack surface condition and $H \to \infty$ corresponds to the conducting crack surface.

The mechanical conditions can be expressed as

$$\sigma_{xy}(x, -h_a) = \sigma_y(x, -h_a) = 0, \quad (|x| < \infty) \tag{6.65a}$$

$$\sigma_{xy}(x, h_b) = \sigma_y(x, h_b) = 0, \quad (|x| < \infty) \tag{6.65b}$$

$$\sigma_{xy}(x, 0) = \sigma_y(x, 0) = 0, \quad (|x| < 1) \tag{6.66}$$

$$\sigma_{xy}(x, 0^+) = \sigma_{xy}(x, 0^-), \quad (|x| \geq 1) \tag{6.67a}$$

$$\sigma_y(x, 0^+) = \sigma_y(x, 0^-), \quad (|x| \geq 1) \tag{6.67b}$$

$$u(x, 0^+) = u(x, 0^-), \quad (|x| \geq 1) \tag{6.67c}$$

$$v(x, 0^+) = v(x, 0^-), \quad (|x| \geq 1) \tag{6.67d}$$

Similar to the derivations for the temperature field as discussed in Sect. 6.1, the temperature field of the cracked strip under thermal impact can be readily obtained. Then we can investigate the thermoelastic field. Details of the derivations can be found in [38].

6.3.2 *Thermal Stresses*

Substituting the temperature expressions [26], the governing equation for the Airy function U^* (6.26) becomes:

$$\begin{aligned}\nabla^2\nabla^2 U^* = &\int_{-\infty}^{\infty} (\xi^2 - m^2)D(\xi)\{\exp(my) - \exp[m(2h_b - y)]\} \\ &\exp(-ix\xi)d\xi - q^2 W(y,p) \quad (y \geq 0)\end{aligned} \tag{6.68a}$$

$$\begin{aligned}\nabla^2\nabla^2 U^* = &\int_{-\infty}^{\infty} r_{ab}(\xi^2 - m^2)D(\xi)[\exp[m(2h_a + y)] - \exp(-my)] \\ &\exp(-ix\xi)d\xi - q^2 W(y,p) \quad (y \leq 0)\end{aligned} \tag{6.68b}$$

The general solution of Eq. (6.68) satisfying the regularity condition at infinity can be expressed as

$$
\begin{aligned}
U^* = & \int_{-\infty}^{\infty} [(A_1 + A_2 y)\exp(|\xi|y) + (A_3 + A_4 y)\exp(-|\xi|y)]\exp(-ix\xi)d\xi \\
& + \int_{-\infty}^{\infty} [C_{11}(\xi)\exp(my) + C_{12}(\xi)\exp(-my)]\exp(-ix\xi)d\xi \\
& - W(y,p)/q^2 \quad (y \geq 0)
\end{aligned} \tag{6.69a}
$$

$$
\begin{aligned}
U^* = & \int_{-\infty}^{\infty} [(B_1 + B_2 y)\exp(|\xi|y) + (B_3 + B_4 y)\exp(-|\xi|y)]\exp(-ix\xi)d\xi \\
& + \int_{-\infty}^{\infty} [C_{21}(\xi)\exp(my) + C_{22}(\xi)\exp(-my)]\exp(-ix\xi)d\xi \\
& - W(y,p)/q^2 \quad (y \leq 0)
\end{aligned} \tag{6.69b}
$$

where A_i and B_i $(i = 1, 2, 3, 4)$ are unknowns to be determined, and

$$C_{11}(\xi) = -D(\xi)/(m^2 - \xi^2) \tag{6.70a}$$

$$C_{12}(\xi) = \exp(2mh_b)D(\xi)/(m^2 - \xi^2) \tag{6.70b}$$

$$C_{21}(\xi) = r_{ab}\exp(2mh_a)D(\xi)/(\xi^2 - m^2) \tag{6.70c}$$

$$C_{22}(\xi) = -\exp(-2mh_a)C_{21}(\xi) = r_{ab}D(\xi)/(m^2 - \xi^2) \tag{6.70d}$$

The stresses in Laplace domain can be obtained by substituting Eq. (6.69) into Eq. (6.56) as

$$
\begin{aligned}
\sigma_y^* & = \frac{\partial U}{\partial x^2} \\
& = \int_{-\infty}^{\infty} -\xi^2[(A_1 + A_2 y)\exp(|\xi|y) + (A_3 + A_4 y)\exp(-|\xi|y)]\exp(-i\xi x)d\xi_{(y \geqslant 0)} \\
& \quad - \int_{-\infty}^{\infty} \xi^2[C_{11}\exp(my) + C_{12}\exp(-my)]\exp(-i\xi x)d\xi,
\end{aligned} \tag{6.71a}
$$

$$\sigma_y^* = \frac{\partial U}{\partial x^2}$$
$$= \int_{-\infty}^{\infty} -\xi^2[(B_1 + B_2 y)\exp(|\xi|y) + (B_3 + B_4 y)\exp(-|\xi|y)]\exp(-i\xi x)d\xi_{(y\leqslant 0)}$$
$$- \int_{-\infty}^{\infty} \xi^2[C_{21}\exp(my) + C_{22}\exp(-my)]\exp(-i\xi x)d\xi, \quad (y\leqslant 0) \tag{6.71b}$$

$$\sigma_x^* = \int_{-\infty}^{\infty} \left\{\left[2A_2|\xi| + \xi^2(A_1 + A_2 y)\right]\exp(|\xi|y) + \left[\xi^2(A_3 + A_4 y) - 2A_4|\xi|\right]\exp(-|\xi|y)\right\}$$
$$\exp(-i\xi x)d\xi + \int_{-\infty}^{\infty} m^2[C_{11}\exp(my) + C_{12}\exp(-my)]\exp(-i\xi x)d\xi - W(y,p)$$
$$(y\geqslant 0) \tag{6.71c}$$

$$\sigma_x^* = \int_{-\infty}^{\infty} \left\{\left[2B_2|\xi| + \xi^2(B_1 + B_2 y)\right]\exp(|\xi|y) + \left[\xi^2(B_3 + B_4 y) - 2B_4|\xi|\right]\exp(-|\xi|y)\right\}$$
$$\exp(-i\xi x)d\xi + \int_{-\infty}^{\infty} m^2[C_{21}\exp(my) + C_{22}\exp(-my)]\exp(-i\xi x)d\xi - W(y,p)$$
$$(y\leqslant 0) \tag{6.71d}$$

$$\sigma_{xy}^* = \int_{-\infty}^{\infty} i\xi[[A_2 + |\xi|(A_1 + A_2 y)]\exp(|\xi|y) + [A_4 - |\xi|(A_3 + A_4 y)]\exp(-|\xi|y)]$$
$$\exp(-i\xi x)d\xi + \int_{-\infty}^{\infty} i\xi m[C_{11}\exp(my) - C_{12}\exp(-my)]\exp(-i\xi x)d\xi$$
$$(y\geqslant 0) \tag{6.71e}$$

$$\sigma_{xy}^* = \int_{-\infty}^{\infty} i\xi[[B_2 + |\xi|(B_1 + B_2 y)]\exp(|\xi|y) + [B_4 - |\xi|(B_3 + B_4 y)]\exp(-|\xi|y)]$$
$$\exp(-i\xi x)d\xi + \int_{-\infty}^{\infty} i\xi m[C_{21}\exp(my) - C_{22}\exp(-my)]\exp(-i\xi x)d\xi$$
$$(y\leqslant 0) \tag{6.71f}$$

Denote the jumps of displacements across the line $y = 0$ by $\langle u \rangle$ and $\langle v \rangle$,

$$\begin{aligned} \langle u \rangle &= u(x, 0^+) - u(x, 0^-) \\ \langle v \rangle &= v(x, 0^+) - v(x, 0^-) \end{aligned} \tag{6.72}$$

Following the procedure in [14] and introducing two dislocation density functions $f_j(x)$ $(j = 1, 2)$ as

$$f_1(x) = \frac{\partial \langle u \rangle}{\partial x} \tag{6.73a}$$

$$f_2(x) = \frac{\partial \langle v \rangle}{\partial x} \tag{6.73b}$$

By applying the boundary conditions (6.65)–(6.67), it can be shown that $f_i(x)$ $(i = 1, 2)$ satisfy the following singular integral equations:

$$\int_{-1}^{1} f_1(t) \left[\frac{1}{t - x} + M_{11}(x, t) \right] dt + \int_{-1}^{1} f_2(t) M_{12}(x, t) dt = \pi L_1(x) \tag{6.74a}$$

$$\int_{-1}^{1} f_1(t) M_{21}(x, t) dt + \int_{-1}^{1} f_2(t) \left[\frac{1}{t - x} + M_{22}(x, t) \right] dt = \pi L_2(x) \tag{6.74b}$$

where

$$L_1(x) = -8 \int_{0}^{\infty} \xi \cdot l_1(\xi) \sin(x\xi) d\xi \tag{6.75a}$$

$$L_2(x) = 8 \int_{-\infty}^{\infty} \xi^2 \cdot l_2(\xi) \cos(x\xi) d\xi \tag{6.75b}$$

$$M_{11}(x, t) = \int_{0}^{\infty} [1 + 4\Re_{11}(\xi)] \sin[\xi(x - t)] d\xi \tag{6.75c}$$

$$M_{12}(x, t) = -4 \int_{0}^{\infty} \Re_{12}(\xi) \cos[\xi(t - x)] d\xi \tag{6.75d}$$

$$M_{21}(x,t) = -4\int_0^{\infty} \Re_{21}(\xi)\cos[\xi(t-x)]d\xi \tag{6.75e}$$

$$M_{22}(x,t) = \int_0^{\infty} [1 - 4\Re_{22}(\xi)]\sin[\xi(x-t)]d\xi \tag{6.75f}$$

and the functions $l_1(\xi)$, $l_2(\xi)$ and $\Re_{ij}(\xi)$ $(i,j=1,2)$ are defined as given in [37].

The functions $f_j(x)$ $(j=1,2)$ also satisfy the singled-value equations

$$\int_{-1}^{1} f_j(x)dx = 0, \quad (j=1,2) \tag{6.76a}$$

$$f_j(x) = 0, \quad (j=1,2) \quad |x| \geq 1 \tag{6.76b}$$

The solution of the integral Eq. (6.76) of $f_j(x)$ $(j=1,2)$ can be expressed as follows

$$f_j(t) = \frac{F_j(t)}{\sqrt{1-t^2}}, \quad (j=1,2) \tag{6.77}$$

Using the Lobatto-Chebyshev method [50, 51], we can reduce the integral equations to the following algebraic equations:

$$\begin{aligned}\sum_{i=1}^{n} w_i\left[\frac{1}{(t_i - x_k)} + M_{11}(x_k,t_i)\right]F_1(t_i)\\ + \sum_{i=1}^{n} w_i M_{12}(x_k,t_i)F_2(t_i) = L_1(x_k)\end{aligned} \tag{6.78a}$$

$$\sum_{i=1}^{n} w_i F_1(t_i) = 0 \tag{6.78b}$$

$$\begin{aligned}\sum_{i=1}^{n} w_i M_{21}(x_k,t_i)F_1(t_i)\\ + \sum_{i=1}^{n} w_i\left[\frac{1}{(t_i - x_k)} + M_{22}(x_k,t_i)\right]F_2(t_i) = L_2(x_k)\end{aligned} \tag{6.78c}$$

$$\sum_{i=1}^{n} w_i F_2(t_i) = 0 \tag{6.78d}$$

where,

$$t_i = \cos\frac{(i-1)\pi}{n-1}, i = 1,2,\ldots,n; x_k = \cos\frac{(2k-1)\pi}{2(n-1)}, k = 1,2,\ldots,n-1. \tag{6.79a}$$

$$w_i = \frac{\pi}{2(n-1)}, i = 1,n; \quad w_i = \frac{\pi}{n-1}, i = 2,3,\ldots,n-1. \tag{6.79b}$$

6.3.3 Asymptotic Stress Field Near Crack Tip

By considering the asymptotic nature of the integrands in Eq. (6.71) for large values of the integration variable, ξ, following the procedure given in references [14, 50] and using the asymptotic formula [35]:

$$\begin{aligned}\int_{-1}^{1} \frac{F_j(t)}{\sqrt{1-t^2}} e^{i\xi t} dt = \sqrt{\frac{\pi}{2|\xi|}} \Big\{ & F_j(1)\exp\left[i\left(\xi - \frac{\pi\xi}{4|\xi|}\right)\right] \\ & + F_j(-1)\exp\left[-i\left(\xi - \frac{\pi\xi}{4|\xi|}\right)\right] + \mathrm{O}\left(\frac{1}{|\xi|}\right)\Big\},\end{aligned} \tag{6.80a}$$

$$\begin{aligned}&\int_0^\infty x^{\mu-1}\exp(-sx)\left\{\begin{matrix}\sin\\ \cos\end{matrix}\right\}(\beta x)dx \\ &= \frac{\Gamma(\mu)}{\left(s^2+\beta^2\right)^{\mu/2}}\left\{\begin{matrix}\sin\\ \cos\end{matrix}\right\}\left(\mu\tan^{-1}\left(\frac{\beta}{s}\right)\right),\end{aligned} \qquad s > 0, \mu > 0, \tag{6.80b}$$

the singular stresses near the crack tip in Laplace domain can be obtained as

$$\begin{aligned}
\sigma_y^*(r,\theta,p) = \frac{1}{\sqrt{2\pi r}}\Big\{ & K_1^*(p)\cos\left(\frac{\theta}{2}\right)\left[1+\sin\left(\frac{\theta}{2}\right)\sin\left(\frac{3\theta}{2}\right)\right] \\
& + K_2^*(p)\sin\left(\frac{\theta}{2}\right)\cos\left(\frac{\theta}{2}\right)\cos\left(\frac{3\theta}{2}\right)\Big\} \\
\sigma_{xy}^*(r,\theta,p) = \frac{1}{\sqrt{2\pi r}}\Big\{ & K_1^*(p)\sin\left(\frac{\theta}{2}\right)\cos\left(\frac{\theta}{2}\right)\cos\left(\frac{3\theta}{2}\right) \\
& + K_2^*(p)\cos\left(\frac{\theta}{2}\right)\left[1-\sin\left(\frac{\theta}{2}\right)\sin\left(\frac{3\theta}{2}\right)\right]\Big\} \\
\sigma_x^*(r,\theta,p) = \frac{1}{\sqrt{2\pi r}}\Big\{ & K_1^*(p)\cos\left(\frac{\theta}{2}\right)\left[1-\sin\left(\frac{\theta}{2}\right)\sin\left(\frac{3\theta}{2}\right)\right] \\
& - K_2^*(p)\sin\left(\frac{\theta}{2}\right)\left[2+\cos\left(\frac{\theta}{2}\right)\cos\left(\frac{3\theta}{2}\right)\right]\Big\}
\end{aligned} \tag{6.81}$$

where (r, θ) are the polar coordinates measured from the crack tip defined by

$$r^2 = (x-1)^2 + y^2, \quad \tan(\theta) = y/(x-1) \tag{6.82}$$

and the dimensionless stress intensity factors (SIFs) $K_1^*(p)$ and $K_2^*(p)$ are

$$\begin{aligned} K_1^*(p) &= \lim_{r\to 0} \sqrt{2r}\sigma_y(r,0,p) = -\frac{\sqrt{\pi}F_2(1,p)}{4} \\ K_2^*(p) &= \lim_{r\to 0} \sqrt{2r}\sigma_{xy}(r,0,p) = -\frac{\sqrt{\pi}F_1(1,p)}{4} \end{aligned} \tag{6.83}$$

By applying the inverse Laplace transform to (6.83), the crack-tip stress fields in the time domain can be obtained as

$$\begin{aligned} \sigma_y(r,\theta,t) &= \frac{1}{\sqrt{2\pi r}}\left\{K_1(t)\cos\left(\frac{\theta}{2}\right)\left[1+\sin\left(\frac{\theta}{2}\right)\sin\left(\frac{3\theta}{2}\right)\right]\right. \\ &\quad \left. + K_2(t)\sin\left(\frac{\theta}{2}\right)\cos\left(\frac{\theta}{2}\right)\cos\left(\frac{3\theta}{2}\right)\right\} \\ \sigma_{xy}(r,\theta,t) &= \frac{1}{\sqrt{2\pi r}}\left\{K_1(t)\sin\left(\frac{\theta}{2}\right)\cos\left(\frac{\theta}{2}\right)\cos\left(\frac{3\theta}{2}\right)\right. \\ &\quad \left. + K_2(t)\cos\left(\frac{\theta}{2}\right)\left[1-\sin\left(\frac{\theta}{2}\right)\sin\left(\frac{3\theta}{2}\right)\right]\right\} \\ \sigma_x(r,\theta,t) &= \frac{1}{\sqrt{2\pi r}}\left\{K_1(t)\cos\left(\frac{\theta}{2}\right)\left[1-\sin\left(\frac{\theta}{2}\right)\sin\left(\frac{3\theta}{2}\right)\right]\right. \\ &\quad \left. - K_2(t)\sin\left(\frac{\theta}{2}\right)\left[2+\cos\left(\frac{\theta}{2}\right)\cos\left(\frac{3\theta}{2}\right)\right]\right\} \end{aligned} \tag{6.84}$$

where the SIFs in the time domain, $K_1(t)$ and $K_2(t)$ are given by

$$\begin{aligned} K_1(t) &= L^{-1}(K_1^*(p)) \\ K_2(t) &= L^{-1}(K_2^*(p)) \end{aligned} \tag{6.85}$$

From Eq. (6.81) we can see that the dynamic stress field exhibits the same form as the static stress field around the crack tip. The dynamic effect is merely introduced by the SIFs, which are time-dependent as shown in Eq. (6.85).

The principal stresses near the crack tip can be expressed as

$$\begin{matrix}\sigma_1 \\ \sigma_2\end{matrix} = \frac{\sigma_x+\sigma_y}{2} \pm \sqrt{\left(\frac{\sigma_x-\sigma_y}{2}\right)^2 + \sigma_{xy}^2} \tag{6.86}$$

The normalized principal stresses can be defined as

$$\begin{aligned} P_1 &= \sigma_1\sqrt{2\pi r} \\ &= K_1(t)\cos\left(\frac{\theta}{2}\right) - K_2(t)\sin\left(\frac{\theta}{2}\right) \\ &\quad + \frac{1}{2}\sqrt{K_1^2(t)\sin^2(\theta) + K_2^2(t)\left[4 - 3\sin^2(\theta)\right] + 2K_1(t)K_2(t)\sin(2\theta)} \end{aligned} \tag{6.87a}$$

$$\begin{aligned} P_2 &= \sigma_2\sqrt{2\pi r} \\ &= K_1(t)\cos\left(\frac{\theta}{2}\right) - K_2(t)\sin\left(\frac{\theta}{2}\right) \\ &\quad - \frac{1}{2}\sqrt{K_1^2(t)\sin^2(\theta) + K_2^2(t)\left[4 - 3\sin^2(\theta)\right] + 2K_1(t)K_2(t)\sin(2\theta)} \end{aligned} \tag{6.87b}$$

6.3.4 Numerical Results and Discussions

The temperature field in the time domain can be obtained through the same approach as discussed in Sect. 6.1, which exhibits the similar thermal wave behavior. For simplicity, details are omitted here. By considering the thermal loading boundary conditions to be $T_a = 2, T_b = 1$ and the geometric size of the strip to be $2h_a/c = h_b/c = 2$, the stress field around the crack can be calculated. The dynamic SIFs can be obtained from (6.83) and (6.85) once the algebraic Eq. (6.78) are solved, and the numerical inversion of Laplace transform is performed.

The effect of the thermal conductivity, H, on the dynamic SIFs is shown in Fig. 6.7 for the symmetric case $h_a/c = h_b/c = 1$ and $R = 0$. It can be seen that K_1 vanishes due to the geometric symmetry of the cracked strip. The magnitudes of K_2 reduce as the thermal conductivity H increases, which means the crack disturbance on the stress field decreases as the crack faces become more conducting. For perfectly conducting crack case, there is no any disturbance around the crack and no any stress concentration.

The variation of dynamic SIFs versus time is shown in Fig. 6.8 for different geometric size when $H = 0$ and $R = 0.5$. The SIFs increase with time and reach their peak values then oscillate for some time to get their steady value. The magnitudes of the peak values increase as the size ratio h_b/h_a increases. The limiting case $h_b \to \infty$ corresponds to the cracked half-plane problem and the results are in agreement with [27].

The variation of peak-values of dynamic SIFs versus h_b for different R when $h_a = c$ and $H = 0$ is shown in Fig. 6.9. The magnitudes of the peak-values of dynamic SIFs increase as the size h_b increases from 1 to about 3.5 and then get steady values. We can see that for bigger relaxation time R corresponds to bigger magnitude of the SIFs.

The dynamic effect on the stress field around the crack tip can be well expressed by studying the variation of the principal stress versus time. From Eq. (6.84), it can be seen that the stress intensity factors introduce the dynamic effect to the stresses

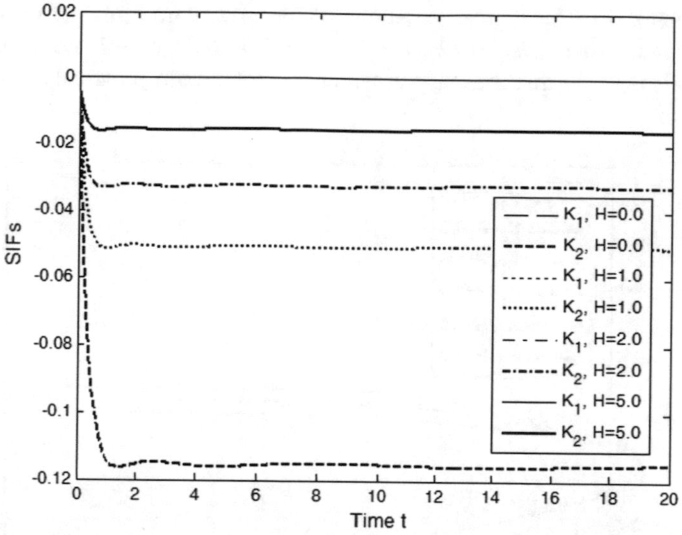

Fig. 6.7 Dynamic SIFs versus time for different thermal conductivity H when $h_a/c = h_b/c = 1$ and $R = 0$ [26]

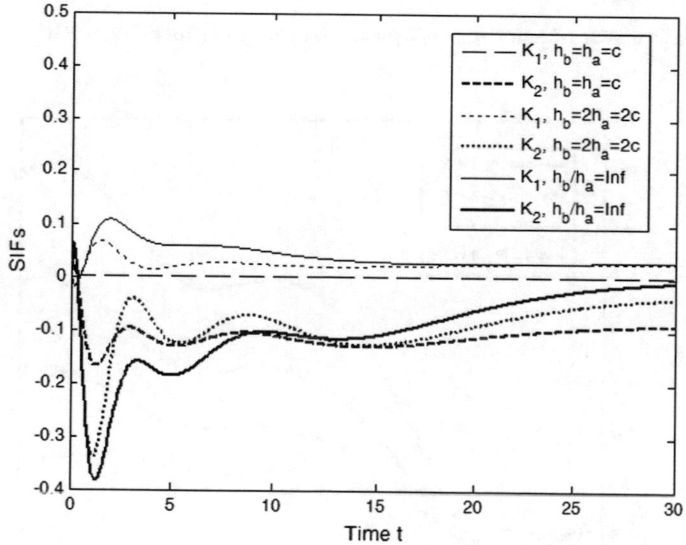

Fig. 6.8 Dynamic SIFs versus time for different geometric size h_a, h_b when $H = 0$ and $R = 0.5$ [26]

near the crack tip. The angular variations of the maximum principal stresses P_1 at different time when $2h_a/c = h_b/c = 2$, $R = 0.5$ and $H = 0$ are displayed in Fig. 6.10. It is noted that the angle at which the maximum principal stress appears

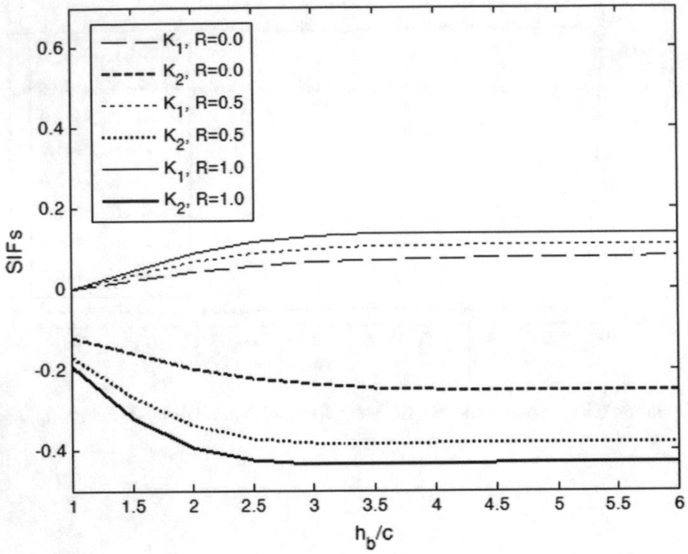

Fig. 6.9 The variation of peak-values of dynamic SIFs versus h_b for different R when $h_a = c$ and $H = 0$ [26]

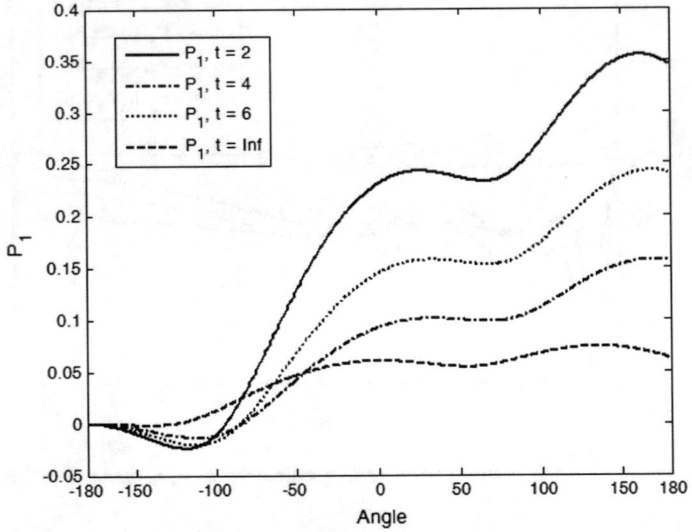

Fig. 6.10 The variation of principal stresses P_1 with angle at different time when $2h_a/c = h_b/c = 2$, $R = 0.5$ and $H = 0$ [26]

varies with time, which implies that the crack propagation direction may change under dynamic thermal loading. The magnitudes of both principal stresses are larger in the early loading period than the steady state values which indicate that the crack is more likely to propagate under dynamic thermal loading than the static thermal loading case.

6.4 Thermal Stresses in a Circumferentially Cracked Hollow Cylinder Based on Memory-Dependent Heat Conduction

The transient, thermoelastic fracture behavior of circumferentially cracked cylinders has been investigated by many researchers. Nied and Erdogan [41] solved the transient, thermal stress problem of an internally cracked hollow cylinder under a sudden cooling load, where the temperature field, parallel to the crack surfaces, was assumed to be independent of the presence of the crack. Nabavi and Ghajar [52] obtained thermal stress intensity factors (SIFs) of a circumferentially cracked, hollow cylinder under steady thermo-mechanical loading using the weight function method. The range of maximum values of SIFs of a circumferential crack in a finite-length, thick-walled cylinder with rotation-restrained edges under thermal striping was obtained in [53]. It is worth mentioning and these studies are all based on the Fourier's law.

As discussed in Chap. 1, Fourier's law leads to inaccurate results when dealing with extremely low temperature, very short time duration or high heat flux [54]. To better predict the heat transfer in solids, a number of non-Fourier heat conduction models, such as the Cattaneo and Vernotte (CV) model [55, 56], the inertial theory [57], the dual-phase lag (DPL) model [58], or the Green and Naghdi (GN) model [59, 60], etc. Employing these models, the mode I fracture problem of circumferential cracks under various crack surface loadings in a solid cylinder has been extensively studied [61–66].

The existence of the anomalous diffusion of a material leads to the abnormal heat conduction [67, 68], whereas the diffusion exhibits path dependence, memory-related behaviour. The standard, integer-order, time derivative is defined by the local limit, which cannot describe the memory-dependent process. The fractional time derivative is essentially a differential-integral convolution operator, and the integral term in its definition reflects the history-dependence of the system. It is very suitable for describing the anomalous heat conduction under extreme conditions [69–73]. The memory-dependent, fractional heat conduction model can be regarded the extension of the CV model with memory-dependent effect. Recently, Wang and Li [74] proposed a memory-dependent derivative (MDD), which is defined in an integral form of a common derivative with a kernel function on a slipping interval. This model performs better than the fractional one in reflecting the memory effect. Subsequently, Yu et al. [75] first established a new, memory-dependent, generalized thermoelasticity based upon MDD. More applications on the MDD-based thermoelasticity model can be found in [76–78]. Here, the transient, thermoelastic response of a hollow cylinder with a circumferential crack under thermal shock is presented using the MDD based, memory-dependent, heat conduction model [79]. The results are compared with the results based on the CV model to illustrate the unique feature of the memory-dependent model in thermoelastic analysis.

6.4.1 Problem Formulation

Consider an infinitely long, hollow cylinder with an initial temperature, T_0. It contains a circumferential crack in the $z = 0$ plane with z being the axis of the cylinder, as shown in Fig. 6.11. The crack occupies the region $c < r < d$, while the inner and outer radii of the cylinder are r_i and r_o, respectively. Suppose the inner surface of the cylinder suffers a sudden, thermal shock, T_i, and the outer surface is insulated. In the present analysis, we only consider that the temperature change alters stress distribution, whereas the elastic deformation does not affect the temperature distribution. As the crack faces are parallel to the direction of heat

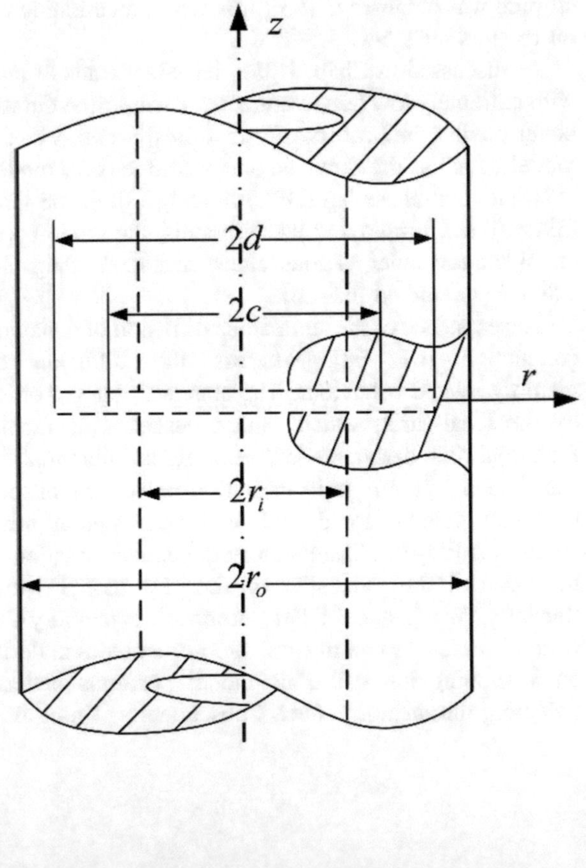

Fig. 6.11 The geometry of a hollow cylinder containing a circumferential crack [79]

conduction, the effect of crack of zero height is negligible. Therefore, the temperature change is independent of z. The procedure to solve the thermoelastic, crack problem usually includes three steps, as illustrated in Sect. 5.1. First, the transient distribution of temperature in a crack-free, hollow cylinder is obtained by solving the heat conduction equation. Then this temperature distribution is employed to determine the thermostress field. Finally, the opposite of the transient, thermal stress obtained in the preceding step is applied on the crack surfaces to solve an elastic, crack problem.

6.4.2 Thermal Axial Stress in an Un-cracked Hollow Cylinder

According to Wang and Li [75] the mth order MDD of function f has the following form

$$D_{\omega}^{m}f(t)=\frac{1}{\omega}\int_{t-\omega}^{t}K(t-\xi)f^{(m)}(\xi)d\xi \tag{6.88}$$

in which, the kernel function, $K(t-\xi)$, and the time delay, ω, can be chosen to reflect the real behavior of the material. Here, the following kernel function is adopted,

$$K(t-\xi)=1-\frac{2b}{\omega}(t-\xi)+\frac{a^2(t-\xi)^2}{\omega^2}=\begin{cases}1 & a=b=0\\ 1-\frac{(t-\xi)}{\omega} & a=0\,b=\frac{1}{2}\\ \left(1-\frac{t-\xi}{\omega}\right)^2 & a=b=1\end{cases} \tag{6.89}$$

where a and b are constants. Employing the first order MDD into the rate of heat flux, a new, memory-dependent CV model can be obtained as [75]

$$\boldsymbol{q}+\tau D_w\boldsymbol{q}=-k\nabla T \tag{6.90}$$

For an axially symmetric problem of an isotropic elastic material in the cylindrical coordinates (r,θ,z), substituting Eq. (6.90) into (6.2), the generalized heat conduction equation with MDD:

$$k\nabla^2T=(1+\tau D_{\omega})\rho_E C\dot{T} \tag{6.91}$$

where $\nabla^2=\frac{1}{r}\frac{\partial}{\partial r}\left(r\frac{\partial}{\partial r}\right)+\frac{\partial^2}{\partial z^2}$ is the Laplace operator, and ρ_F is the density of the materials. It is worthing noting that for the infinitely long hollow cylinder, $\partial\theta/\partial z=0$.

The initial and boundary conditions have the forms:

$$T(r,t) = 0 \quad \dot{T}(r,t) = 0 \quad t = 0 \tag{6.92a}$$

$$T(r,t) = H(t)(T_i - T_0) \quad r = r_i \tag{6.92b}$$

$$\partial T(r,t)/\partial r = 0 \quad r = r_o \tag{6.92c}$$

where $H(t)$ is the Heaviside step function.

For convenience, introduce the following non-dimensional variables:

$$r' = \frac{1}{r_0} r, (t', \tau', \omega') = \frac{k}{\rho C r_o^2}(t, \tau, \omega), T' = \frac{T}{T_i - T_o} \tag{6.93}$$

the governing Eq. (6.91), boundary conditions (6.92b) and (6.92c) can be rewritten as:

$$\frac{\partial^2 T'}{\partial r'^2} + \frac{1}{r'}\frac{\partial T'}{\partial r'} = (1 + \tau' D_{\omega'})\dot{T}' \tag{6.94}$$

$$T'(r', t') = H(t') \quad r' = r_{in} \tag{6.95a}$$

$$\frac{\partial T'(r', t')}{\partial r'} = 0 \quad r' = 1 \tag{6.95b}$$

where $r_{in} = r_i / r_o$.

Applying Laplace transform to both sides of Eqs. (6.94) and (6.95), we have

$$\frac{\partial^2 \overline{T'}(r', s)}{\partial r'^2} + \frac{1}{r'}\frac{\partial \overline{T'}(r', s)}{\partial r'} = (1 + G)s\overline{T'}(r', s) \tag{6.96}$$

$$\overline{T'}(r_{in}, s) = \frac{1}{s} \tag{6.97a}$$

$$\frac{\partial \overline{T'}(1, s)}{\partial r'} = 0 \tag{6.97b}$$

where

$$G = \frac{\tau'}{\omega'}\left[(1 - \exp(-s\omega'))\left(1 - \frac{2b}{\omega' s} + \frac{2a^2}{\omega'^2 s^2}\right) - \left(a^2 - 2b + \frac{2a^2}{\omega' s}\right)\exp(-s\omega')\right].$$

Considering initial conditions (6.92a), the solution of Eq. (6.96) can be expressed as:

$$\overline{T'}(r', s) = A_1 I_0\left(\sqrt{\lambda} r'\right) + A_2 K_0\left(\sqrt{\lambda} r'\right) \tag{6.98}$$

where $\lambda = (1 + G)s$, I_n and K_n represent the nth-order, modified Bessel functions of the first and second kinds, respectively.

The unknown coefficients A_1 and A_2 can be determined by the boundary conditions (6.97a) and (6.97b):

$$A_1 = \frac{K_1\left(\sqrt{\lambda}\right)}{s\left[I_0\left(\sqrt{\lambda} r_{in}\right) K_1\left(\sqrt{\lambda}\right) + I_1\left(\sqrt{\lambda}\right) K_0\left(\sqrt{\lambda} r_{in}\right)\right]}$$

$$A_2 = \frac{I_1\left(\sqrt{\lambda}\right)}{s\left[I_0\left(\sqrt{\lambda} r_{in}\right) K_1\left(\sqrt{\lambda}\right) + I_1\left(\sqrt{\lambda}\right) K_0\left(\sqrt{\lambda} r_{in}\right)\right]}$$

As there is no axial force over any cross section of the cylinder, we have:

$$\int_{r_i}^{r_o} r\sigma_{zz} dr = 0 \tag{6.99}$$

where the axial stress component σ_{zz} in an un-cracked hollow cylinder is governed by the following equation:

$$\sigma_{zz} = \frac{2E\alpha_t}{(1 - \upsilon)\left(r_o^2 - r_i^2\right)} \int_{r_i}^{r_o} r\delta T dr - \frac{E\alpha_t}{(1 - \upsilon)} \delta T \tag{6.100}$$

where E, υ, and α_t are Young's modulus, Poisson's ratio, and the coefficient of linear thermal expansion, respectively. By introducing the following non-dimensional axial stress:

$$\sigma'_{zz} = \frac{\sigma_{zz}(1 - \upsilon)}{E\alpha_t (T_i - T_0)} \tag{6.101}$$

and then the non-dimensional, thermal axial stress in the Laplace domain can be rewritten as:

$$\bar{\sigma}'_{zz} = \frac{2}{1 - r_{in}^2} \int_{r_{in}}^{1} r' \overline{T'} dr' - \overline{T'} \tag{6.102}$$

Substituting Eq. (6.98) into Eq. (6.102) leads to:

$$\begin{aligned}\bar{\sigma}'_{zz} = & \frac{2}{\sqrt{\lambda}(1-r_{in}^2)}\left\{A_1\left[I_1\left(\sqrt{\lambda}\right) - r_{in}I_1\left(\sqrt{\lambda}r_{in}\right)\right] + A_2\left[r_{in}K_1\left(\sqrt{\lambda}r_{in}\right) - K_1\left(\sqrt{\lambda}\right)\right]\right\} \\ & - A_1I_0\left(\sqrt{\lambda}r'\right) - A_2K_0\left(\sqrt{\lambda}r'\right) \triangleq \bar{p}(r', s)\end{aligned} \tag{6.103}$$

Thus far, the temperature and thermal axial stress in the Laplace domain have been solved.

6.4.3 Thermal Stress in the Axial Direction

With the thermal stress in an un-cracked hollow cylinder in hand, in this section we will determine the transient thermal stress in a circumferentially cracked hollow cylinder. For simplicity, the following non-dimensional quantities are introduced:

$$\begin{aligned}&\left(u'_r, u'_z\right) = \frac{1-\upsilon}{\alpha_t(1+\upsilon)(T_i - T_0)r_o}(u_r, u_z) \quad \left(\sigma'_{rr}, \sigma'_{rz}\right) = \frac{1-\upsilon}{\alpha_t E(T_i - T_0)}(\sigma_{rr}, \sigma_{rz}) \\ &(z', c', d') = \frac{1}{r_o}(z, c, d)\end{aligned}$$

where u_r and u_z are the displacements in the radial and axial directions, respectively. σ_{rr} and σ_{rz} denote the normal and shear stress components.

By superposition, the perturbation problem for the cracked cylinder can be formulated with the crack surface tractions equal to the negative of the thermal axial stress $\bar{p}(r', s)$ given in Eq. (6.103). As the problem is symmetric about plane $z = 0$, the non-dimensional, boundary conditions in the Laplace domain for the mode I, crack problem can be written as:

$$\bar{\sigma}'_{rr}(1, z', s) = 0 \quad \bar{\sigma}'_{rz}(1, z', s) = 0 \quad 0 \le z' \le \infty \tag{6.104a}$$

$$\bar{\sigma}'_{rr}(r_{in}, z', s) = 0 \quad \bar{\sigma}'_{rz}(r_{in}, z', s) = 0 \quad 0 \le z' \le \infty \tag{6.104b}$$

$$\bar{\sigma}'_{rz}(r', 0, s) = 0 \quad r_{in} \le r' \le 1 \tag{6.104c}$$

$$\bar{u}'_z(r', 0, s) = 0 \quad r_{in} \le r' < c', d' < r' \le 1 \tag{6.104d}$$

$$\bar{\sigma}'_{zz}(r', 0, s) = -\bar{p}(r', s) \quad c' < r' < d' \tag{6.104e}$$

The axisymmetric problem for an isotropic cylinder can be solved by introducing the Love potential function $\Phi(r, z, t)$. The non-dimensional governing equation, displacement components, and stress components are expressed in terms of $\Phi(r, z, t)$ as:

$$\nabla^2\nabla^2\overline{\Phi}' = 0 \tag{6.105}$$

$$\bar{u}'_r = -\frac{\partial^2\overline{\Phi}'}{\partial r'\partial z'} \tag{6.106a}$$

$$\bar{u}'_z = 2(1-\upsilon)\nabla^2\overline{\Phi}' - \frac{\partial^2\overline{\Phi}'}{\partial z'^2} \tag{6.106b}$$

$$\bar{\sigma}'_{zz} = \frac{\partial}{\partial z'}\left((2-\upsilon)\nabla^2\overline{\Phi}' - \frac{\partial^2\overline{\Phi}'}{\partial z'^2}\right) \tag{6.106c}$$

$$\bar{\sigma}'_{rr} = \frac{\partial}{\partial z'}\left(\upsilon\nabla^2\overline{\Phi}' - \frac{\partial^2\overline{\Phi}'}{\partial r'^2}\right) \tag{6.106d}$$

$$\bar{\sigma}'_{rz} = \frac{\partial}{\partial r'}\left((1-\upsilon)\nabla^2\overline{\Phi}' - \frac{\partial^2\overline{\Phi}'}{\partial z'^2}\right) \tag{6.106e}$$

in which $\overline{\Phi}'$ is a non-dimensional, Love potential function in the Laplace domain

$$\overline{\Phi}'(r', z', s) = \frac{\overline{\Phi}(r, z, s)(1-\upsilon)}{\alpha_t E(T_i - T_0)r_o^3} \tag{6.107}$$

As the stress vanishes at $z \to \pm\infty$, the bi-harmonic Eq. (6.105) can be solved by using the Fourier transform and the Hankel transform as:

$$\begin{aligned}\overline{\Phi}' = {} & \frac{2}{\pi}\int_0^\infty [C_1 I_0(\zeta r') + C_2 K_0(\zeta r') + r' C_3 I_1(\zeta r') + r' C_4 K_1(\zeta r')]\sin(\zeta z')d\zeta \\ & + \int_0^\infty \eta(C_5 + C_6 z')J_0(\eta r')\exp(-\eta z')d\eta\end{aligned} \tag{6.108}$$

where $C_i\,(i = 1, 2, \ldots, 6)$ are unknown functions of s to be determined from the boundary conditions (6.105). Substituting Eq. (6.108) into Eq. (6.106), the displacement and stress components can be expressed as:

$$\bar{u}'_r(r', z', s) = -\frac{2}{\pi}\int_0^\infty \zeta^2[C_3 r' I_0(\zeta r') - C_4 r' K_0(\zeta r') + C_1 I_1(\zeta r') - C_2 K_1(\zeta r')] \times \cos(\zeta z')d\zeta + \int_0^\infty \eta^2[C_6 - \eta(C_5 + C_6 z')]J_1(\eta r')\exp(-\eta z')d\eta \tag{6.109a}$$

$$\begin{aligned}\bar{u}'_z(r', z', s) = \frac{2}{\pi}\int_0^\infty \zeta\{[\zeta C_1 + 4(1-\upsilon)C_3]I_0(\zeta r') + [\zeta C_2 - 4(1-\upsilon)C_4] \\ \times K_0(\zeta r') + \zeta C_3 r' I_1(\zeta r') + \zeta C_4 r' K_1(\zeta r')\}\sin(\zeta z')d\zeta \\ + \int_0^\infty \eta^2[2(2\upsilon - 1)C_6 - \eta(C_5 + C_6 z')]J_0(\eta r')\exp(-\eta z')d\eta\end{aligned} \tag{6.109b}$$

$$\begin{aligned}\bar{\sigma}'_{rr}(r', z', s) = \frac{2}{\pi}\int_0^\infty \zeta^2\{[-\zeta C_1 + (2\upsilon - 1)C_3]I_0(\zeta r') + [-\zeta C_2 - (2\upsilon - 1)C_4] \\ \times K_0(\zeta r') + \left(\frac{C_1}{r'} - \zeta C_3 r'\right)I_1(\zeta r') - \left(\frac{C_2}{r'} + \zeta C_4 r'\right)K_1(\zeta r')\Big\} \\ \times \cos(\zeta z')d\zeta - \int_0^\infty \frac{\eta^2}{r'}[C_6 - \eta(C_5 + C_6 z')]J_1(\eta r')\exp(-\eta z')d\eta \\ + \int_0^\infty \eta^3[(1 + 2\upsilon)C_6 - \eta(C_5 + C_6 z')]J_0(\eta r')\exp(-\eta z')d\eta\end{aligned} \tag{6.109c}$$

$$\begin{aligned}\bar{\sigma}'_{zz}(r', z', s) = \frac{2}{\pi}\int_0^\infty \zeta^2\{[\zeta C_1 + (4 - 2\upsilon)C_3]I_0(\zeta r') + [\zeta C_2 - (4 - 2\upsilon)C_4] \\ \times K_0(\zeta r') + \zeta C_3 r' I_1(\zeta r') + \zeta C_4 r' K_1(\zeta r')\}\cos(\zeta z')d\zeta \\ + \int_0^\infty \eta^3[(1 - 2\upsilon)C_6 + \eta(C_5 + C_6 z')]J_0(\eta r')\exp(-\eta z')d\eta\end{aligned} \tag{6.109d}$$

$$\bar{\sigma}'_{rz}(r',z',s) = \frac{2}{\pi}\int_0^\infty \zeta^2\{\zeta C_3 r' I_0(\zeta r') - \zeta C_4 r' K_0(\zeta r') + [\zeta C_1 + 2(1-\upsilon)C_3] \\ \times I_1(\zeta r') - [\zeta C_2 - 2(1-\upsilon)C_4]K_1(\zeta r')\}\sin(\zeta z')d\zeta \\ + \int_0^\infty \eta^3[-2\upsilon C_6 + \eta(C_5 + C_6 z')]J_1(\eta r')\exp(-\eta z')d\eta \tag{6.109e}$$

In order to determine the unknown coefficients, a dislocation density function can be defined as:

$$\frac{d}{dr'}\bar{u}'_z(r',0,s) = \phi(r',s) \tag{6.110}$$

In order to determine the unknown function $\phi(r',s)$, the following singular integral equation can be obtained by substituting Eq. (6.109d) into (6.106c):

$$\int_{c'}^{d'} \phi(x',s)\left[\frac{1}{x'-r'} + L(r',x') + 2x'\int_0^\infty M(r',x',\zeta)d\zeta\right]dx' \\ = -2\pi(1-\upsilon)\bar{p}(r',s) \quad c' < r' < d' \tag{6.111}$$

where

$$M(r',x',\zeta) = \frac{\zeta^2}{\Delta}\{\zeta I_0(\zeta r')\Delta_1 + \zeta K_0(\zeta r')\Delta_2 + [\zeta r' I_1(\zeta r') - (2\upsilon - 4) \\ \times I_0(\zeta r')]\Delta_3 + [\zeta r' K_1(\zeta r') + (2\upsilon - 4)K_0(\zeta r')]\Delta_4\} \tag{6.112}$$

$$L(r',x') = \frac{m(r',x') - 1}{x' - r'} + \frac{m(r',x')}{x' + r'} \tag{6.113}$$

and

$$m(r',x') = \begin{cases} E\left(\frac{r'}{x'}\right) & r' < x' \\ \frac{r'}{x'}E\left(\frac{x'}{r'}\right) + \frac{x'^2 - r'^2}{x'r'}K\left(\frac{x'}{r'}\right) & r' > x' \end{cases} \tag{6.114}$$

In the above equation, $K()$ and $E()$ are the complete elliptic integrals of the first and second kinds, respectively. $\Delta_i\ (i = 1,2,3,4)$ can be found in [79].

From the displacement boundary condition (6.104d) and the definition (6.110), it is clear that the integral equation must be solved under the following single-value condition:

$$\int_{c'}^{d'} \phi(r', s)dr' = 0 \tag{6.115}$$

Introduce the following normalized parameters:

$$r' = \frac{d' - c'}{2}\rho + \frac{d' + c'}{2}, \quad x' = \frac{d' - c'}{2}\rho_0 + \frac{d' + c'}{2}$$

where $-1 < \rho,\ \rho_0 < 1$, the singular integral Eq. (6.111) and the single-value condition (6.115) could be rewritten as:

$$\int_{-1}^{1} \varphi(\rho_0, s)\left[\frac{1}{\rho_0 - \rho} + L'(\rho, \rho_0) + 2x' \int_{0}^{\infty} M'(\rho, \rho_0, \zeta)d\zeta\right] d\rho_0 \\ = -2\pi(1 - \upsilon)\bar{p}'(\rho, s) \tag{6.116}$$

$$\int_{-1}^{1} \varphi(\rho_0, s)d\rho_0 = 0 \tag{6.117}$$

in which

$$\varphi(\rho_0, s) = \phi(x', s) \tag{6.118a}$$

$$L'(\rho, \rho_0) = \frac{d' - c'}{2} L(r', x') \tag{6.118b}$$

$$M'(\rho, \rho_0, \zeta) = \frac{d' - c'}{2} M(r', x', \zeta) \tag{6.118c}$$

$$\bar{p}'(\rho, s) = \bar{p}(r', s) \tag{6.118d}$$

Using the numerical quadrature formulas mentioned in [80], the fundamental solution of Eq. (6.116) may be expressed as:

$$\varphi(\rho_0, s) = f(\rho_0, s)(1 + \rho_0)^{\beta_1}(1 - \rho_0)^{\beta_2} \tag{6.119}$$

where $f(\rho_0, s)$ is a bounded function in the interval $-1 < \rho_0 < 1$.

6.4.4 Stress Intensity Factors

In this section, the SIFs for an embedded crack, an outer edge crack, and an inner edge crack are defined, respectively for further numerical calculation.

6.4.4.1 Embedded Crack ($r_{in} < c' < d' < 1$)

For an embedded crack, the Cauchy kernel $1/(\rho_0 - \rho)$ is the dominant kernel and the power-law exponents are taken to be $\beta_1 = \beta_2 = -0.5$ due to the thermal stresses at the crack tips $r' = c', d'$ exhibiting singular behavior. In this case, the unknown function $f(\rho_0, s)$ in Eq. (6.119) is determined using the numerical technique described in [52] and the approximate expression reads:

$$\int_{-1}^{1} \frac{f(\rho_0, s)}{(\rho_0 - \rho)\sqrt{1 - \rho_0^2}} d\rho_0 = \sum_{j=0}^{n} \frac{\pi}{n} \gamma_j \frac{f(\rho_{0j}, s)}{\rho_{0j} - \rho_l} \tag{6.120}$$

Thus, Eqs. (6.116) and (6.117) can be approximated by the following system of $n + 1$ linear algebraic equations at $n + 1$ unknown discrete points of $f(\rho_{0j}, s)$

$$\sum_{j=0}^{n} \frac{\gamma_j f(\rho_{0j}, s)}{n} \left[\frac{1}{\rho_{0j} - \rho_l} + L'(\rho_l, \rho_{0j}) + 2x' \int_0^{\infty} M'(\rho_l, \rho_{0j}, \zeta) d\zeta \right] = -2(1 - \upsilon)\bar{p}'(\rho_l, s) \quad l = 1, 2, \ldots, n \tag{6.121}$$

$$\sum_{j=0}^{n} \frac{\gamma_j f(\rho_{0j}, s)}{n} = 0 \tag{6.122}$$

where

$$\begin{aligned} &\rho_{0j} = \cos\left(\tfrac{j}{n}\pi\right) \qquad j = 0, 1, \ldots, n \\ &\rho_l = \cos\left(\tfrac{2l-1}{2n}\pi\right) \qquad l = 1, 2, \ldots, n \\ &\gamma_0 = \gamma_n = 0.5, \qquad \gamma_1 = \gamma_2 = \cdots = \gamma_{n-1} = 1. \end{aligned}$$

After solving the integral equation, the SIFs in the Laplace domain are defined as:

$$\bar{k}_c(s) = \lim_{r \to c} \sqrt{2(c - r)}\bar{\sigma}_{zz}(r, 0, s), \quad \text{inner crack tip} \tag{6.123a}$$

$$\bar{k}_d(s) = \lim_{r \to d} \sqrt{2(r - d)}\bar{\sigma}_{zz}(r, 0, s), \quad \text{outer crack tip} \tag{6.123b}$$

By introducing the normalized SIFs:

$$k'(s) = \frac{(1-\upsilon)k(s)}{\alpha_t E(T_i - T_0)\sqrt{r_o}} \tag{6.124}$$

Equations (6.123a) and (6.123b) can be rewritten as:

$$\bar{k}'_{c'}(s) = \frac{f(-1,s)}{2(1-\upsilon)}\sqrt{\frac{d'-c'}{2}} \tag{6.125a}$$

$$\bar{k}'_{d'}(s) = -\frac{f(1,s)}{2(1-\upsilon)}\sqrt{\frac{d'-c'}{2}} \tag{6.125b}$$

where $f(-1,s)$ and $f(1,s)$ can be obtained from $f(\rho_{0j},s)$ using the method described in [52].

6.4.4.2 Outer Edge Crack ($r_{in} < c' < d' = 1$)

For an outer edge crack, the single-value condition (6.117) is not valid anymore since the surface crack has only one crack tip $r' = c'$, i.e. the crack tip $r' = d'$ disappears. Special attention should be paid to the integral Eq. (6.116). The asymptotic analysis of the integrand in (6.112) for large values of ζ indicates that the kernel $M(r', x', \zeta)$ may be expressed as the sum of two parts as follows:

$$M(r', x', \zeta) = M_\infty(r', x', \zeta) + \Delta M(r', x', \zeta) \tag{6.126}$$

where $\Delta M(r', x', \zeta)$ is nonsingular in the corresponding interval and $M_\infty(r', x', \zeta)$ becomes singular as r' and x' approach the end point $d' = 1$. After some manipulations the asymptotic expressions for the integrand and the singular part of the kernel $M(r', x', \zeta)$ are found to be:

$$\begin{aligned} M_\infty(r', x', \zeta) = & -\frac{1}{2\sqrt{r'x'}}\left[2\zeta^2(1-r')(1-x') - 3\zeta(1-x')\right. \\ & \left. -\zeta(1-r') + 2\right]\exp^{-\zeta(2-r'-x')} \end{aligned} \tag{6.127}$$

Thus, one can get:

$$\begin{aligned} \int_0^\infty M_\infty(r', x', \zeta)d\zeta = & \frac{1}{2\sqrt{r'x'}}\left[\frac{1}{2-r'-x'} - \frac{6(1-r')}{(2-r'-x')^2} + \frac{4(1-r')^2}{(2-r'-x')^3}\right] \\ & \triangleq W(r', x') \end{aligned} \tag{6.128}$$

As ζ approaches infinity, $\Delta M(r', x', \zeta)$ convergence rapidly. Therefore, Eq. (6.121) can be rewritten as:

$$\int_{-1}^{1} \varphi(\rho_0, s)\left[\frac{1}{\rho_0 - \rho} + L'(\rho, \rho_0) + 2x'W'(\rho, \rho_0) + 2x'\int_{0}^{\infty} [M'(\rho, \rho_0, \zeta) - M'_{\infty}(\rho, \rho_0, \zeta)]d\zeta\right]d\rho_0 = -2\pi(1 - \upsilon)\bar{p}'(\rho, s) \tag{6.129}$$

in which

$$M'_{\infty}(\rho, \rho_0, \zeta) = \frac{d' - c'}{2} M_{\infty}(r', x', \zeta)$$
$$W'(\rho, \rho_0) = \frac{d' - c'}{2} W(r', x')$$

For an outer surface crack, the singularity indices must satisfy $\beta_1 = -\beta_2 = -0.5$. Applying the Gauss-Jacobi quadrature formulas [81]:

$$\int_{-1}^{1} \sqrt{\frac{1 - \rho_0}{1 + \rho_0}} \frac{f(\rho_0, s)}{\rho_0 - \rho} d\rho_0 = \sum_{j=1}^{n} \frac{2\pi(1 - \rho_{0j})}{2n + 1} \frac{f(\rho_{0j}, s)}{\rho_{0j} - \rho_l} \tag{6.130}$$

One derives a system of linear algebraic equations at n collocation points to solve Eq. (6.130) as:

$$\sum_{j=1}^{n} \frac{1 - \rho_{0j}}{2n + 1} f(\rho_{0j}, s)\left[\frac{1}{\rho_{0j} - \rho_l} + L'(\rho_l, \rho_{0j}) + 2x'_j W'(\rho_l, \rho_{0j}) + 2x'_j \times \int_{0}^{\infty} [M'(\rho_l, \rho_{0j}, \zeta) - M'_{\infty}(\rho_l, \rho_{0j}, \zeta)]d\zeta\right] = -(1 - \upsilon)\bar{p}'(\rho_l, s) \tag{6.131}$$

where

$$\rho_{0j} = \cos\left(\frac{2j}{2n + 1}\pi\right) \quad j = 1, 2, \ldots, n$$
$$\rho_l = \cos\left(\frac{2l\ 1}{2n + 1}\pi\right) \quad l = 1, 2, \ldots, n$$

Similar to the embedded crack, the non-dimensional SIF at the crack tip of the outer edge crack in the Laplace domain can be expressed as:

$$\bar{k}'_{c'}(s) = \frac{f(-1,s)}{2(1-\upsilon)}\sqrt{2(1-c')} \tag{6.132}$$

where $f(-1,s)$ can be obtained from $f(\rho_{0j},s)$ using the interpolation formulas.

6.4.4.3 Inner Edge Crack ($r_{in} = c' < d' < 1$)

For an inner edge crack, the similar method for the outer edge crack is adopted here. In this case, the singularity indices must satisfy $-\beta_1 = \beta_2 = -0.5$. As the integral of $M(r',x',\zeta)$ is singular when r' and x' approach $c' = r_{in}$ simultaneously, the asymptotic analysis of $M(r',x',\zeta)$ leads to:

$$\begin{aligned} M_\infty(r',x',\zeta) = \frac{1}{2\sqrt{r'x'}}\big[&2\zeta^2(r'-r_{in})(x'-r_{in}) - 3\zeta(x'-r_{in}) \\ &- \zeta(r'-r_{in}) + 2\big]\exp^{-\zeta(r'+x'-2r_{in})} \end{aligned} \tag{6.133}$$

and

$$\begin{aligned} \int_0^\infty M_\infty(r',x',\zeta)d\zeta = \frac{1}{2\sqrt{r'x'}}\Bigg[&\frac{-1}{r'+x'-2r_{in}} + \frac{6(r'-r_{in})}{(r'+x'-2r_{in})^2} \\ &- \frac{4(r'-r_{in})^2}{(r'+x'-2r_{in})^3}\Bigg] \triangleq W(r',x') \end{aligned} \tag{6.134}$$

Considering the quadrature formulas [31]:

$$\int_{-1}^{1}\sqrt{\frac{1+\rho_0}{1-\rho_0}}\frac{f(\rho_0,s)}{\rho_0-\rho}d\rho_0 = \sum_{j=1}^{n}\frac{2\pi(1+\rho_{0j})}{2n+1}\frac{f(\rho_{0j},s)}{\rho_{0j}-\rho_l} \tag{6.135}$$

One can have the following system of algebraic equations:

$$\begin{aligned} \sum_{j=1}^{n}\frac{1+\rho_{0j}}{2n+1}f(\rho_{0j},s)\Bigg[&\frac{1}{\rho_{0j}-\rho_l} + L'(\rho_l,\rho_{0j}) + 2x'_jW'(\rho_l,\rho_{0j}) + 2x'_j \\ &\times\int_0^\infty\left[M'(\rho_l,\rho_{0j},\zeta) - M'_\infty(\rho_l,\rho_{0j},\zeta)\right]d\zeta\Bigg] = -(1-\upsilon)\bar{p}'(\rho_l,s) \end{aligned} \tag{6.136}$$

where

$$\rho_{0j} = \cos\left(\frac{2j-1}{2n+1}\pi\right) \quad j = 1, 2, \ldots, n$$
$$\rho_l = \cos\left(\frac{2l}{2n+1}\pi\right) \quad l = 1, 2, \ldots, n$$

The non-dimensional SIF at the crack tip of the inner edge crack in the Laplace domain can be expressed as:

$$\bar{k}'_{d'}(s) = -\frac{f(1,s)}{2(1-\upsilon)}\sqrt{2(d'-c')} \tag{6.137}$$

where $f(1,s)$ can be obtained from $f(\rho_{0j}, s)$ through interpolation.

Thus far, all the results in the Laplace domain have been obtained. To obtain the solutions in the time domain, numerical inversion of Laplace transform is needed.

6.4.5 Results and Discussion

In this section, our focuses are placed on the effects of time delay, kernel function and crack geometry on the transient temperature field, thermal axial stress, and the SIFs for three different kinds of cracks in a hollow cylinder. Here, it is worth noting that the apostrophes of dimensionless quantities have been left out for brevity. In the calculation process, the non-dimensional inner and outer radius of the hollow cylinder is chosen to be $r_i = 0.5$ and $r_o = 1$, non-dimensional relaxation time $\tau = 0.1$ and Poisson's ratio $\upsilon = 0.32$, unless otherwise specified. A numerical algorithm for inverse Laplace transform proposed by Brancik [82], with a further improvement using a quotient-difference algorithm in [83], is employed in the calculation.

Figure 6.12 shows the variation of transient temperature over the thickness of the cylinder at time $t = 0.05$ for various values of time delay. Clearly, the larger the time delay, the smoother the distribution of the temperature through the thickness of the cylinder, similar to the findings in [75].

Figure 6.13 illustrates the effects of time delay on the history of SIFs at the inner and outer crack tips of a hollow cylinder with the embedded crack lying in the region $0.7 < r < 0.9$ when $K(t-\xi) = 1$. As expected, the maximum SIF at the inner crack tip is larger than that at the outer crack tip, indicating the crack growth may start at the inner tip. Furthermore, a larger value of ω will result in a lower maximum SIF and a shorter duration for the SIF to reach the steady value.

Figure 6.14 displays the effects of time delay and kernel function on the transient SIF history near the crack tip for an outer edge crack in a hollow cylinder, in which the crack depth is set to be $l/h = 0.4$. Taking the time delay into account leads to a lower SIF, and an increase in ω decreases the magnitude of SIF.

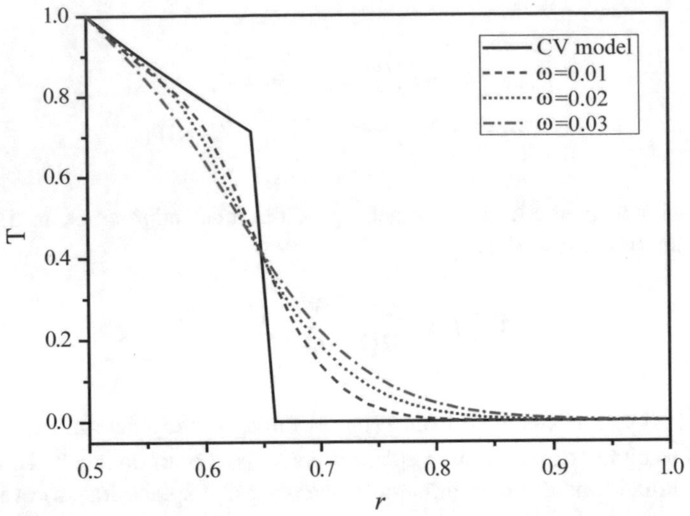

Fig. 6.12 Effects of time delay on the temperature distribution in a hollow cylinder without crack when heated inside at $t = 0.05$ for $K(t - \xi) = 1$ [79]

In an analogous manner, the transient SIF history near the crack tip for an inner edge crack in a hollow cylinder when cooled inside is depicted in Fig. 6.15. Similar effects of time delay and kernel function on the SIFs as reported for the embedded crack shown in Figs. 6.13 can be observed for the inner edge crack. Obviously, when the kernel function is determined, the maximum SIF with a particular crack depth decreases with the increase of ω. Moreover, for given values of ω and $K(t - \xi)$, the maximum SIF will increase first for shallow edge cracks and then decrease as the crack depth increases. For the same time delay, the maximum SIF is different under different kernel functions.

6.5 Transient Thermal Stress Analysis of a Cracked Half-Plane of Functionally Graded Materials

Extensive research has been accumulated on functionally graded materials (FGMs) due to their increasing application in heat engineering, such as high temperature chambers, heat exchanger tubes, thermoelectric generators, gas turbines etc. [84–90]. Compared to homogenous composite materials, FGMs possess gradual changes in composition and microstructure with spatially continuous variations in physical and mechanical properties. Invented as a thermal shield to sustain very high temperature gradients in thin-wall structures [84], one primary advantage of FGMs is their excellent performance in improving bonding strength and reducing residual and thermal stresses [37]. Fracture of a cracked FGM may occur when the

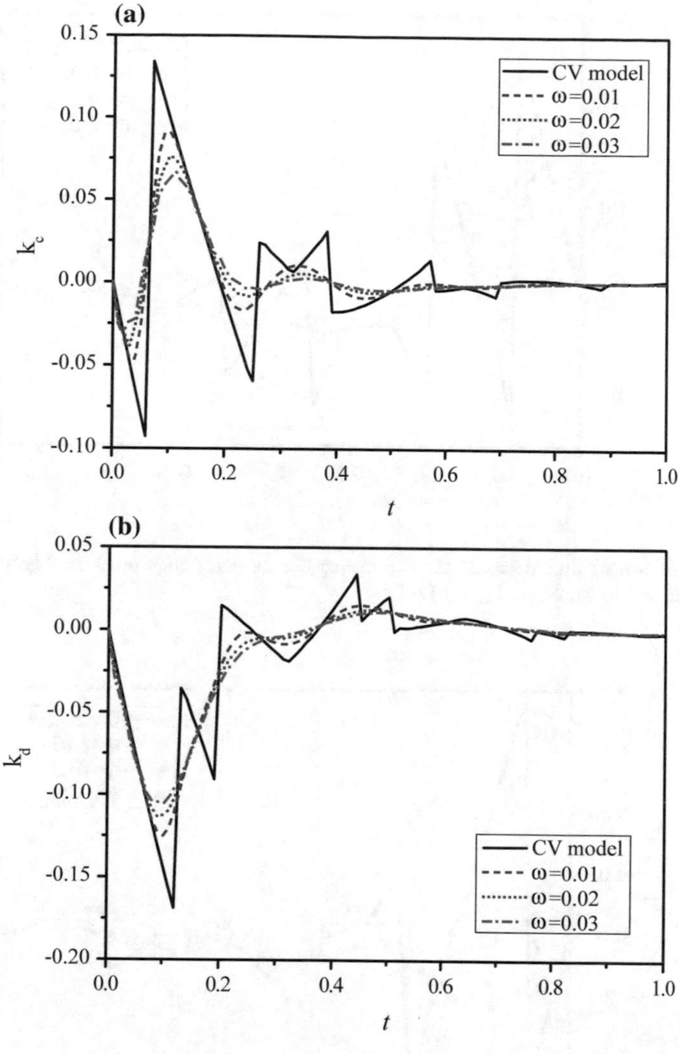

Fig. 6.13 Effects of time delay on the SIF history at **a** the inner crack tip $r_c = 0.7$, **b** the outer crack tip $r_d = 0.9$ for an embedded crack in a hollow cylinder when cooled inside for $K(t - \xi) = 1$ [79]

crack propagation is induced by an external thermal shock. To investigate the thermal stress concentration around cracks under high temperature, numerous studies have been reported on crack problems in FGMs under thermal loading [26, 91–93].

However, almost all the analyses of heat conduction in FGMs are based on the classical Fourier's Law. Although Fourier's Law is practical in many engineering applications, it is incapable of dealing with heat conduction in micro or nano scales,

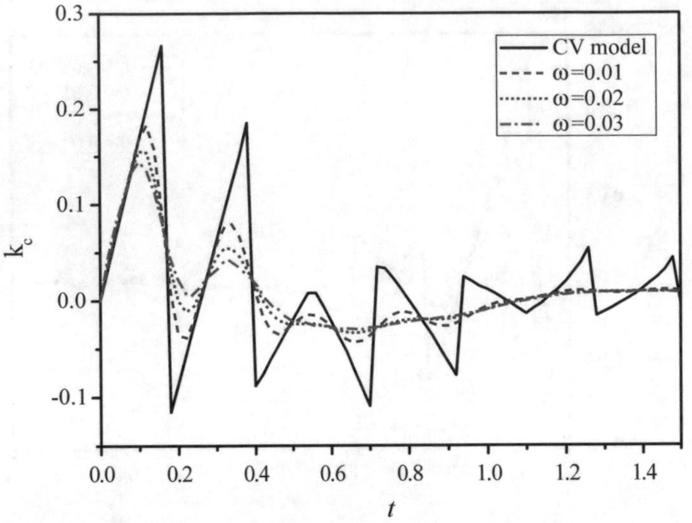

Fig. 6.14 Effects of time delay on the SIF history for an outer edge crack in a hollow cylinder when heated inside for $K(t-\xi)=1$ [79]

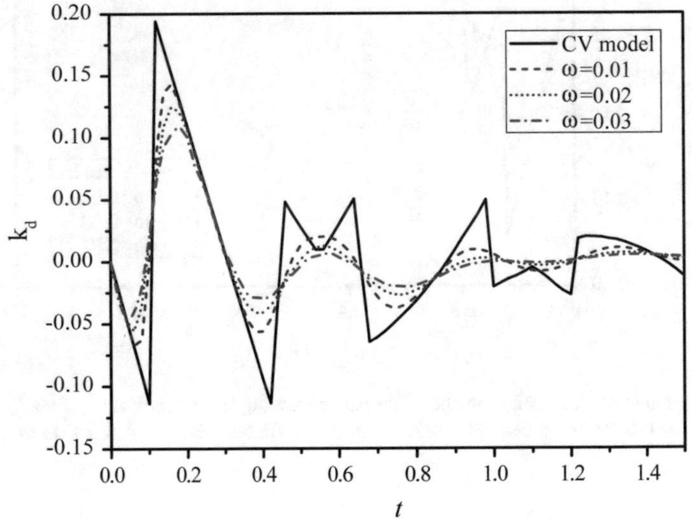

Fig. 6.15 Effects of time delay on the SIF history for an inner edge crack in a hollow cylinder when cooled inside for $K(t-\xi)=1$ [79]

or the extremely low or high temperature conditions [94]. In these situations, the measured results showed significant discrepancy with the temperature predicted by Fourier's Law [6, 7], attributed to the finite thermal wave speed, as discussed before. To compensate the effect of this non-Fourier heat conduction effect, a so-called hyperbolic heat conduction equation is proposed [55, 56] by simply introducing the time lag of thermal wave propagation into the heat conduction equation. Due to its simplicity, the hyperbolic heat conduction model has been widely used in the theoretical analysis of thermomechanical problems. For crack problems, application of the hyperbolic heat conduction model on homogeneous material has increasingly been found in the literature [27, 93–96]. Thermal stress analysis of cracked homogeneous materials based on non-Fourier heat conduction can be found in [97, 98], among others.

The transient crack problem in FGMs under thermal loading conduction model has only been investigated by Eshraghi et al. [99] assuming the circumferential crack does not disturb the temperature field. In this section, we build a thermo-elastic, analytical model for a FGM half-plane containing a crack under a thermal shock impact using the hyperbolic heat conduction theory. The crack is parallel to the free surface and assumed to be thermally insulated, so its disturbance to the temperature field could not be neglected. Employing the theoretical framework developed earlier in this chapter, the problem is solved to illustrate the effect of nn-Fourier heat conduction on the transient thermoelastic response of the cracked, FGM half-plane.

6.5.1 Formulation of the Problem and Basic Equations

As shown in Fig. 6.16, assume a crack of length $2c$ parallel to the free surface is located in a nonhomogeneous, functionally graded half-plane subjected to a thermal shock impact $T_0 H(t)$ on the free surface at time $t = 0$, where $H(t)$ is the Heaviside function. The crack is assumed to be fully thermally insulated so the redistribution of temperature field must to be taken into considerations. At time $t = 0$, the temperature of the entire half-plane is initially a constant, without loss of generality, which is set as zero. For simplicity, inertia effects and body forces are neglected.

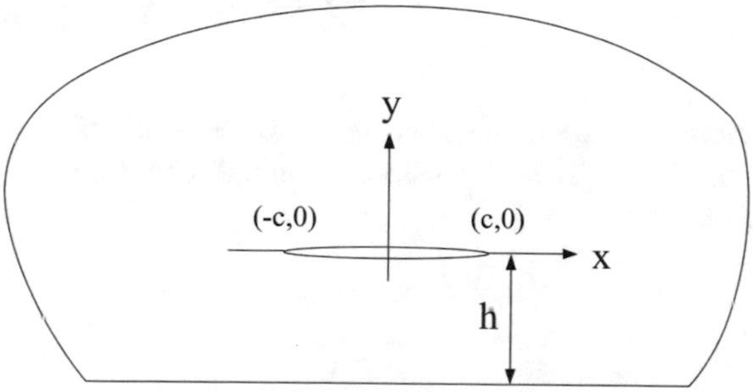

Fig. 6.16 Crack geometry and coordinates [100]

The heterogenous properties of the FGM are assumed to vary exponentially with y-coordinate, expressed as:

$$\begin{aligned} E &= E_0 \exp(\beta y), \\ \upsilon &= \upsilon_0 (1+\varepsilon y) \exp(\beta y), \\ \alpha &= \alpha_0 \exp(\gamma y), \\ k &= k_0 \exp(\delta y), \\ \kappa &= \kappa_0, \end{aligned} \tag{6.138}$$

where β is graded material constant for Young's modulus and Poisson's ratio, ε is graded material constant for Poisson's ratio, γ is graded material constant for thermal expansion coefficient and δ is graded material constant for thermal conductivity; E, υ, α, k and κ are the Young's modulus, Poisson's ratio, thermal expansion coefficient, heat conductivity and thermal diffusivity, respectively.

6.5.1.1 Heat Conduction Equations

For the hyperbolic the heat conduction model, when the inner heat source is negligible, the governing equation of the temperature field in FGMs can be obtained as:

$$\nabla^2 T + \delta \frac{\partial T}{\partial y} = \frac{1}{\kappa}\frac{\partial T}{\partial t} + \frac{\tau}{\kappa}\frac{\partial^2 T}{\partial t^2}. \tag{6.139}$$

By introducing the following dimensionless variables,

$$\overline{T} = T/T_0, \bar{t} = t/(c^2/\kappa), (\bar{x}, \bar{y}, \bar{h}) = (x, y, h)/c, \bar{\delta} = \delta \cdot c,$$

Eq. (6.139) is converted into the following dimensionless form:

$$\nabla^2 T + \delta \frac{\partial T}{\partial y} = \frac{\partial T}{\partial t} + \frac{\kappa\tau}{c^2}\frac{\partial^2 T}{\partial t^2}. \tag{6.140}$$

Here and after, the hats of the dimensionless variables have been omitted for simplicity. The dimensionless initial and boundary conditions for temperature field are:

$$
\begin{aligned}
&T = 0, \quad \frac{\partial T}{\partial t} = 0, \quad (t = 0), \\
&T(x, -h) = 1, \quad (t > 0, |x| < \infty), \\
&T = 0, \quad (y \to \infty), \\
&\frac{\partial T}{\partial y} = 0, \quad (y = 0, |x| \leq 1), \\
&T(x, 0^+) = T(x, 0^-), \quad (|x| > 1), \\
&\frac{\partial T(x, 0^+)}{\partial y} = \frac{\partial T(x, 0^-)}{\partial y}, \quad (|x| > 1).
\end{aligned}
\tag{6.141}
$$

6.5.1.2 Thermal Stress Field Equations

In the following, we assume the functionally graded half plane is under plane stress condition; i.e. $\sigma_{zz} = \sigma_{zx} = \sigma_{zy} = 0$. Without considering the body force and inertia effect, the equilibrium equations are:

$$
\frac{\partial \sigma_x}{\partial x} + \frac{\partial \sigma_{xy}}{\partial y} = 0, \quad \frac{\partial \sigma_{xy}}{\partial x} + \frac{\partial \sigma_y}{\partial y} = 0,
\tag{6.142}
$$

and the strain-displacement relations are:

$$
\varepsilon_x = \frac{\partial u}{\partial x}, \varepsilon_y = \frac{\partial v}{\partial y}, \varepsilon_{xy} = \frac{1}{2}\left(\frac{\partial u}{\partial y} + \frac{\partial v}{\partial x}\right),
\tag{6.143}
$$

The compatibility equation is:

$$
\frac{\partial^2 \varepsilon_x}{\partial y^2} + \frac{\partial^2 \varepsilon_y}{\partial x^2} = 2\frac{\partial^2 \varepsilon_{xy}}{\partial x \partial y},
\tag{6.144}
$$

while the thermoelastic constitutive equations are:

$$
\begin{aligned}
\varepsilon_x &= \frac{1}{E}(\sigma_x - \upsilon\sigma_y) + \alpha T, \\
\varepsilon_y &= \frac{1}{E}(\sigma_y - \upsilon\sigma_x) + \alpha T, \\
\varepsilon_{xy} &= \frac{1 + \upsilon}{E}\sigma_{xy}.
\end{aligned}
\tag{6.145}
$$

Again, employing Airy's function U, the governing equation of elastic stress field in FGMs can be obtained as:

$$\begin{aligned} & \nabla^2\nabla^2 U - 2\beta\frac{\partial}{\partial y}(\nabla^2 U) + \beta^2\frac{\partial^2 U}{\partial y^2} \\ & + E_0\alpha_0 \exp((\beta+\gamma)y)(\nabla^2 T + 2\gamma\frac{\partial T}{\partial y} + \gamma^2 T) = 0. \end{aligned} \tag{6.146}$$

By introducing the other following dimensionless variables as follows,

$$\begin{aligned} & \overline{\sigma_{ij}} = \sigma_{ij}/(E_0\alpha_0 T_0), \overline{U} = U/(E_0\alpha_0 T_0 c^2), \\ & (\overline{u}, \overline{v}) = (u, v)/(c\alpha_0 T_0), \overline{\varepsilon_{ij}} = \varepsilon_{ij}/(\alpha_0 T_0), \\ & (\overline{\beta}, \overline{\varepsilon}, \overline{\gamma}) = (\beta, \varepsilon, \gamma)\cdot c, \end{aligned}$$

the governing equations can be reduced to dimensionless forms:

$$\begin{aligned} & \nabla^2\nabla^2 U - 2\beta\frac{\partial}{\partial y}(\nabla^2 U) + \beta^2\frac{\partial^2 U}{\partial y^2} \\ & + \exp((\beta+\gamma)y)(\nabla^2 T + 2\gamma\frac{\partial T}{\partial y} + \gamma^2 T) = 0. \end{aligned} \tag{6.147}$$

Similarly, the hat of the dimensionless variables is omitted for simplicity. And the boundary conditions for mechanical conditions are:

$$\begin{aligned} \sigma_{xy}(x, -h) &= \sigma_y(x, -h) = 0, \quad (|x| < \infty), \\ \sigma_{xy}(x, 0) &= \sigma_y(x, 0) = 0, \quad (|x| \le 1), \\ \sigma_{xy}(x, 0^+) &= \sigma_{xy}(x, 0^-), \quad (|x| > 1), \\ \sigma_y(x, 0^+) &= \sigma_y(x, 0^-), \quad (|x| > 1), \\ u(x, 0^+) &= u(x, 0^-), \quad (|x| > 1), \\ v(x, 0^+) &= v(x, 0^-), \quad (|x| > 1). \end{aligned} \tag{6.148}$$

6.5.2 Solution of the Temperature Field

The Laplace transform is employed against time variable, thus the governing Eq. (6.140) and the corresponding boundary conditions can be transformed to:

$$\nabla^2 T^* + \delta \frac{\partial T^*}{\partial y} = pT^* + \frac{\kappa\tau}{c^2} p^2 T^*, \tag{6.149}$$

$$\begin{aligned}
&T^*(x, -h) = 1/p, \quad (|x| < \infty),\\
&T^* = 0, \quad (y \to \infty),\\
&\frac{\partial T^*}{\partial y} = 0, \quad (y = 0, |x| \le 1),\\
&T^*(x, 0^+) = T^*(x, 0^-), \quad (|x| > 1),\\
&\frac{\partial T^*(x, 0^+)}{\partial y} = \frac{\partial T^*(x, 0^-)}{\partial y}, \quad (|x| > 1).
\end{aligned} \tag{6.150}$$

Here and after, the superscript * denotes the variables in the Laplace domain, and p is the Laplace transform variable.

Applying Fourier transform to (6.149), the solution of temperature field subjected to the boundary conditions (6.151) in the Laplace domain can be obtained as:

$$\begin{aligned}
T^*(x, y, p) &= \int_{-\infty}^{\infty} D(\xi, p) \exp(-m_2 y - ix\xi) d\xi + \frac{1}{p} \exp(-q(y+h)), \quad y > 0,\\
T^*(x, y, p) &= \int_{-\infty}^{\infty} \frac{m_2 D(\xi, p)}{m_1 - m_2 \exp(-2mh)} \{1 - \exp(-2m(h+y))\} \exp(-m_1 y - ix\xi) d\xi\\
&\quad + \frac{1}{p} \exp(-q(y+h)), \quad y < 0,
\end{aligned} \tag{6.151}$$

where $m_1 = \frac{\delta}{2} - m, m_2 = \frac{\delta}{2} + m, m = \sqrt{p + \xi^2 + \frac{\delta^2}{4} + Bp^2}, q = \frac{\delta}{2} + \sqrt{p + \frac{\delta^2}{4} + Bp^2}$, $B = \frac{\kappa\tau}{c^2}$, and $D(\xi, p)$ is unknown and will be determined by the following density function:

$$\phi^*(x, p) = \frac{\partial T^*(x, 0^+, p)}{\partial x} - \frac{\partial T^*(x, 0^-, p)}{\partial x}. \tag{6.152}$$

Incorporating Eqs. (6.152) and (6.151), and employing Fourier inverse transform, we have

$$D(\xi, p) = -\frac{i[m_1 - m_2 \exp(-2mh)]}{4\pi\xi m} \int_{1}^{1} \phi^*(\tau, p) \exp(i\xi\tau) d\tau. \tag{6.153}$$

Then from the continuity condition of (6.151), it is clear that

$$\int_{-1}^{1} \phi^*(x,p)dx = 0, \tag{6.154}$$

$$\phi^*(x,p) = 0, (|x| > 1). \tag{6.155}$$

Substituting Eq. (6.15) into the temperature distribution (6.151), by using the boundary condition on the crack faces in (6.150), the following singular integral equation is obtained:

$$\int_{-1}^{1} \phi^*(\tau,p)[\frac{1}{\tau - x} + k^*(x,\tau,p)]d\tau = \frac{2\pi q}{p}\exp(-qh), \quad |x| \le 1, \tag{6.156}$$

and the kernel function is given as:

$$k^*(x,\tau,p) = \int_0^{\infty} \left\{1 + \frac{m_2[m_1 - m_2\exp(-2mh)]}{m\xi}\right\} \sin[(x-\tau)\xi]d\xi. \tag{6.157}$$

The numerical technique in [101] is employed to solve the integral Eqs. (6.156) and (6.154), and the following algebraic equation is obtained:

$$\sum_{k=1}^{n} \frac{1}{n} F^*(\tau_k,p)\left[\frac{1}{\tau_k - x_r} + k^*(x_r,\tau_k,p)\right] = \frac{2\pi q}{p}\exp(-qh), \quad |x| \le 1, \tag{6.158a}$$

$$\sum_{k=1}^{n} \frac{\pi}{n} F^*(\tau_k,p) = 0. \tag{6.158b}$$

where $\tau_k = \cos\frac{(2k-1)\pi}{2n}, k = 1,2,\ldots,n;\ x_r = \cos\frac{r\pi}{n}, r = 1,2,\ldots,n-1$ and

$$F^*(x,p) = \frac{\phi^*(x,p)}{\sqrt{1-x^2}}, |x| \le 1. \tag{6.159}$$

Once the integral equations are solved, the temperature field in the Laplace domain can be obtained. The numerical technique in [36] is again used for the Laplace inverse transform, thus the temperature field in the time domain is obtained.

6.5.3 Solution of Thermal Stress Field

Once the temperature field in the Laplace domain is obtained, the general solution of Eq. (6.147) satisfying the regular condition at infinity can be obtained as:

$$
\begin{aligned}
U^*(x,y,p) &= \int_{-\infty}^{\infty} (B_1 + B_2 y)\exp(-s_2 y - ix\xi)d\xi \\
&\quad - \int_{-\infty}^{\infty} C_1 \exp[(\beta+\gamma-m_2)y - ix\xi]d\xi, \quad y>0, \\
U^*(x,y,p) &= \int_{-\infty}^{\infty} \{(A_1 + A_2 y) + (A_3 + A_4 y)\exp(-2sy)\}\exp(-s_1 y - ix\xi)d\xi \\
&\quad - \int_{-\infty}^{\infty} \{C_{21} + C_{22}\exp(-2\mu y)\exp[(\beta+\gamma-m_1)y - ix\xi]d\xi, \quad y<0,
\end{aligned}
\tag{6.160}
$$

where $A_1, A_2, A_3, A_4, B_1, B_2$ can be derived from the boundary conditions (6.150), and

$$
s_1 = -\frac{\beta}{2} - s, s_1 = -\frac{\beta}{2} + s, s = \sqrt{\xi^2 + \frac{\beta^2}{4}},
$$

and

$$
\begin{aligned}
C_1(\xi,p) &= [(\beta+\gamma-m_2)(\gamma-m_2) - \xi^2]^{-2}[\gamma^2 + p - (2\gamma-\delta)m_2]D(\xi,p), \\
C_{21}(\xi,p) &= [(\beta+\gamma-m_1)(\gamma-m_1) - \xi^2]^{-2}[\gamma^2 + p - (2\gamma-\delta)m_1]\frac{m_2 D(\xi,p)}{m_1 - m_2\exp(-2mh)}, \\
C_{22}(\xi,p) &= [(\beta+\gamma-m_2)(\gamma-m_2) - \xi^2]^{-2}[(2\gamma-\delta)m_2 - \gamma^2 - p]\frac{m_2 D(\xi,p)\exp(-2mh)}{m_1 - m_2\exp(-2mh)}.
\end{aligned}
\tag{6.161}
$$

Then the plane stresses in the Laplace domain can be obtained directly from Eq. (6.160) by taking some derivatives according to the definition of Airy's function. Similarly, to solve the displacement field, two dislocation density functions are introduced here:

$$
\psi_1^*(x,p) = \frac{\partial[u^*(x,p)]}{\partial x}, \quad \psi_2^*(x,p) = \frac{\partial[v^*(x,p)]}{\partial x}, \tag{6.162}
$$

where $[u^*(x,p)]$, and $[v^*(x,p)]$ are the displacement jumps across the crack faces. Considering the mechanical boundary condition on crack faces in Eq. (6.148), the following singular integral equations can be obtained as:

$$\int_{-1}^{1}\sum_{j=1}^{2}[\frac{\delta_{ij}}{\tau - x}+k_{ij}(x,\tau)]\psi_j^*(\tau,p)d\tau = 4\pi W_i^*(x,p), i=1,2, \quad -1\leq x\leq 1, \tag{6.163}$$

with

$$\int_{-1}^{1}\psi_i^*(x,p)dx = 0, \quad i=1,2. \tag{6.164}$$

The Fredholm-type kernels are given by:

$$\begin{aligned}
k_{11}(x,\tau) &= \int_0^\infty [1-4\xi f_{11}(\xi)]\sin[(x-\tau)\xi]d\xi,\\
k_{22}(x,\tau) &= \int_0^\infty [1-4\xi^2 f_{22}(\xi)]\sin[(x-\tau)\xi]d\xi,\\
k_{12}(x,\tau) &= \int_0^\infty -4\xi f_{12}(\xi)\cos[(x-\tau)\xi]d\xi,\\
k_{21}(x,\tau) &= \int_0^\infty -4\xi^2 f_{21}(\xi)\cos[(x-\tau)\xi]d\xi,
\end{aligned} \tag{6.165}$$

and

$$\begin{aligned}
W_1^*(x,p) &= 2\int_0^\infty \xi w_1^*(\xi,p)\sin(x\xi)d\xi,\\
W_2^*(x,p) &= -2\int_0^\infty \xi^2 w_2^*(\xi,p)\cos(x\xi)d\xi,\\
w_1^*(\xi,p) &= -\frac{h_{11}(\beta g_1+2g_2)+2sh_{12}(s_2g_1-g_2)}{8s^3}-g_3,\\
w_2^*(\xi,p) &= -\frac{h_{21}(\beta g_1+2g_2)+2sh_{22}(s_2g_1-g_2)}{8s^3}-g_4,
\end{aligned} \tag{6.166}$$

where the expressions of $f_{ij}(\xi), h_{ij}(\xi)$ $(i,j = 1,2)$ and $g_i(\xi)$ $(i = 1,2,3,4)$ can be found in [100]. The solutions of the above integral equations can be expressed as:

$$\psi_i^*(x,p) = \frac{G_i^*(x,p)}{\sqrt{1-x^2}}, \quad (i = 1,2), |x| \leq 1. \tag{6.167}$$

Using the Lobatto-Chebyshev method [102], the above singular integral equations can be transformed to algebraic equations:

$$\begin{aligned}
&\sum_{i=1}^{n} A_i[\frac{1}{\tau_i - x_k} + k_{11}(x_k,\tau_i)]G_1^*(\tau_i,p) + \sum_{i=1}^{n} A_i k_{12}(x_k,\tau_i)G_2^*(\tau_i,p) = 4\pi W_1^*(x_k,p),\\
&\sum_{i=1}^{n} A_i G_1^*(\tau_i,p) = 0,\\
&\sum_{i=1}^{n} A_i k_{21}(x_k,\tau_i)G_1^*(\tau_i,p) + \sum_{i=1}^{n} A_i[\frac{1}{\tau_i - x_k} + k_{22}(x_k,\tau_i)]G_2^*(\tau_i,p) = 4\pi W_2^*(x_k,p),\\
&\sum_{i=1}^{n} A_i G_2^*(\tau_i,p) = 0,
\end{aligned} \tag{6.168}$$

where

$$\begin{aligned}
\tau_i &= \cos\frac{(i-1)\pi}{n-1}, i = 1,2,\ldots n,\\
x_k &= \cos\frac{(2k-1)\pi}{2(n-1)}, k = 1,2,\ldots n-1,\\
A_i &= \frac{\pi}{2(n-1)}, i = 1,n; A_i = \frac{\pi}{n-1}, i = 2,3,\ldots n-1.
\end{aligned}$$

From reference [88], the stress intensity factors (SIFs) in the Laplace domain can be obtained as:

$$K_I^*(p) = -\frac{\sqrt{\pi}}{4}G_2^*(1,p), K_{II}^*(p) = -\frac{\sqrt{\pi}}{4}G_1^*(1,p). \tag{6.169}$$

The dynamic stress intensity factors in the time domain can be obtained by the Laplace inverse transform via Eq. (6.39),

$$\begin{aligned}
K_I(t) &= \frac{1}{2\pi i}\int_{Br} -\frac{\sqrt{\pi}}{4}G_2^*(1,p)\exp(pt)dp,\\
K_{II}(t) &= \frac{1}{2\pi i}\int_{Br} -\frac{\sqrt{\pi}}{4}G_1^*(1,p)\exp(pt)dp,
\end{aligned} \tag{6.170}$$

where "Br" stands for the Bromwich path. In the following section, the numerical algorithm of Laplace inverse transform proposed by Miller and Guy [36] will be used to obtain the SIFs in the time domain.

6.5.4 Numerical Results and Discussion

The temperature field distribution in the time domain can be obtained after taking the inverse Laplace transform of Eq. (6.151). Since the crack is assumed to be thermally insulated, the existence of crack parallel to the free surface will disturb the temperature field. At the beginning, the temperature variations of the mid-points of crack faces versus dimensionless time are investigated under the influence of $B = \frac{\kappa\tau}{c^2}$, which plays a vital role in the hyperbolic heat conduction theory. From Ref. [33], the thermal relaxation time for nonhomogeneous FGMs could be up to the order of 10 s. If we take the typical crack size as 1 mm, the parameter B can be up to 10 according the experiment results in [33], which is much larger than that in homogenous materials, such as metals.

To give a better illustration of the transient dynamic stress field around the crack tips, the variation of cleavage stresses defined by the following equation:

$$\sigma_\theta = \frac{1}{\sqrt{2\pi r}}\left\{K_I[\frac{3}{4}\cos(\frac{\theta}{2}) + \frac{1}{4}\cos(\frac{3\theta}{2})] + K_{II}[-\frac{3}{4}\sin(\frac{\theta}{2}) - \frac{3}{4}\sin(\frac{3\theta}{2})]\right\} \tag{6.171}$$

are plotted against angle $\theta \in (-180°,\ 180°)$, as shown in Fig. 6.17. Two different time instants, $t = 3$, $t = 15$ are considered when $B = 1, B = 10$. Clearly the cleavage stresses reach their maximum at the same angle, which means the direction of the possible crack propagation will always be the same at different time instants, independent of the thermal relaxation time (Fig. 6.17).

In FGMs, the nonhomogeneous material constants play a vital role as they affect the SIFs significantly according to the literature. As a result, the parametric investigations are conducted under the framework of hyperbolic heat conduction theory when $B = 0.5$. From the singular integral equation in thermal stress field, the Poisson's ratio would have no influence on the stress intensity factors, only the material constants $\beta,\ \gamma,\ \delta$ will affect the SIFs. As δ, β play a rather more dominant role than γ in the thermoelastic response of the cracked structure [100], we present only the stress intensity factors history at various values of β in Fig. 6.18 and their peak values versus the gradient parameters $\delta \in (-2, 2)$ and $\beta \in (-2, 2)$ in Fig. 6.19, respectively. It is noted the negative values of K_I indicate crack faces would be under compression (Figs. 6.18 and 6.19).

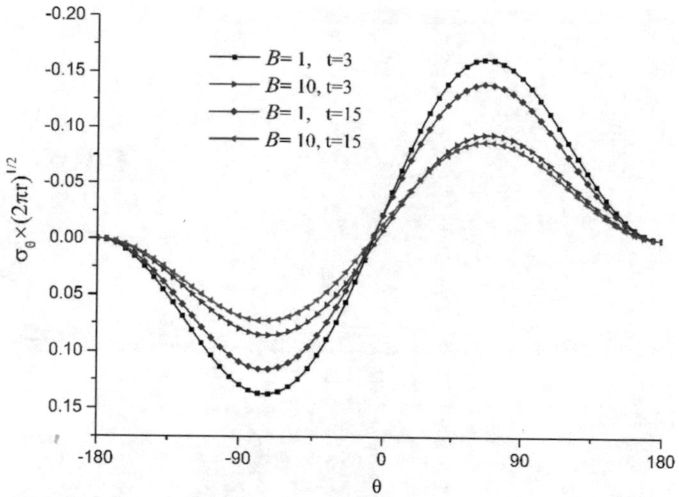

Fig. 6.17 Variation of the cleavage stress versus angle $\theta \in (-180°,\ 180°)$ at time $t = 3, t = 15$ when $B = 1, B = 10$ [100]

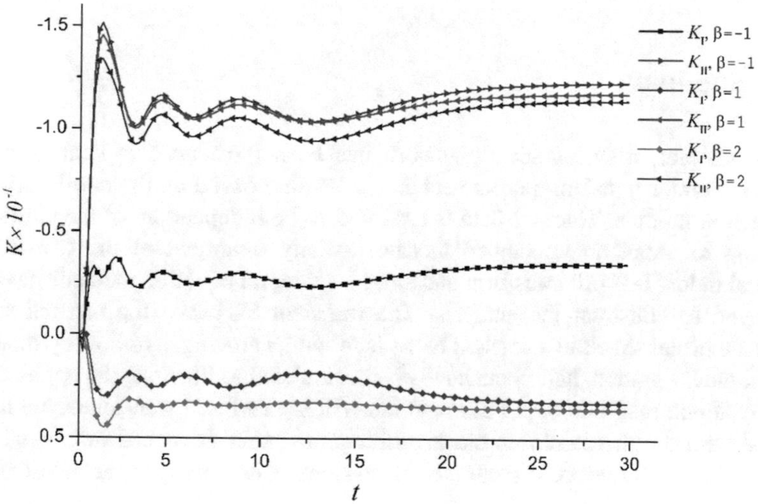

Fig. 6.18 The effect of the gradient parameter β on the SIFs when $\delta = 1, \gamma = 0.1, B = 0.5$ [100]

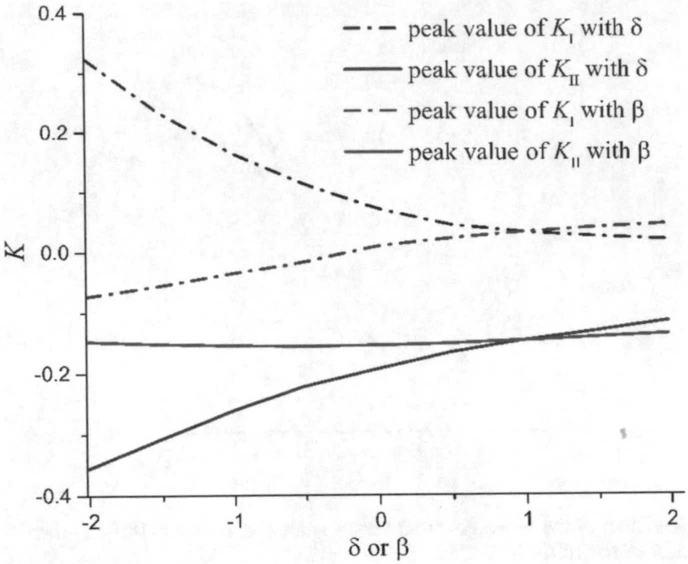

Fig. 6.19 The variation of peak values of stress intensity factors with: (1) the gradient parameter $\beta \in (-2, 2)$ when $\delta = 1, \gamma = 0.1$. (2) the gradient parameter $\delta \in (-2, 2)$ when $\beta = 1, \gamma = 0.1$ [100]

6.6 Summary

In this chapter, a systematic framework has been introduced to deal with crack problems under transient thermomechanical loading based on the non-Fourier heat conduction models. Thermal field is assumed to be independent of the elastic field allowing to adopt an uncoupled thermoelasticity treatment of the two different physical fields. Integral transform and singular integral equation methods have been employed to construct the analysis. The transient SIFs and the thermal stresses under a thermal shock in a cracked half-plane with a coating, a hollow cylinder, and a functionally graded, half plane have been calculated to illustrate the application of the developed methodology. Future works will see further extension of the method to deal with the thermoelastic crack problems of other advanced functional materials, such as piezoelectric materials, nano-composites, and magnetoelectroelastic materials.

References

1. Zhang MY, Cheng GJ (2011) Pulsed laser coating of bioceramic/metallic nanoparticles on metal implants: multiphysics simulation and experiments. IEEE Trans Nanobiosci 99:1
2. Babaei MH, Chen ZT (2010) Transient hyperbolic heat conduction in a functionally gradient hollow cylinder. J Thermophys Heat Transf 24(2):325

3. Roy S, Vasudeva Murthy AS, Kudenatti RB (2009) A numerical method for the hyperbolic-heat conduction equation based on multiple scale technique. Appl Numer Math 59:1419
4. Al-Nimr MA (1997) Heat transfer mechanisms during short-duration laser heating of thin metal films. Int J Thermophys 18:1257
5. Naji M, Al-Nimr M, Darabseh T (2007) Thermal stress investigation in unidirectional composites under the hyperbolic energy model. Int J Solids Struct 44:5111
6. Maurer MJ, Thompson HA (1973) Non-Fourier effects at high heat flux. ASME J Heat Transf 95:284
7. Babaei MH, Chen ZT (2008) Hyperbolic heat conduction in a functionally graded hollow sphere. Int J Thermophys 29:1457
8. Ozisik M, Tzou DY (1994) On the wave theory in heat conduction. ASME J Heat Transf 116:526
9. Tzou DY (1989) The effects of thermal shock waves on the crack initiation around a moving heat source. Eng Fract Mech 34:1109
10. Al-Khairy RT, Al-Ofey ZM (2009) Analytical solution of the hyperbolic heat conduction equation for moving semi-infinite medium under the effect of time-dependent laser heat source. J Appl Math 2009:1
11. Sih GC (1965) Heat conduction in the infinite medium with lines of discontinuities. ASME J Heat Transf 87:283
12. Tzou DY (1990) The singular behavior of the temperature gradient in the vicinity of a macrocrack tip. Int J Heat Mass Transf 33(12):2625
13. Tzou DY (1992) Characteristics of thermal and flow behavior in the vicinity of discontinuities. Int J Heat Mass Transf 35(2):481
14. Jin Z-H, Noda N (1993) An internal crack parallel to the boundary of a nonhomogeneous half plane under thermal loading. Int J Eng Sci 31:793
15. Noda N, Jin Z-H (1993) Thermal stress intensity factors for a crack in a strip of a functionally gradient material. Int J Solids Struct 30:1039
16. Erdogan F, Wu BH (1996) Crack problems in FGM layers under thermal stresses. J Therm Stresses 19:237–265
17. Itou S (2004) Thermal stresses around a crack in the nonhomogeneous interfacial layer between two dissimilar elastic half-planes. Int J Solids Struct 41:923
18. El-Borgi S, Erdogan F, Hidri L (2004) A partially insulated embedded crack in an infinite functionally graded medium under thermo-mechanical loading. Int J Eng Sci 42:371–393
19. Wang BL, Mai Y-W (2005) A periodic array of cracks in functional graded materials subjected to thermo-mechanical loading. Int J Eng Sci 43:432
20. Ueda S (2008) Transient thermoelectroelastic response of a functionally graded piezoelectric strip with a penny-shaped crack. Eng Fract Mech 75:1204
21. Zhou YT, Li X, Yu DH (2010) A partially insulated interface crack between a graded orthotropic coating and a homogeneous orthotropic substrate under heat flux supply. Int J Solids Struct 47:768
22. Qin QH (2000) General solutions for thermopiezoelectrics with various holes under thermal loading. Int J Solids Struct 37:5561
23. Gao CF, Noda N (2004) Thermal-induced interfacial cracking of magnetoelectroelastic materials. Int J Eng Sci 42:1347
24. Manson JJ, Rosakis AJ (1993) The effects of hyperbolic heat conduction around a dynamically propagating crack tip. Mech Mater 15:263–278
25. Tzou DY (1990) Thermal shock waves induced by a moving crack. ASME J Heat Transf 112:21
26. Hu KQ, Chen ZT (2012) Thermoelastic analysis of a partially insulated crack in a strip under thermal impact loading using the hyperbolic heat conduction theory. Int J Eng Sci 51:144–160
27. Chen ZT, Hu KQ (2012) Thermo-elastic analysis of a cracked half-plane under a thermal shock impact using the hyperbolic heat conduction theory. J Therm Stresses 35:342–362

28. Chen W-H, Huang C-C (1992) On the singularity of temperature gradient near an inclined crack terminating at bimaterial interface. Int J Fract 58:319–324
29. Achenbach JD (1973) Wave propagation in elastic solids. North-Holland, Amsterdam
30. Carslaw HS, Jaeger JC (1990) Conduction of heat in solids. Clarendon Press, Oxford
31. Lewandowska M, Malinowski L (1998) Hyperbolic heat conduction in the semi-infinite body with the heat source which capacity linearly depends on temperature. Heat Mass Transf 33:389
32. Ali YM, Zhang LC (2005) Relativistic heat conduction. Int J Heat Mass Transf 48:2397
33. Kaminski W (1990) Hyperbolic heat conduction equation for materials with a nonhomogeneous inner structure. ASME J Heat Transf 112(3):555
34. Chen ZT, Hu KQ (2012) Hyperbolic heat conduction in a cracked thermoelastic half-plane bonded to a coating. Int J Thermophys 33:895–912
35. Gradshteyn IS, Ryzhic IM (1965) Tables of integrals, series and products. Academic Press, New York
36. Miller MK, Guy WT (1966) Numerical inversion of the Laplace transform by use of Jacobi polynomials. SIAM J Numer Anal 3:624
37. Kovalenko AD (1969) Thermoelasticity: basic theory and applications. Noordhoff, Groningen
38. Williams ML (1959) The stress around a fault or crack in dissimilar media. Bull Seismol Soc Am 49:199–204
39. Sih GC (1962) On singular character of thermal stress near a crack tip. ASME J Appl Mech 51:587–589
40. Tsai YM (1984) Orthotropic thermalelastic problem of uniform heat flow disturbed by a central crack. J Compos Mater 18:122–131
41. Nied HF, Erdogan F (1983) Transient thermal stress problem for a circumferentially cracked hollow cylinder. J Therm Stress 6(1):1–14
42. Noda N, Matsunaga Y, Nyuko H (1986) Stress intensity factor for transient thermal stresses in an infinite elastic body with external crack. J Therm Stresses 9:119–131
43. Chen TC, Weng CI (1991) Coupled transient thermo-elastic response in an edge-cracked plate. Eng Fract Mech 39:915–925
44. Atkinson C, Craster RV (1992) Fracture in fully coupled dynamic thermoelasticity. J Mech Phys Solids 40:1415–1432
45. Sternberg E, Chakravorty JG (1959) On inertia effects in a transient thermoelastic problem. ASME J Appl Mech 26:503–508
46. Boley BA, Weiner JH (1985) Theory of thermal stresses. Wiley, New York
47. Noda N, Matsunaga Y, Nyuko H (1990) Coupled thermoelastic problem of an infinite solid containing a penny-shaped crack. Int J Eng Sci 28:347–353
48. Arpaci VS (1966) Conduction heat transfer. Addison-Wesley Pub. Co., Reading
49. Choi HJ, Thangjitham S (1993) Thermal-induced interlaminar crack-tip singularities in laminated anisotropic composites. Int J Fract 60:327–347
50. Delale F, Erdogan F (1979) Effect of transverse shear and material orthotropy in a cracked spherical cap. Int J Solids Struct 15:907–926
51. Theocaris P, Ioakimidis N (1977) Numerical integration methods for the solution of singular integral equations. Quart Appl Math 35:173–183
52. Nabavi SM, Ghajar R (2010) Analysis of thermal stress intensity factors for cracked cylinders using weight function method. Int J Eng Sci 48(12):1811–1823
53. Meshii T, Watanabe K (2004) Stress intensity factor of a circumferential crack in a thick-walled cylinder under thermal striping. J Press Vessel Technol Trans ASME 126(2):157–162
54. Wang L (1994) Generalized Fourier law. Int J Heat Mass Transf 37:2627–2634
55. Cattaneo C (1958) A form of heat conduction equation which eliminates the paradox of instantaneous propagation. Comp Rend 247(4):431–433
56. Vernotte P (1958) Paradoxes in the continuous theory of the heat conduction. Comp Rend 246:3154–3155

57. Joseph DD, Preziosi L (1989) Heat waves. Rev Mod Phys 61:41–73
58. Tzou DY (1995) A unified field approach for heat conduction from macro to micro scales. J Heat Transf Trans ASME 117:8–16
59. Mallik SH, Kanoria M (2008) A two dimensional problem for a transversely isotropic generalized thermoelastic thick plate with spatially varying heat source. Eur J Mech A/Solids 27:607–621
60. Taheri H, Fariborz SJ, Eslami MR (2005) Thermoelastic analysis of an annulus using the Green-Naghdi model. J Therm Stress 28:911–927
61. Keer LM, Freedmann JM, Watts HA (1977) Infinite tensile cylinder with circumferential edge crack. Lett Appl Eng Sci 5:129–139
62. Erdol R, Erdogan F (1975) A thick-walled cylinder with an axisymmetric internal or edge crack. J Appl Mech Trans ASME 45:281–286
63. Aydin L, Secil Altundag Artem H (2008) Axisymmetric crack problem of thick-walled cylinder with loadings on crack surfaces. Eng Fract Mech 75(6):1294–1309
64. Fu JW, Chen ZT, Qian LF, Xu YD (2014) Non-Fourier thermoelastic behavior of a hollow cylinder with an embedded or edge circumferential crack. Eng Fract Mech 128:103–120
65. Chen LM, Fu JW, Qian LF (2015) On the non-Fourier thermal fracture of an edge-cracked cylindrical bar. Theor Appl Fract Mech 80:218–225
66. Guo SL, Wang BL, Zhang C (2016) Thermal shock fracture mechanics of a cracked solid based on the dual-phase-lag heat conduction theory considering inertia effect. Theor Appl Fract Mech 86:309–316
67. Scher H, Montroll EW (1975) Anomalous transit-time dispersion in amorphous solid. Phys Rev B 12:2455–2477
68. Koch DL, Brady JF (1988) Anomalous diffusion in heterogeneous porous media. Phys Rev Fluids 31:965–973
69. Li B, Wang J (2003) Anomalous heat conduction and anomalous diffusion in one-dimensional systems. Phys Rev Lett 91:044301–1–044301–4
70. Povstenko YZ (2004) Fractional heat conduction equation and associated thermal stress. J Therm Stress 28(1):83–102
71. Sherief HH, El-Sayed AMA, Abd El-Latief AM (2010) Fractional order theory of thermoelasticity. Int J Solids Struct 47(2):269–275
72. Youssef HM (2010) Theory of fractional order generalized thermoelasticity. J Heat Transf Trans ASME 132(6):061301–1–7
73. Ezzat MA, El-Karamany AS, Ezzat SM (2012) Two-temperature theory in magneto-thermoelasticity with fractional order dual-phase-lag heat transfer. Nucl Eng Des 252:267–277
74. Wang JL, Li HF (2011) Surpassing the fractional derivative: concept of the memory-dependent derivative. Comput Math Appl 62(3):1562–1567
75. Yu YJ, Tian XG, Lu TJ (2013) Fractional order generalized electro-magneto-thermo-elasticity. Eur J Mech A/Solids 42:188–202
76. Yu YJ, Hu W, Tian XG (2014) A novel generalized thermoelasticity model based on memory-dependent derivative. Int J Eng Sci 81:123–134
77. Ezzat MA, El-Karamany AS, El-Bary AA (2014) Generalized thermo-viscoelasticity with memory-dependent derivatives. Int J Mech Sci 89:470–475
78. Ezzat MA, El-Karamany AS, El-Bary AA (2016) Electro-thermoelasticity theory with memory-dependent derivative heat transfer. Int J Eng Sci 99:22–38
79. Xue Z, Chen Z, Tian X (2018) Transient thermal stress analysis for a circumferentially cracked hollow cylinder based on a memory-dependent heat conduction model. Theoret Appl Fract Mech 96:123–133
80. El-Karamany AS, Ezzat MA (2016) Thermoelastic diffusion with memory-dependent derivative. J Therm Stress 39(9):1035–1050
81. Erdogan F, Gupta GD, Cook TS (1973) Numerical solution of singular integral equations. In: Sih GC (ed) Mechanics of fracture, methods of analysis and solutions of crack problems, vol 1. Noordhoff, Leyden, Netherlands, pp 368–425

82. Brancik L (1999) Programs for fast numerical inversion of Laplace transforms in MATLAB language environment. In: Proceedings of the 7th conference MATLAB'99, Czech Republic, Prague, pp 27–39
83. Brancık L (2001) Utilization of quotient-difference algorithm in FFT-based numerical ILT method. In: Proceedings of the 11th international Czech-Slovak SCIENTIfiC CONFERENCE RADIOELEKTRONIKA, Czech Republic, Brno, pp 352–355
84. Erdogan F (1995) Fracture mechanics of functionally graded materials. Compos Eng 5:753–770
85. Singh AK, Siddhartha (2018) A novel technique for manufacturing polypropylene based functionally graded materials. Int Polym Process 33:197–205
86. Singh AK, Vashishtha S (2018) Mechanical and tribological peculiarity of nano-TiO_2-augmented, polyester-based homogeneous nanocomposites and their functionally graded materials. Adv Polym Technol 37:679–696
87. Parameswaran V, Shukla A (1998) Dynamic fracture of a functionally gradient material having discrete property variation. J Mater Sci 33:3303–3311
88. Jin ZH, Noda N (1994) Transient thermal stress intensity factors for a crack in a semi-infinite plate of a functionally gradient material. Int J Solids Struct 31:203–218
89. Erdogan F, Wu BH (1996) Crack problems in FGM layers under thermal stresses. J Therm Stress 19:237–265
90. Zhou YT, Li X, Qin JQ (2007) Transient thermal stress analysis of orthotropic functionally graded materials with a crack. J Therm Stress 30:1211–1231
91. Rao BN, Kuna M (2010) Interaction integrals for thermal fracture of functionally graded piezoelectric materials. Eng Fract Mech 77:37–50
92. Chen ZT, Hu KQ (2014) Thermoelastic analysis of a cracked substrate bonded to a coating using the hyperbolic heat conduction theory. J Therm Stress 37:270–291
93. Chang DM, Wang BL (2012) Transient thermal fracture and crack growth behavior in brittle media based on non-Fourier heat conduction. Eng Fract Mech 94:29–36
94. Wang BL, Li JE (2013) Thermal shock resistance of solids associated with hyperbolic heat conduction theory. Proc R Soc A 469:20120754
95. Fu J, Chen ZT, Qian L, Hu K (2014) Transient thermoelastic analysis of a solid cylinder containing a circumferential crack using the C-V heat conduction model. J Therm Stress 37:1324–1345
96. Wang BL (2013) Transient thermal cracking associated with non-classical heat conduction in cylindrical coordinate system. Acta Mech Sin 29:211–218
97. Zhang XY, Li XF (2017) Transient thermal stress intensity factors for a circumferential crack in a hollow cylinder based on generalized fractional heat conduction. Int J Therm Sci 121:336–347
98. Keles I, Conker C (2011) Transient hyperbolic heat conduction in thick-walled FGM cylinders and spheres with exponentially-varying properties. Eur J Mech A Solids 30(3):449–455
99. Eshraghi I, Soltani N, Dag S (2018) Hyperbolic heat conduction based weight function method for thermal fracture of functionally graded hollow cylinders. Int J Pres Ves Pip 165:249–262
100. Yang W, Chen Z (2019) Investigation of the thermoelastic problem in cracked semi-infinite FGM under thermal shock using hyperbolic heat conduction theory. J Therm Stresses 42:993–1010. https://doi.org/10.1080/01495739.2019.1590170
101. Chen EP, Sih GC (1977) Mechanics of fracture, elastodynamic crack problems, vol 4. Noordhoff International Publishers
102. Erdogan F (1975) Complex function technique. In: Eringen AC (ed) Continuum physics, vol 2. Academic Press, New York, pp 523–603

Chapter 7
Future Perspectives

7.1 Heat Conduction Theories

Non-Fourier heat conduction theories have seen wide range of applications in both theoretical and applied science in recent years. However, a better understanding of the mechanism of non-Fourier heat conduction and its impact on multiphysical responses of advanced materials is still needed, in particular with respect to the industrial applications. A few points deserve further exploration before these theories can be applied in a comparable fashion as the classical Fourier heat conduction:

- Selection of appropriate non-Fourier heat conduction theories for particular applications. Non-Fourier heat conduction theories of various forms show their advantages in dealing with transient, time-dependent, heat processes as well as generated thermal wave. For instant, for transient processes on metals with short phase lags in the order of less than 100 ps, a single phase lag model such as C-V can be the best choice to address the thermal wave effect. On the other hand, for biological materials such as mammal skins where thermal wave travels at a relatively low speed, and the phase lag is relatively large (in an order from milliseconds up to seconds), a memory-based fractional differential model can be ideal to capture the history of heat conduction and its effect on the thermo-mechanical behavior of advanced materials.
- Application to nanoscale materials. Advances in ultrafast, laser-assisted manufacturing have enabled the fabrication of miniaturized, nano/microscale devices for applications in electronics, optics, medicine, and energy applications, where a non-equilibrium heat transfer model incorporating the size-dependent multi-physical properties is required to evaluate temperature rise during laser-assisted manufacturing. In these applications a non-local, non-Fourier heat conduction, e.g. NL C-V or NL TPL, should be considered for thermo-mechanical analysis. To also avoid the field singularity around wavefronts, a fractional version of non-local, non-Fourier heat conduction, e.g. NL FTPL, can be considered for ultrafast heat transfer in nanomaterials.

Z. T. Chen and A. H. Akbarzadeh, *Advanced Thermal Stress Analysis of Smart Materials and Structures*, Structural Integrity 10,
https://doi.org/10.1007/978-3-030-25201-4_7

- Determination of appropriate phase lags for different non-Fourier theories. The existing non-Fourier heat conduction theories have their specific advantages in particular applications from advanced manufacturing to medicine. However, determination of the thermal phase lags for various types of advanced materials is still a major challenge and requires dedicated experimentation to first capture the thermal wave, and then correlate it to phase-lags and microstructural features of advanced materials through hybrid, experimental-multiscale modelling approaches.

7.2 Application in Advanced Manufacturing Technologies

Application of non-Fourier theories in advanced manufacturing technologies, such as laser-based additive manufacturing and 3D printing of high-melting temperature materials such as metals and ceramics, will be more frequently seen in the academic and industrial community due to the recent paradigm shift in manufacturing and design caused by advances in additive manufacturing (3D printing). Non-Fourier heat conduction provides a more viable tool to capture the transient, thermomechanical behavior of advanced materials and offers a more conservative, but accurate prediction of the integrity of structures under thermal shocks and/or high-strain rate mechanical deformation. In particular, with the development of soft machines and robotics in biomedical engineering and advanced manufacturing, transient heat processes with non-negligible thermal phase lags will be encountered more often where non-Fourier heat conduction will turn out to be the best choice for thermomechanical analysis in the design and analysis of advanced material and structures.